# Materials Chemistry
# at High Temperatures

*Volume 1*
## Characterization

# Materials Chemistry at High Temperatures

## Volume 1

## Characterization

*Editor*

## *John W. Hastie*

*National Institute of Standards and Technology*
*Gaithersburg, MD*

**Humana Press • Clifton, New Jersey**

Based on papers presented at the Sixth International Conference
on High Temperatures—Chemistry of Inorganic Materials, held April 3–7, 1989 ·
at The National Institute of Standards and Technology, Gaithersburg, Maryland
Sponsored by: The National Institute of Standards and Technology
and The International Union of Pure and Applied Chemistry
With the Participation of: The Air Force Office of Scientific Research

ISBN: 0-89603-186-1

Many of these papers were selected from the Proceedings of the Sixth International Conference on High Temperatures—Chemistry of Inorganic Materials, Originally published in *High Temperature Science, An International Journal,* vols. 26–28. The plenary papers were originally published by IUPAC in *Pure and Applied Chemistry,* **62,** No. 1, January 1990.

# Introduction

## Conference Overview and the Role of Chemistry in High-Temperature Materials Science and Technology

LEO BREWER

*Department of Chemistry, University of California, and Materials and Chemical Sciences Division, Lawrence Berkeley Laboratory, 1 Cyclotron Road, Berkeley, CA 94720*

I don't want to compete with the fascinating historic account that John Drowart gave us, but I would like to go through the history of high-temperature symposia. I hope I don't get the reaction that I get from some of my classes when I say, "Remember when such-and-such happened during the War?" And I get this blank look, and one of the students will say, "I wasn't born until after the Korean War." Nevertheless, during World War II, many people in the high-temperature field had their first initiation. But there was one handicap. Owing to security measures, they were not able to interact with one another. Following the War, it was recognized that the high-temperature field was going to expand to meet the demands for materials with unique properties. To meet the demands for new fabrication techniques, it was important to establish better communications among various people. High-temperature symposia were established at that time and have continued very frequently, and I'd like to point out why they are especially important for this field.

One problem is that it is not easy to work at high temperatures. There are many unusual aspects of high-temperature science that often are not recognized by those who have not had extensive experience in the field. It's very important to be acquainted with the instrumentation and techniques that might be useful. One of the purposes then of these symposia was to acquaint people with the new techniques that had developed. This conference has been very successful in that regard. We've had both poster and oral presentations that were clearly presentations of new techniques. They emphasize the value of these techniques that weren't even in use at the time following the War, for example, the lasers. One of the areas in which high-temperature chemistry is particularly handicapped is the problem of containers. How do you contain these materials without contamination? The use of fast-pulse heating and levitation are certainly techniques that are going to be greatly expanded to overcome this. But there are some other problems that require adequate communication between people in the field. Another important aspect of these high-temperature symposia are not only the formal presentations; it is very easy to get together during coffee breaks, at meals, and during social hours to get acquainted with other people in the field and with what facilities they have. As we saw in many of the presentations here, there is collaboration between people at various institutions. Because it is often very difficult to have these very specialized pieces of equipment in all the laboratories, collaboration in use of facilities is very important.

There are some other aspects of high temperature chemistry that make these symposia particularly important. Fritz Franzen mentioned my definition of high-temperature chemistry. At what temperature should research be labeled as high temperature? Everyone is accustomed to the behavior of materials at room temperature and usually assume that they can extrapolate to higher temperatures. However, in many systems, quite unexpected behavior is observed upon raising the temperature. Such behavior marks the boundary of high-temperature science. So the definition of high-temperature science is "when things start behaving the way you don't expect them to behave," and I usually cite the case of boron chemistry where its high-temperature behavior starts below room temperature. So we do have to recognize that we have things happening that people don't normally expect or believe. An example of this came after the War when people heard that there were people interested in studying gases at high temperatures. The usual reaction was, "Oh, that's a waste of time. If you go up to high temperatures, everything will decompose to atoms and there won't be anything to study." But what they didn't recognize is that these gases are usually in contact with condensed phases. One can show through simple thermodynamics that the higher you go in temperature, the more complex the vapor becomes. It becomes more complex with respect to the number of different of species that are significant, as well as the size and complexity of the molecules that are there. Even today, it frustrates people that this is so. The lack of recognition of the complexity of high-temperature vapors has handicapped many industrial processes where unexpected reactions were taking place at high temperatures. On setting up new pilot plants, they've really gone astray because things have not behaved at high temperatures as they would at room temperature. The clarification of the role of complex gaseous species in high-temperature systems has not been an easy chore and much more work is needed to characterize the structures and properties of these species. The presentations at this conference illustrate the recognition of the importance of this problem. Both equilibrium species, as well as clusters formed by expansion of vapors, have been covered. Over 30 papers and posters deal with the problem of characterization of complex species in high-temperature gases.

This complexity of high-temperature systems requires a very broad outlook on the part of investigators of these systems. One must consider a wide variety of species to gain an adequate understanding of high-temperature behavior. High-temperature science is going to be important, but it is going to be important in unexpected directions. I think the high-temperature superconductor is an illustration of why we must maintain a broad outlook on how we are going to use high-temperature science, because we can't anticipate the leads that will come up and the new directions that will arrive. Students often come in to talk to me and ask, "In what direction should I concentrate? What's going to be the important direction? I have a mobile in my office that has seven hands pointing in different directions. I tell them, "Those are the directions you should be interested in." You want a broad base, because you can never anticipate what things will be important, and you should be prepared to aim for the new things that are coming up in the future. High-temperature research is difficult. One not only has to consider the possibility of many unusual species, but the experimental conditions are difficult to control. A common difficulty is the matter of containment without excessive contamination. One of the areas that was covered very extensively in this symposium is the problem of interactions between vapors and solids. Clearly, not only in erosion and corrosion, but also in preparation of materials, this is a very important topic, and we had 50 presentations at this symposium. This was very successful in presenting the topics that are important for this field.

Understanding of the kinetics of reactions at interfaces is particularly important and is an area that needs much more work to develop a broader understanding of high-temperature reactions, whose rates vary rapidly with temperature. The emphasis on thermodynamics is that we have a foundation that ties it together, whereas in kinetics, it is very diverse and they are really handicapped there. Obviously, from many of the presentations, the lack of understanding in kinetics is a critical barrier that is going to be a very difficult obstacle to overcome. There will certainly have to be more work in those directions. Even the problem of adequately characterizing the temperature can be a difficult job. The complexity of the systems and the difficulty of making reliable measurements set severe limits on the amount of information that can be obtained by direct experiment. But present day technology requires a great deal of information about high-temperature behavior to prepare materials by high-temperature processes and improve the efficiency of energy utilization and other high-temperature operations.

It's obviously very important to get data. Because of the fact that the behavior is unexpected, we need measurements of what really goes on at high temperatures. But if one considers the various combinations of the elements just in binary and ternary systems, we are up to astronomical numbers. And as we go to higher components, it is clear that nobody is going to be able to measure all these properties that we would like to know, particularly considering the difficulties in many cases of doing these measurements at high temperatures. So it becomes very important to pick decisive measurements, not just to pick anything at random and measure it. You must be systematic in that when you work with a phase diagram, you not only want to pick one phase, but pick related phases and how they relate to one another. It is particularly important to develop predictive models since it is not possible to hope to get all the data one needs. In other words, the measurements can be increased in value many fold if they will help improve our ability to predict reactions that haven't been measured, or, in some cases, can't be conveniently measured. So it is very important then to do measurements, perhaps for a sequence of elements in the periodic table, to give one an idea of what the trends are so that you can get some insight into the chemical bonding and then develop some better ways of predicting. We had, at this symposium, 40 presentations of various types of predictive models that will help improve our ability.

One of the areas that is most important for high-temperature science is thermodynamics. Thermodynamics is a most powerful tool in allowing you to take measurements that are made at low temperatures and use those to predict what will happen at high temperatures. One of my regrets is that there has been such a decline in room temperature calorimetry, because that has been a very important source of information for high temperatures. If you can determine the enthalpy of formation, even if you do it at room temperature, determine the entropies from spectroscopic data or by using the third law, and estimate high-temperature heat capacities, you can often make much more accurate predictions of high-temperature behavior than you could hope to measure. However, the importance of thermodynamics is well-recognized by the high-temperature community. This is illustrated by the more than 50 papers and posters that deal with characterization of the thermodynamics of systems of importance for high-temperature chemistry or deal with the important job of providing data banks.

The availability of data banks is most important. The lack of needed information is a serious problem for all disciplines, but it is a particularly critical problem for high-temperature scientists. Even information that may be available is not always readily accessible. What we need, and there have been attempts, is a large-scale effort to establish data bases. At the end of World War I, a substantial effort was made and produced a large amount of data, but it was not maintained. Then, the Bureau of Standards started an effort in this direction, but they were never adequately funded. One time when their funds were being reduced, when I was in Washington on other business, I had an appointment with the person in charge of their budget. He agreed to talk to me and I found out that he knew nothing whatever about what was going on in the Bureau. He called in some of the people under him that did know something about the Bureau and they discussed it. Their reaction was, ''We don't think that the work is worth supporting because we never hear from industry that it is of any use to them." Well, these people in industry, of course, get these tables and they think that there is nothing to putting these numbers together, and they never point out to anyone how important it is to have these data.

What are we going to do? We want a centrally coordinated organization to provide a consistent data base. But it is impractical. We're never going to get the funding for it. One other problem is that we are never going to get enough qualified people together in one place to do all the things that would have to be done. So we have to compromise. I wish to discuss the responsibility of scientists in the matter of making information more readily available. Each one of us goes through the literature for data that is important for your particular measurement. What one has to recognize is that, after that effort has been made, it is important to make a little more effort to put those data in a form where they can be published so that other people can use them. What we have to do is really recruit everybody in the field to do their part in the preparation of these data bases.

There are some disadvantages. We are not going to have the consistency that we would like to have. But I don't think that we are going to be able to ever get one group that will put these all together and get them consistent. It will be useful to take these data, though they may be not quite consistent, and make some little shifts and get them on a consistent basis. But I think that to meet this demand for these critical data that people will need, we all have to get in here and try to contribute. The Bureau of Standards has had a program where they've tried to stimulate work of this type. I don't know whether future budgets will supply much money for it, but it is important to have some coordinating agency that will at least tie together these evaluations so that it will at least have a similar format for presenting the information.

It is impractical for most scientists and engineers to retrieve, critically evaluate, and put into usable form any significant fraction of the information they need. The retrieval of information from the literature is often a serious bottleneck. It is not only the tedious and time-consuming work of finding papers dispersed in many journals and published in many languages that stands in the way of utilizing the exponentially growing mass of information. A more serious obstacle is the need for a broad experience in the field to be able to evaluate critically the reliability of the measurements. A large fraction of the data in the primary literature is unreliable. This is particularly true of high-temperature literature because of the difficulty of obtaining measurements under extreme conditions. It requires a great deal of experience to extract reliable information from the literature. Most people are not familiar enough with the experimental or theoretical techniques to be able to critically evaluate the various, often conflicting, values. Reliable data are even more important when no measurements exist for the compounds of interest, and one must use various predictive correlation models to obtain the needed information. If the input data for the model are in error, the predictions can be greatly in error.

Bob Lamoreaux and I found many examples where a large fraction of the published data were seriously in error in the review of the thermodynamic data and phase diagrams for the binary systems of molybdenum that was published in *Atomic Energy Review Special Issue No.7* by the International Atomic Energy Agency. As an example, we found 13 papers all using the same technique, high-temperature solid-electrolyte cells, to determine the Gibbs energy of formation. We critically evaluated those papers: what sort of containers they used; how pure the materials were; did they use techniques like the simple technique of varying the gas flow through their system to see if degassing was affecting the potential and various other techniques of that type; many of the papers gave no indication that they had done that. So that is an example where it is easy to take measurements, but the measurements can often be useless. Of the 13 papers, we found only three that we felt were accurate. It is very important not only to review all the data, but to apply appropriate weighting factors that can vary by many orders of magnitude. One of the other examples was molybdenum diiodide, where we had to reject every publication of measured values, because we did have some predictive models as to what would be reasonable bonding energies for $MoI_2$. All these measurements used transpiration experiments that indicated they were getting $MoI_2$. But from our previous experience, and it was possible to confirm this later, we recognized that they were working in quartz tubes that would yield enough oxygen so that they were measuring $MoOI_2$, not $MoI_2$. In some cases, you have to throw away all of the data. And finally, one other example, the chromium–molybdenum system, where there have been seven determinations of the liquidus and solidus. We went through and reviewed again the purities of materials, the containers they used, and previous work by the same investigators to see whether there was similar consistency as far as doing careful experiments. We concluded that we would reject six of them, and only take one of them.

Thus, it is clear that it is most important to have data banks available that have been prepared by persons with adequate background to critically evaluate the data. A critical evaluation is not an easy chore. One must often reject most or all of the reported results. One must review carefully the experimental procedures used and examine the publication or previous publications by the same authors to determine whether they have been taking the effort to pin down systematic errors. It is necessary to utilize any reliable predictive models to determine if the experimental data are reasonable even within an order of magnitude. It is most important to give the reader some feeling for the possible uncertainty of the final accepted values. The $\pm$ values assigned should not be just the statistical uncertainty (which is a measure of the scatter around the average). The uncertainty cited should include an estimate of possible systematic errors. It is particularly important to not arrive at the recommended value by just averaging all the reported results; that would just mix the reasonably correct results with those in serious error.

There is a tremendous economic saving if each user of data need not take the time to search the literature independently. There is an even greater saving if the user is directed to the most reliable data and is not misled by the erroneous values. The wrong choice between divergent values could make the difference between success and failure of a proposed process. Each of us has to critically evaluate data of interest to our particular subfield. It is important to make the results of our evaluations generally available to others who would not be as expert in our subfield to arrive at a reliable evaluation.

Because of the difficulty of obtaining reliable data for high-temperature systems, there are two approaches to compilations of thermodynamic data. One is the compilation in which the data are critically evaluated in terms of the experimental methods used and possible systematic errors and evaluation of other results from the same laboratory compared to laboratories with established reliability. Such compilations can be carried out independently of any models to establish what appear to be the best experimental thermodynamic results. Another type of compilation is carried out in the context of some model where the data are examined with respect to the expectations of the model. An example where this is now routinely done is in the treatment of aqueous electrolyte data in terms of the Debye-Hückel theory. In this instance, the model is of very great importance because its greatest validity is in the extremely dilute range where it is particularly difficult to obtain reliable experimental data.

Another example is the use of pair-interaction theory as a basis for use of a second virial-coefficient equation of state for extrapolation of gaseous thermodynamic data to the low pressure gaseous standard state. A similar pair-interaction theory is often used as the basis for a Gibbs energy of dilution equation for solutions corresponding to a linear deviation from the Henry's Law slope. In using models of these types in compilations, one must be cautious about the implications of some of the assumptions of the model. In evaluating uncertainty limits for the compiled results, one must include uncertainties introduced by the model. As an example, one would not use an equation of state based on just a second virial coefficient for an ionized gas just as a linear deviation from Henry's Law is not applicable to an aqueous electrolyte. In the extrapolation of the thermodynamic data to the infinitely dilute or solute standard state, one must consider which systems might have long-range forces that would cause substantial deviations from Henry's Law even at the most dilute concentrations, which can be studied experimentally. There is also the problem that the component being assumed may be dissociated at the low concentrations, such as to make Henry's Law inoperable even in the limit of infinite dilution. It is important to use bonding models as a guide that warns when difficulties can be expected and provides for reformulation of the thermodynamic data in terms of a component that is actually the major species at the lowest concentrations. These are standard thermodynamic problems, but it is very important that the role of these essential models be clearly defined by the compiler.

Models are of value to the compiler aside from the problem of extrapolating to the infinitely dilute standard state or for carrying out the Gibbs-Duhem integration to infinite dilution. For example, it is sometimes convenient to express solution data in terms of the solubility parameter of the regular solution theory as a valuable means of tabulating the data more compactly. However, one must consider the implications of entropy assumptions that are implicit in the model. All of these models are useful in compilation efforts, but the possible influence of the model upon the resulting values must be clearly spelled out.

Much valuable data have been reported at this conference. These data are not only of value for the specific systems studied, but when used to develop or improve predictive models, the value of the data are increased many fold. At least 40 of the papers and posters at this conference were tied to models that were used to interpret and evaluate the measurements or the measurements were used to check existing models or develop new models that can yield useful data.

There are several other important areas covered in this conference. A most common problem in high-temperature processing is the understanding of interactions between condensed and gaseous phases. Over 50 of the presentations at this conference have dealt with various aspects of the problems of contact between condensed and gaseous phases. It is clear that the high-temperature community is addressing that problem. Because of the difficulty of obtaining reliable measurements at high temperatures, it is most important to have a variety of techniques available. Almost 30 presentations dealt with new techniques or improvements of previous techniques. The laser is becoming a widely-used tool for characterizing high-temperature systems.

For those of us who promoted high-temperature conferences in the past, the success of the present conference is a pleasant experience. It has been a wonderful opportunity for high-temperature workers around the world to get together and share new developments that will be useful to all. One can always ask for expansion of research in the high-temperature field to meet the many industrial demands for new materials, but we must face the budgetary restrictions that have been with us and might get worse. There is one area where we must press strongly for increased support. It is most important to have an expansion of the programs to develop data bases. Any money invested in these programs will be paid off many fold by the savings of time for the users who would not have to dig through the literature and who would not be misled by erroneous data that would be screened by the experienced people compiling the data bases. The other area that should be strongly supported is the development of chemical bonding models that would allow predictions of thermodynamic data for the multitude of systems where data do not exist. There are many fruitful opportunities in that direction. Of course, neither the data base compilations nor the development of models can be carried out adequately without the continued influx of new data measurements.

## ACKNOWLEDGMENT

This work was supported by the Director, Office of Energy Research, Office of Basic Energy Sciences, Materials Sciences Division of the United States Department of Energy under Contract DE-AC03-76SF00098.

*Editors' Note: This account was given orally on behalf of IUPAC and is not available in document form.

# Preface

The role of chemistry in materials processing and performance, at high temperatures, is of the utmost importance in today's technology. This is particularly the case for advanced inorganic materials, including the elements, their compounds, alloys, ceramics, composites—and specialized forms, such as films, coatings, powders, slags, and fluxes. The key combination of inorganic chemistry with high temperatures, known as the field of High Temperature Chemistry, has grown dramatically during the past 50 years. A number of books, symposia proceedings, and journals representing this field have appeared during this period. However, much of the early research emphasis and reports have been on the formation, characterization, and thermodynamic properties of high temperature species because of the intrinsic molecular nature of materials at high temperatures. In response to the need for improved processing and performance of materials, increased attention has recently been given to the physicochemical interplay between high temperature molecular species and solid or liquid interfaces and phases. New measurement techniques, models, and databases have been developed and applied to these heterogeneous systems. Even the previously hypothetical intermediate state of critical nuclei, or molecular clusters is now amenable to detailed experimental (and theoretical) investigation. These, and related aspects of high temperature and materials science, are discussed here by prominent international experts. The purpose of these two volumes, then, is to represent the current status of fundamental research in the various disciplines that deal with the scientifically and technologically important interplay of high temperature and materials chemistry. In the first volume, two major chapters are offered that deal with (1) Advances in Measurement Techniques and (2) Thermochemistry and Models. The second volume contains major chapters on (3) Processing and Synthesis and (4) Performance Under Extreme Environments. The special Introduction, by Leo Brewer, discusses the importance of these particular areas to the overall field.

Many of the papers that follow were presented at an international conference, cosponsored by NIST and the IUPAC Commission for High Temperature and Solid State Materials, with additional support from the Air Force Office of Scientific Research. Other papers presented at the conference, but not included here, appear in volume 28 of *High Temperature Science, An International Journal.*

*John W. Hastie*
National Institute of Standards and Technology
(formerly the National Bureau of Standards)
Gaithersburg, MD

# TABLE OF CONTENTS

### Physicochemical Methods

### CHAPTER TWO: THERMOCHEMISTRY AND MODELS

#### Databases and Phase Equilibria Models
*Chair: Malcolm Rand*

*Advances in Measurement Techniques*

# Optical Second Harmonic Generation as a Probe of Properties and Processes at Surfaces and Interfaces

Richard J. M. Anderson* and John C. Hamilton

*Physical Science Department, Sandia National Laboratory, Livermore, California 94551-0969*

## ABSTRACT

The technological significance of surfaces and interfaces hardly can be overstated. Whether the effects of surface processes are primary to the technology, as in heterogeneous catalysis and tribology, or limiting of the application of bulk material properties, such as corrosion of structural metals and ceramic bonding, the chemistry and physics of interfaces are critical to successful long-term operation of systems that are dependent on specific materials. The necessity to understand beneficial and deleterious changes at these interfaces, coupled with their relative inaccessibility, requires the utilization of every analytical technique that is surface specific. An important, new diagnostic tool of this type is surface optical second harmonic generation (SHG).

The coherent interaction of intense laser beams to generate new wavelengths is determined by material-specific nonlinear optical susceptibilities. The lowest order nonlinear response, the second order susceptibility, which mediates the bilinear combination of optical frequencies, is nonzero only in those regions of space that lack a center of symmetry. At the interface of two bulk centrosymmetric media (all gases and liquids, nearly all metals, many oxides, and chalcogenides), the second order susceptibility, therefore, endures only at the discontinuously thin surface separating the materials. The pairwise coupling to form sum and difference frequencies is dependent on this local second order susceptibility, which is a function of the electronic and vibrational structures of the interface. As a chemi-

*Author to whom all correspondence and reprint orders should be addressed.

cal reaction or other process changes the electronic states of the interface, or a change in bonding alters the vibrational frequencies of interface species, characteristic resonances between photon energies of the incident or generated frequencies, and transitions connecting electronic and vibrational states, will evolve and change the second order susceptibility. Thus, the intensities of these new coherent beams can be used *in situ* and in real time to follow the evolution of these interface properties.

The practical application of these techniques, with special attention to second harmonic generation and sum frequency generation, will be discussed in detail. Specific examples will be given for adsorbed gases on metals, high temperature segregation of bulk impurities to the metal surface, surface states of dielectrics and ceramics, ceramic–metal interactions, and other topics.

**Index Entries:** Interfaces; surfaces; optical susceptibilities.

# INTRODUCTION

Optical second harmonic generation (SHG) from surfaces is a process in which a fraction of the light incident on a material interface is converted upon reflection to light at twice the incident frequency. In addition to polarizing sample electrons at the fundamental frequency, $\omega$, of the incident laser beam, the strong electric fields induce a nonlinear source polarization, $P_{nls}$, at $2\omega$, which reradiates a coherent beam of light at the second harmonic frequency. Using a point dipole approximation, $P_{nls}$ is zero for bulk centrosymmetric media because of symmetry considerations. Only at interfaces between dissimilar materials is the symmetry broken, giving rise to a nonlinear source polarization. Thus, interfaces make a major contribution to the total second harmonic radiation detected from centrosymmetric materials. This interface sensitivity is a major advantage of nonlinear optical spectroscopies over linear optical techniques, such as Raman spectroscopy, which average over the penetration depth of the laser. The experiment is conceptually quite straightforward. The light generated at the second harmonic frequency is colinear with the reflected pump beam at the fundamental, but it can be readily filtered and its (normalized) intensity detected for analysis.

Material property that bilinearly couples the incident fields to form the nonlinear polarization, the second order susceptibility, $\chi^{(2)}$, is a third rank tensor that must transform as the surface symmetry of the sample. Furthermore, the susceptibility shows maxima at resonances between the photon energies and transitions connecting occupied and vacant surfazce electronic states. Thus, at an ordered interface surface phenomena that modify the energy levels or spatial symmetry can be probed by measuring the subsequent change in the second harmonic (SH) intensity. We will illustrate these basic properties by showing the effect that monolayer adsorbates hydrogen and carbon monoxide have on the SH intensity from a Ni(111) surface, then demonstrate their application to the problem of high temperature segregation of a dilute bulk impurity to the surface of an alloy, the (110) face of an Fe–18Cr–3Mo single crystal.

Hydrogen adsorbed on Ni(111) has been extensively studied using conventional surface probes, including thermal desorption (3–5), low energy electron diffraction (LEED) (3), angle resolved photoemission (10), and work function measurements (3). It is a relatively simple system showing two distinct thermal desorption peaks corresponding to $\beta_1$ and $\beta_2$ adsorption states of hydrogen (3–5). A half monolayer of hydrogen in the $\beta_2$ state is observed to undergo an order–disorder transformation at about 270 K (3). Multiple-scattering calculations show that hydrogen is adsorbed in the threefold hollow sites and that, below 270 K, a graphite-like array of hydrogen exists on the surface (3).

We have measured the SH intensity as a function of both azimuthal orientation (rotation about the surface normal) and surface hydrogen coverage, from which we have inferred the changes in the isotropic and anisotropic components of the second-order susceptibility associated with both the $\beta_1$ and $\beta_2$ states of hydrogen. Although previous experiments (1) have shown a difference in susceptibilities for the top site and bridge site of CO on Rh(111), this is the first case where susceptibilities for different adsorption *states* associated with the same adsorption site have been measured. We find that the 3m symmetry of the Ni(111) surface is manifest by a threefold symmetry in the SH intensity, which persists at coverages and temperatures at which the LEED pattern shows both ordered and disordered hydrogen overlayers, demonstrating that SHG probes local symmetry, not long-range ordering. We find that second harmonic generation is dramatically reduced by hydrogen adsorption, almost disappearing at saturation coverage. This indicates that the major contribution to the second-order susceptibility comes from the surface, suggesting that resonances with electronic transitions between surface states are responsible for the large SH intensity from clean Ni(111).

The adsorption system CO on Ni(111) is one of considerable complexity involving three distinct adsorption sites for molecularly-adsorbed CO. For room temperature adsorption, the saturation coverage is $\theta = 0.5$. At temperatures below 270 K, additional CO is adsorbed with a saturation coverage of $\theta = 0.57$ and a $(\sqrt{7}/2 \times 7/2)R19°$ LEED structure (12). Vibrational spectroscopies indicate population of terminal, twofold and threefold-bridged sites depending on coverage and temperature (13). Threefold and twofold sites are both occupied at low coverage, with a decrease in occupancy of threefold sites at $\theta > 0.07$. For $0.3 < \theta < 0.5$, terminal-bonded CO is observed in addition to CO at the other two sites. For $\theta = 0.5$, only twofold CO is observed and a $c(4 \times 2)$ LEED pattern is observed (14,15). The clean surface has $C_{3v}$ symmetry, which results in a threefold symmetric SH intensity as a function of azimuthal rotation about the surface normal. The symmetry of the CO overlayer alone is

$C_{4v}$. By putting the overlayer in the twofold-bridge sites, as shown, and considering the Ni surface and the CO overlayer, the rotational symmetry is removed completely ($C_1$ symmetry). Since the adsorbate changes the overall point group symmetry of the surface region responsible for second harmonic generation, this is a particularly interesting text case.

## EXPERIMENTAL

The sample was mounted in a UHV chamber ($2 \times 10^{-10}$ torr base pressure) on a manipulator with liquid nitrogen cooling, resistive heating, and azimuthal rotation over $\approx 180°$. Cleanliness, surface ordering, and azimuthal orientation were confirmed with Auger spectroscopy and LEED. The 700 nm $p$-polarized beam (20 mJ/cm$^2$ in a 7-ns pulse at 10 Hz) from a Nd:YAG-pumped dye laser was incident on the crystal surface at a 60° angle to the surface normal. $P$-polarized second harmonic radiation was measured with a filter, phototube, and gated integrator and divided on a shot-by-shot basis by a reference SH signal generated in bulk quartz.

For $p$-polarized excitation and detection, the intensity of the second harmonic generated from a surface of 3m symmetry can be written (2)

$$I(2\omega) \propto |\chi^{(2)}|^2 I^2(\omega) = |A + B\cos(3\psi)|^2 I^2(\omega) \qquad (1)$$

Here, $\chi^{(2)}$ is the second-order susceptibility, and $A$ and $B$ are linear combinations of the susceptibility components. $A$ and $B$ depend on the angle of incidence and, in general involve both bulk and surface contributions. $\psi$ is the orientation of the plane of incidence measured relative to the surface vector <211>. Included in Fig. 1 is the square root of the measured SH intensity for the clean Ni(111) surface over the range of azimuthal angles available to us; this data is fit well by Eq. (1), using A/B = 3.26. It is generally not possible to separate surface and bulk contributions to the susceptibility for a clean surface (6,7). However, it is possible to study relative contributions from surface and bulk by chemical means since the bulk contribution will not be affected by adsorption of gases or other surface chemical changes. At azimuthal angles $\psi$ = 0, 30, 60, and 90°, we have measured SH intensity from Ni(111) during adsorption and thermal desorption of hydrogen and carbon monoxide. Hydrogen adsorption took place at sample temperatures ranging from 225 to 200 K. After $4.35 \times 10^{17}$ mol/cm$^2$ exposure, corresponding to essentially saturation coverage (4), hydrogen was removed from the chamber, and thermal desorption was initiated using a heating rate of 0.26 K/s. CO adsorption was effected at 298 K using gas pressures in the range $5 \times 10^{-9}$ to $5 \times 10^{-8}$ torr. Thermal desorption was driven by a heating rate of 1 K/s.

Figures 2 and 3 show the data from hydrogen adsorption and desorption experiments, respectively. The data are all plotted on a common intensity scale, allowing direct comparison of adsorption and desorption data. Adsorption of a full monolayer of hydrogen almost completely

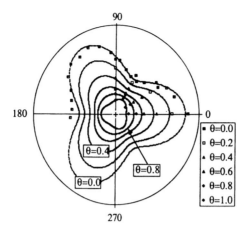

Fig. 1. Continuous curves show the calculated square root of the SH intensity (proportional to $\chi^{(2)}$ as a function of azimuthal angle, $\psi$, for hydrogen coverages, $\theta = 0.0, 0.2, 0.4, 0.6, 0.8,$ and $1.0$, using Eq. (3) and the susceptibility coefficients from Table 1. Discrete data points are shown for the clean surface (outer curve) and at inferred coverages from the adsorption/desorption data.

quenched the second harmonic intensity at all orientations, indicating that the dominant contribution comes from the surface susceptibility. For identical, optically noninteracting oscillators, a linear relationship between the components of $\chi^{(2)}$ and coverage will exist (1). Thus, for a single site, one expects

$$\chi^{(2)} \alpha \ 1 \ - \ C\theta/\theta_s \tag{2}$$

where $C$ is a constant, $\theta$ is the coverage and $\theta_s$ is the saturation coverage for the site. For two nonequivalent states, we generalize Eqs. (1) and (2).

$$\chi^{(2)} \alpha \ A \ (1 \ - C_{1A}\theta_1/\theta_{1s} \ - \ C_{2A}\theta_2/\theta_{2s})$$
$$+ \ B \ \cos(3\psi) \ (1 \ - C_{1B}\theta_1/\theta_{1s} \ - \ C_{2B}\theta_2/\theta_{2s}) \tag{3}$$

where $\theta_1$ and $\theta_2$ are the absolute coverages in the $\beta_1$ and $\beta_2$ states, and $\theta_{1s}$ and $\theta_{2s}$ are the saturation coverages for the $\beta_1$ and $\beta_2$ states. The $C_{ij}$ are constants giving the linear decrease in the components of the second order susceptibility caused by hydrogen adsorbed in the $\beta_i$ state.

The fractional coverage of hydrogen in each state during absorption was calculated using one-site Langmuirian kinetics (5). To explain the observed data, we found it necessary to assume independent one-site Langmuirian kinetics for both states, an approximation ultimately justified by the large ratio of the sticking coefficients. In this absorption model

$$d\theta_1/d\epsilon \ = \ 2S_1^o(1 \ - \ \theta_1/\theta_{1s}); \quad d\theta_2/d\epsilon \ = \ 2S_2^o(1 \ - \ \theta_2/\theta_{2s}) \tag{4}$$

where $S_i^o$ is the sticking coefficient for a hydrogen molecule in an empty $\beta_1$ site, $\theta_{is}$ is the saturation coverage of hydrogen for the $\beta_i$ state, and $\epsilon$ is the hydrogen exposure. We have used $\theta_{is} = \theta_{2s} = 0.93 \times 10^{15}$ sites/cm$^2$

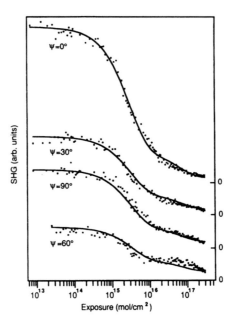

Fig. 2.    Optical second harmonic intensity at four different azimuthal angles as a function of hydrogen exposure on Ni(111). The fits to data are based on Eqs. (3) and 4 using values from Table 1.

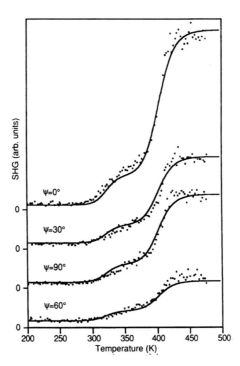

Fig. 3.    Optical second harmonic intensity at four different azimuthal angles during thermal desorption of hydrogen from Ni(111). The fits to data are based on Eqs. (3) and (5) using values from Table 1.

(3–5). The evolution of second harmonic intensity during thermal desorption was modeled using the Polanyi-Wigner differential equation (8). For a state, $i$, on the surface, the rate of desorption is

$$d\theta_i/dt = v_i\theta_1{}^{n_i}\exp(-E_i/RT) \tag{5}$$

where $v_i$ is a frequency factor, $n_i$ is the reaction order for desorption from the ith state, and $E_i$ is the desorption activation energy. For hydrogen on nickel, second-order desorption kinetics for both states has been demonstrated previously (3,4). The adsorption and desorption data at all four azimuths were fit collectively by varying A/B, $S_1^o$, $S_2^o$, $E_1$, $E_2$, $C_{1A}$, $C_{2A}$, $C_{1B}$, and $C_{2B}$ of Eqs. (3), (4), and (5) using a least-square fitting routine. The fits to the data are shown in Figs. 2 and 3 and the values of the parameters are listed in Table 1. Although A/B and the susceptibility coefficients, $C_{ij}$, are generally complex, varying phases of these parameters in no way improved the fit to the data. To minimize the number of adjustable paraments, we used $v_1 = 2 \times 10^{-4}$ cm²/s and $v_2 = 1 \times 10^{-3}$ cm²/s in agreement with literature values (3).

## DISCUSSION

### Hydrogen Adsorption

The good agreement between these results, obtained with SHG, and results using traditional techniques demonstrates the validity of this model and the inferred dependences of $\chi^{(2)}$ on the hydrogen coverage. We calculate an initial sticking probability $S_1^o + S_2^o = 0.084$, in good agreement with literature measurements ranging from 0.02 to 0.15 (3–5). To our knowledge, we are the first to apply an adsorption model for hydrogen on Ni(111) that utilizes independent sticking coefficients for $\beta_1$ and $\beta_2$. Attempts to fit the data of Fig. 2 using a one-site Langmuirian model with a single sticking coefficient (as in Russel et al. (5)) did not reproduce the slow decrease in the SHG signal seen in Fig. 2 at exposures over about 10 Langmuirs. Actually, of course, hydrogen will occupy the $\beta_1$ state only after the $\beta_2$ state is partially filled; thus, the sticking probability for the $\beta_1$ state should be coupled to the fractional coverage in the $\beta_2$ state. In our calculations, the initial sticking coefficient for $\beta_2$ is 24 times larger than $\beta_1$, so the error introduced by ignoring this coupling is insignificant. The calculated coverages of $\beta_1$ and $\beta_2$ reproduce the observation that $\beta_1$ is populated only after $\beta_2$ is largely filled. The activation energy for thermal desorption from the $\beta_1$ state, $E_1 = 18.9$ kcal/mol, is essentially identical to that of Christmann et al. (3), whereas the energy for $\beta_2$, $E_2 = 25.0$ kcal/mol, is in reasonable agreement.

The azimuthal dependence of $\chi^{(2)}$ was calculated as a function of total fractional coverage, $\theta = (\theta_1 + \theta_2)/(\theta_{1s} + \theta_{2s})$, and is shown in Fig. 1. Data points representing $\sqrt{I(2\omega)}$ at appropriate coverages from averaged adsorption/desorption data of Figs. 2 and 3 are also plotted, along with

Table 1
Susceptibility Coefficients (60° Incidence), Initial Sticking Probabilities,
and Desorption Activation Energies used in Eqs. (3), (4),
and (5) to Calculate Second-Order Susceptibility Obtained from Least-Squares Fit
to Data Shown in Figs. 2 and 3[a]

| Isotropic susceptibility coefficients | Anisotropic susceptibility coefficients | Initial sticking probability | Desorption activation energy, kcal/mol | Zero-coverage susceptibility components |
|---|---|---|---|---|
| $C_{2A} = 0.499(7)$ | $C_{2B} = 0.78(2)$ | $S_2^0 = 0.081(5)$ | $E_2 = 25.0(1)$ | $A/B = 3.26(2)$ |
| $C_{1A} = 0.260(7)$ | $C_{1B} = 0.33(3)$ | $S_1^0 = 0.0034(3)$ | $E_1 = 18.9(1)$ | – |

[a]Values in parentheses are standard deviations in the least significant figure.

additional measurements taken for the clean surface at 10° azimuthal increments. The isotropic components of the susceptibility is considerably greater than the anisotropic component at all hydrogen coverages. The change in the susceptibility associated with adsorption of a half monolayer into the $\beta_2$ state is about twice the change in susceptibility associated with subsequent adsorption of a half monolayer into the $\beta_1$ state. Furthermore, the anisotropic portion of the susceptibility decreases more rapidly than the isotropic portion as hydrogen is adsorbed, passing through zero at total fractional coverage 0.84 and reversing the sign of the contribution at full coverage. The calculated angular patterns and discrete data points combine to show graphically the progressive reduction of the total susceptibility and the sign-reversal of the anisotropic component. Based on the quality of the fits shown in Figs. 2 and 3, we believe that this model gives a reasonable representation of the effects of hydrogen adsorption on the isotropic and anisotropic components of the second-order susceptibility.

Given the excellent fit to data obtained with Eq. (3), it appears that the 3m symmetry persists for all hydrogen coverages. This is somewhat surprising since hydrogen occupies disordered sites at coverages significantly below or above $\theta = 0.5$ (7). Since a disordered hydrogen layer would not have 3m symmetry on a scale of several unit cells, we conclude that the nonlinear susceptibility primarily reflects the local symmetry of bonding sites on the surface. This conclusion was confirmed by an experiment in which we attempted to see a change in the SH intensity associated with the order-disorder transformation of hydrogen, which occurs for the $\beta_2$ state at about 270 K. We adsorbed hydrogen at a low temperature until a well-ordered (2 × 2) LEED pattern was obtained. The sample was then warmed and cooled to verify that the order-disorder transformation could be seen using LEED. Next, the same heating and cooling cycle was performed while monitoring the SH intensity. Finally, we verified that the order-disorder transformation was still observable using LEED. At no time did we see any significant change in the second harmonic generation at the surface. We conclude that for this adsorbate/surface system, the *local* point group symmetry of surface and adsorbate plays the major role in determining the form of the second-order suscep-tibility and, thus, the azimuthal symmetry of optical second harmonic generation.

Recent work on Cu(111) attributes the existence of azimuthal an-isotropy to bulk interband transitions at the energy of the second har-monic (2.33 eV) (2). For nickel, interband transitions occur in the vicinity of 1.4 eV, between d band levels in the vicinity of the L symmetry point in the Brillouin zone (9). However, in view of the fact that adsorbed surface hydrogen so markedly affects the magnitude of both isotropic and anisotropic components of the susceptibility, we find it difficult to attribute the presence of anisotropy to bulk interband transitions. Rather, it is more likely that $\chi^{(2)}$ derives strength from a resonance between our incident light at 1.77 eV energy and a transition involving surface states. Studies of Ni(111) using angle-revolved photoemission show an occu-pied sp-derived surface state of symmetry $\Lambda_1$ about 0.25 eV below the Fermi energy ($E_F$) at $\Gamma$ that disperses to about $-1.0$ eV at the K point of the surface Brillouin zone (SBZ) and is dramatically shifted and quenched by the absorption of hydrogen (10). Observed bulk bands are much less strongly affected by hydrogen adsorption. Furthermore, momentum-resolved inverse photoemission experiments on clean Ni(111) show a crystal-induced surface state/resonance about 0.5 eV above $E_F$ at $\Gamma$ that disperses rapidly upward along $\Gamma M$ (11). Since $\chi^{(2)}$ gains oscillator strength from any dipole-allowed (vertical, or k-conserving) transition, we suggest that the 1.77 eV exciting light is resonant with a direct transition between these surface states at some point in the SBZ. Also, at the resonance energy, these surface levels carry the projection of the bulk symmetry onto the (111) surface, which is point group 3m, accounting qualitatively for the threefold rotational symmetry of SH intensity. Tran-sitions involving bulk d bands may be responsible for the residual signal from the hydrogen-saturated surface. An adequate quantitative descrip-tion of the anisotropy for clean and partially covered surfaces must await calculations that include the size of the transition dipole connecting these surface states as a function of the E-field component parallel to the surface.

## CO Adsorption

Since the $C_{3v}$ symmetry of the clean surface is not maintained when CO is adsorbed on Ni(111), there is no *a priori* reason why Eq. 1 should pertain after CO adsorption. Examination of the data shown in Figs. 4 and 5 suggests, however, that the threefold symmetry of the SH intensity is maintained for much of the range of CO coverages. Thus, the data was fit successfully, using Eq. 1, with A and B a function of CO coverage, as described in the next paragraph. We shall consider possible explanations for the persistence of threefold SH intensity later.

Although adsorption of CO on Ni(111) is complex, we shall begin by fitting the data using a one-site Langmuirian adsorption model (12,13).

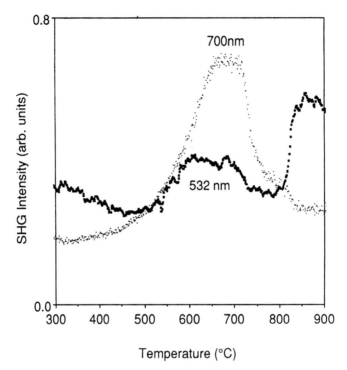

Fig. 4.  Optical second harmonic intensity as a function of carbon monoxide exposure on Ni(111). Data collected at four different azimuthal angles, $\psi = 0$, 30, 90, and 60°, are shown. The fits to data are based on Eqs. (3) and (4) in the text and represent the values of the parameters, A/B, $C_A$, and $C_B$, obtained by least-square fits to the experimental data, as described in the text. The surface temperature during carbon monoxide adsorption was 25°C. The 30 and 90° data is grouped here and in Fig. 3 because these two azimuths are equivalent by symmetry for $C_{3v}$.

The adsorption data of Fig. 4 was fit by Eqs. (3) and (4) using a least-squares fitting routine to vary $S_o$, A, B, $C_A$, and $C_B$. Least-squares fit of the data, assuming Langmuirian adsorption, gives $C_A = 1.05 \times$ exp (i11°), $C_B = 0.87 \times$ exp ($-$i2°), and A/B $= 3.57$, with an initial sticking probability on the order of unity in agreement with thermal desorption measurements made using quadrupole mass spectroscopy. Figure 6 shows a polar graph of Eq. 3 plotted using the constants derived from the least-squares fit. The plotted points are data from the adsorption runs shown in Fig. 4. Also included are data taken at intermediate azimuthal orientations for the clean surface. We see that the azimuthal pattern is little changed since CO is adsorbed until near-saturation coverage is reached. For saturation coverage, the SH intensity is almost independent of azimuthal sample orientation.

It appears that the changes in the nonlinear susceptibility are virtually independent of the site occupied by the carbon monoxide molecules. This may be concluded since the relative occupations of various

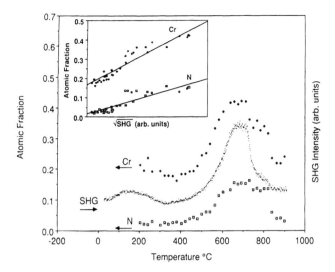

Fig. 5. Optical second harmonic intensity as a function of temperature during thermal desorption of carbon monoxide from Ni(111). The sample heating rate during these desorption runs was 1K/s. Data collected at four different azimuthal angles, y = 0, 30, 90, and 60°, are shown. The fits to data are based on Eqs. (3) and (6) in the text. The coverage dependence of the activation energy for desorption used in these fits is plotted in Fig. 6.

sites vary with total coverage, yet the susceptibility varies linearly with coverage independently of adsorption site distribution.

Perhaps the most surprising aspect of this data is the persistence of threefold symmetry in the SH intensity over the whole range of CO coverages. This is true even though IR data shows occupation of twofold sites lacking $C_{3v}$ symmetry. There are two possible explanations for the persistence of threefold intensity: (1) The threefold symmetry is broken for individual domains, but coherent optical interference between the three domains on the surface results in a threefold symmetric pattern or (2) the electronic states that give rise to the second-order nonlinear susceptibility retain their threefold symmetry even after CO adsorption. This would be plausible if the initial states for resonant SHG were the somewhat localized d-band derived states at the surface. If CO adsorption binds some of this electron density, it is possible that sites remaining free of CO would retain their threefold symmetry, giving rise to the observed SH intensity pattern. We tend to prefer the second explanation based upon the excellent fit to the data using Eqs. (4) and (5); however, the first explanation cannot be ruled out for the experimental data presented here.

Isosteric heats of adsorption have been reported in the literature for CO adsorption on Ni(111) (16). The heat of adsorption is strongly dependent on coverage with a marked decrease at ~0.25 monolayer coverage. The coverage dependence of the heat of adsorption must be included in

CO on Ni(111)

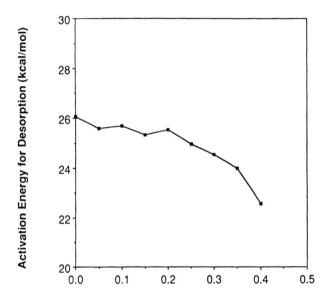

Fig. 6.    Calculated square root of the SHG intensity as a function of azimuthal angle, $\psi$, for carbon monoxide coverages, $\theta = 0.0, 0.1, 0.2, 0.3,$ 0.4, and 0.5. The square root of SHG is proportional to the magnitude of the second-order nonlinear susceptibility, $\chi$. The anisotropy is slightly reduced as carbon monoxide is adsorbed on the surface.

order to model the thermally-activated desorption of CO from Ni(111). For CO on Ni(111), we assumed first order desorption kinetics. The coverage dependence of the activation energy was approximated by discrete values $E(\theta_i)$ for $\theta_i = 0.0, 0.05, 0.10, \ldots, 0.40$ with a linear dependence between these values. Equations (3) and (5) were used with the values of A, B, $C_A$, and $C_B$ from the adsorption fit to calculate the SH intensity. A least-squares fit to the data was used to determine $\upsilon$ and $E(\theta_i)$. The best fit was obtained with $\upsilon = 2.2 \times 10^{12}$. The best fit values for $E(\theta_i)$ are plotted in Fig. 7. The activation energy for desorption is large at low coverages (~26 kcal/mol) and decreases with increasing coverage above about 0.25 monolayer. These desorption activation energies agree reasonably well for $0.1 < \theta < 0.4$ with the isosteric heat of adsorption for CO previously reported in the literature (16). However, this reference reports a large isosteric heat of adsorption for coverages less than 0.1 monolayer, which does not agree with the activation energy determined here for that coverage range. The activation energy for desorption at low coverages should be very dependent on the defect density, which may explain this apparent discrepancy.

The data presented here show the azimuthal and CO coverage dependence of SH intensity generated at the Ni(111) surface using 700 nm excitation. We find that threefold azimuthal symmetry of the SH

CO on Ni(111)

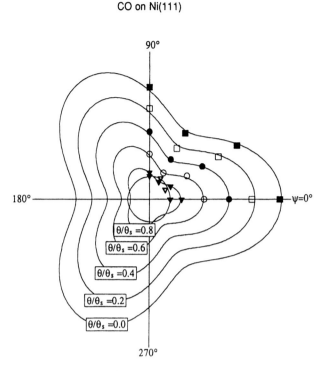

Fig. 7. Activation energy for desorption of carbon monoxide from Ni(111) as a function of coverage. These values were derived from a least squares fit of Eq. (6) to the thermal desorption data using Eq. (3) with coefficients, A/B, $C_A$, and $C_B$ derived from the adsorption data.

intensity persists over a wide range of coverages even though IR experiments show that CO is located predominantly in the twofold sites. We suggest that most of the SH intensity is derived from resonance with a transition involving at least one electronic state localized in the vicinity of unoccupied sites. These states apparently retain threefold symmetry even upon CO adsorption in nearby sites. Thus, the SH intensity can be described by the $C_{3v}$ symmetry of these states. These results emphasize the critical importance of an understanding of the symmetry of the surface and adsorbate electronic wavefunctions that give rise to the second-order nonlinear susceptibility.

## High Temperature Segregation Experiments

The modification of material properties owing to exposure to high temperatures has enormous technological significance. As an application of the SHG technique, we have studied modification of the surface of a stainless steel-like alloy by high temperature segregation. Our sample was an Fe–18Cr–3Mo single crystal oriented to within 0.5° of the (110) plane and mechanically polished with diamond paste to an optically smooth surface. The sample was mounted in an ultrahigh vacuum system with a $2 \times 10^{-10}$ torr base pressure. The sample manipulator was liquid nitrogen cooled and resistively heated. The sample was cleaned using argon-ion bombardment and subsequent annealing for about a minute to 400°C; cleanliness and surface ordering were confirmed with Auger spectroscopy and LEED. Gas exposures were performed using leak valves and ion gage pressure monitoring. For the experiments reported here, the (110) surface was oriented so that the plane defined by the incident and reflected laser beams was the (001) plane of the crystal.

We performed segregation experiments on the Fe–18Cr–3Mo (110) surface using a heating rate of 7.5°C/min. With this heating rate, near equilibrium conditions were attained above ~500° C (verified by measuring rates of approach to equilibrium vs temperature). Figure 8 includes the atomic fractions of chromium and nitrogen determined using Auger spectroscopy on the originally clean Fe–18Cr–3Mo (110) surface during heating from 50 to 950°C. Chromium and nitrogen were enriched on the surface over the temperature range from 550 to 850°C. At temperatures above 850°C, the surface returned to very nearly the bulk composition. Figure 8 also shows the SH intensity measured in an equivalent heating run using 700 nm excitation. The increase in SH intensity between 500 and 800°C is clearly associated with the cosegregation of chromium and nitrogen.

The strong correlation between the SH intensity for 700 nm excitation and the atomic fractions of chromium and nitrogen seen in Fig. 8 suggests that the coverage and the nonlinear susceptibility tensor, $\chi^{(2)}$, are related. For equivalent noninteracting surface sites, the surface nonlinear susceptibility can be written as a linear function of coverage

$$\sqrt{I(2\omega)} \; \alpha \; \chi^{(2)} \; I(\omega) = (A + B\theta/\theta_s) \; I(\omega) \tag{1}$$

Here, $A$ and $B$ are constants, $q$ is an adsorbate surface coverage, $\theta_s$ is a saturation coverage, $I(\omega)$ is the incident laser intensity, and $I(2\omega)$ is the second harmonic intensity. Given this assumption, the coverage is related linearly to the square root of the second harmonic intensity, i.e.

$$\theta = \theta_s \; [\sqrt{I(2\omega)}/I(\omega) - A]/B \tag{2}$$

The insert in Fig. 8 plots the measured atomic fractions of chromium and nitrogen against the square root of the second harmonic intensity. The linear fits to the data indicate that Eq. (2) is a reasonable approximation in this case. Since chromium and nitrogen are probably forming a surface compound, it seems unlikely that all sites should be noninteracting; nonetheless, the strong correlation between chromium/nitrogen atomic fractions and the square root of the second harmonic intensity suggests a nearly linear relationship between the chromium/nitrogen atomic fraction and the surface nonlinear susceptibility at this wavelength. The wavelength dependent studies described below suggest that resonant

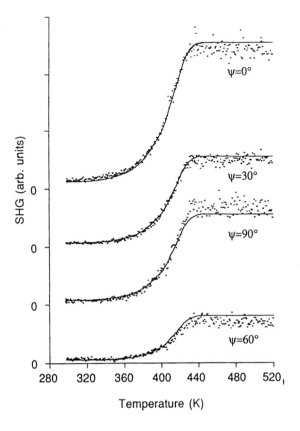

Fig. 8. Second harmonic intensity as a function of temperature during linear heating ramp from room temperature to 950°C at 7.5°C/min. Excitation wavelength was 700 nm. Atomic fractions of chromium and nitrogen in the surface obtained in an equivalent heating run using Auger spectroscopy are also plotted. The insert shows the correlation between the atomic fractions of chromium and nitrogen and the square root of the second harmonic intensity.

electronic transitions in the chromium/nitrogen overlayer are responsible for this quantitative relationship at 700 nm.

Optical second harmonic generation is highly sensitive to electronic structure since the cross-sections for the process are determined by resonant denominators involving electronic transitions between surface states or other electronic levels. Fig. 9 compares the second harmonic intensity measured during the previous experiments using 700 nm excitation with the intensity measured using 532 nm excitation during a temperature ramp identical to Fig. 8. Clearly, there are major differences in SH intensity at these two wavelengths. For 532 nm excitation, the second harmonic generation at 650°C is only a little more than half that for the clean surface at 900°C. This implies that cosegregated chromium/ nitrogen actually decreases the SH intensity relative to a clean surface at

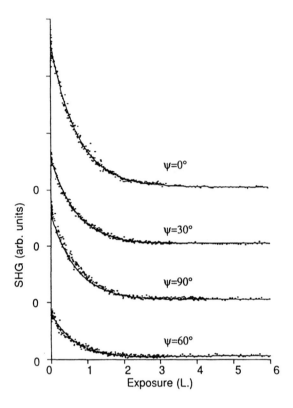

Fig. 9. Second harmonic intensity as a function of temperature during linear heating ramp from room temperature to 950°C at 7.5°C/min. Intensities observed using excitation wavelengths of 532 to 700 nm are plotted.

this wavelength. The SH intensity measured using 700 nm excitation increases considerably with the cosegregation of chromium/nitrogen, as discussed previously. The dramatic dependence of the SH intensity on wavelength indicates that resonant electronic transitions are responsible for the large signal from chromium/nitrogen seen using 700 nm excitation. Another major difference found with 532 nm excitation was the sensitivity of SHG to CO adsorption during the heating from 100 to 400°C. The CO adsorbed during the slow heating at these temperatures was removed by desorption (or dissociation and dissolution into the bulk) at higher temperatures. For 700 nm excitation, the adsorbed CO had little effect on the SH signal. For 532 nm excitation, the adsorbed CO reduced the SH signal dramatically. Since we did not study this effect quantitatively, we choose the surface at 900°C as our clean reference and did not attempt to draw conclusions from the portions of the curve below about 550°C. This strong SH intensity dependence on wavelength could provide species-specific detection using a tunable laser to excite resonant transitions of various surface species.

## SUMMARY

The effect of segregated chromium/nitrogen on the second harmonic generation from an Fe–18Cr–3Mo (110) surface has been observed using 532 and 700 nm excitations. In the case of chromium/nitrogen, resonance occurs for 700 nm excitation, and the square root of the second harmonic intensity is linearly related to the atomic fractions of chromium and nitrogen in the surface layers. This was in direct contrast to the case of adsorbed oxygen (studied but not reported here), for which the second harmonic intensity did not show major resonant effects.

Second harmonic generation is a promising surface diagnostic that will allow surface studies under ambient pressures, and studies of liquid–solid and solid–solid interfaces. Considerable work will be necessary under UHV conditions to allow proper interpretation of the electronic and structural information obtainable using this technique.

## ACKNOWLEDGMENT

This work was supported by the US Department of Energy, Office of Basic Energy Sciences, Division of Materials Sciences, under contract DEAC04-76DP00789

## REFERENCES

1. Tom, H. W. K., Mate, C. M., Zhu, X. D., Crowell, J. E., Heinz, T. F., Somorjai, R. A., and Shen, Y. R., *Phys. Rev. Lett.* **52**, 348 (1984).
2. Tom, H. W. K. and Aumiller, G. D., *Phys. Rev. B.* **33**, 8818 (1986).
3. Christmann, K., Behm, R. J., Ertl, G., Van Hove, M. A., and Weinberg, W. H., *J. Chem. Phys.* **70**, 4168 (1979).
4. Winkler, A. and Rendulic, K. D., *Surf. Sci.* **118**, 19 (1982).
5. Russell, J. N., Jr., Gates, S. M., and Yates, J. T., Jr., *J. Chem. Phys.* **85**, 6792 (1986).
6. Guyot-Sionnest, P., Chen, W., and Shen, Y. R., *Phys. Rev. B.* **33**, 8254 (1986).
7. Sipe, J. E., Moss, D. J., and van Driel, H. M., *Phys. Rev. B.* **35**, 1129 (1987).
8. King, D. A., *Surf. Sci.* **47**, 384 (1975).
9. Shiga, M. and Pells, G. P., *J. Phys. C.* **2**, 1847 (1969).
10. Greuter, F., Strathy, I., Plumer, E. W., and Eberhardt, W., *Phys. Rev. B.* **33**, 736 (1986).
11. Goldmann, A., Donath, M., Altmann, W., and Dose, V., *Phys. Rev. B.* **32**, 837 (1985).
12. Netzer, F. P. and Madey, T. E., *J. Chem. Phys.* **76**, 710 (1982).
13. Surnev, L., Xu, Z., and Yates, J. T., *Surf. Sci.*, in press (1989).
14. Conrad, H., Ertl, G., Kuppers, J., and Latta, E. E., *Surf. Sci.* **57**, 475 (1976).
15. Erley, W., Besocke, K., and Wagner, H., *J. Chem. Phys.* **66**, 5269 (1977).
16. Christmann, K., Schober, O., and Ertl, G., *J. Chem. Phys.* **60**, 4719 (1974).

# Investigations of Materials at High Temperatures Using Raman Spectroscopy

K. F. McCARTY

*Sandia National Laboratories, Livermore, CA 94551-0969*

## ABSTRACT

This paper will give a brief overview of the use of Raman spectroscopy to investigate materials at high temperatures. The first section describes the apparatus and experimental techniques necessary for *in situ*, high-temperature measurements. Following this, specific results are illustrated for three different systems: (1) materials formed from flame-deposition processes, (2) analysis of the oxides formed during the oxidation of metallic alloys, and (3) the high-temperature processing of oxide superconductors.

**Index Entries:** Raman spectroscopy; high temperatures; apparatus and experimental techniques.

## INTRODUCTION

Laser Raman spectroscopy is a powerful technique for studying materials at high temperatures. Materials can be examined in nearly any environment as long as there is optical access for an incident laser beam and the collection of scattered light. This requirement is not overly restrictive and permits *in situ* investigations of materials in environments such as high-temperature furnaces. Conditions can range from high vacuum to high pressure and from inert to highly corrosive atmospheres. Using modern instrumentation, Raman spectra can be obtained rapidly in times ranging from fractions of seconds to a few minutes. This permits the temporal evolution of a system to be examined, such as the growth of oxides on metallic alloys. Finally, a laser beam can be tightly focused, permitting microscopic examination of materials.

When light is incident upon a solid, a very small fraction is scattered at energies different from the incident energy. In Raman spectroscopy, this inelastically-scattered light is analyzed. The dominant mechanism of inelastic light scattering in solids is the interaction of the light with the lattice vibrations of the solid (1). Through this interaction, the Raman technique determines the energies of some fraction of the lattice vibrations of the solid, and these are referred to as the Raman-active phonons. Different chemical phases have distinct "phonon fingerprints" since different phases have distinct distributions of lattice vibrations. Raman spectroscopy exploits these unique fingerprints to determine chemical phase.

## EXPERIMENTAL

The experimental techniques required to perform Raman measurements of materials at high temperatures are somewhat more involved than measurements at ambient temperatures (1) owing to several additional concerns. Specifically, it is necessary to collect light efficiently while maintaining a large working distance between the sample and the optics, and the Raman-scattered radiation must be discriminated against the intense blackbody radiation emitted from high-temperature surfaces. Large working distances are required in order to: not overheat and damage the optics, and analyze materials remotely in equipment, such as furnaces, and vacuum chambers. The apparatus shown schematically in Fig. 1, specifically used to collect Raman spectra of deposits forming on a substrate immersed in the exhaust gases of a seeded burner (2), is a typical experimental setup. A highly-corrected lens (commercially available, although at some expense) allows light collection with an efficiency of f/2.2 while still maintaining a distance of about 20 cm between the laser spot and the front surface of the lens. The collected light is focused through a slit before finally being focused into the spectrometer. The intermediate slit serves to spatially filter the imaged laser spot from the blackbody radiation collected from surrounding regions.

When making measurements of materials at high temperatures, the blackbody radiation emitted by hot surfaces can overwhelm the relatively weak Raman signal. When using the 488- or 514.5-nm lines of an Ar-ion laser, analysis above about 700°C becomes difficult because of this problem. Using laser lines of shorter wavelength, e.g., the 457.9-nm line of an Ar-ion laser, allows measurements at slightly higher temperatures. Analyses at much higher temperature are possible using pulsed lasers and gated detection. For example, copper-vapor lasers produce nominally 20-ns pulses at repetition rates of about 5 kHz. In gated detection, the detector is only active for the duration of the laser pulse. Because of the much higher peak power available from pulsed lasers than CW lasers, the intensity of the (pulsed) Raman signal is increased compared to the

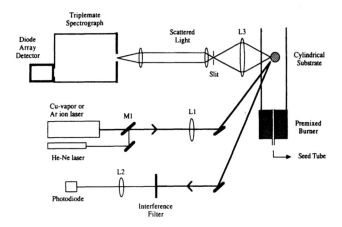

Fig. 1. Schematic illustration of experimental apparatus used for collection of Raman spectra in a high-temperature environment e.g., from deposits forming in the exhaust gas of a seeded flame. Large aperture lenses allow efficient (f/2.2) collection of light at a large working distance (>20 cm), whereas multichannel detection allows real-time measurements. Deposit thickness can be determined by interferometric techniques by monitoring the intensity of a laser beam reflected from the substrate.

(continuous) blackbody radiation by a factor approaching the duty cycle of the laser (about $10^4$ for a copper-vapor laser). Thus, the amount of blackbody radiation detected can be reduced to negligible levels compared to the Raman radiation. Quantitative estimates of the advantage of gated detection are given in ref. (3).

The advantage of pulsed-laser excitation coupled with detection by a gated diode array is illustrated in Fig. 2 for $Na_2SO_4$ being deposited from a flame seeded with $Na_2SO_4$ particles. In this example, the principal background interference was the radiance of the incandescent $Na_2SO_4$ particles and the emission from volatilized sodium. For nongated detection, the peak at 985 cm$^{-1}$, indicative of the sulfate deposit, is barely discernable above the intense background; gating the detector eliminates this background, allowing the signal to be clearly observed.

Applications of Raman spectroscopy to the solid state typically involve measurements of the lattice vibrations. Since the energies of these vibrations are small, on the order of 100 cm$^{-1}$, the inelastically-scattered radiation must be filtered efficiently from the much more intense elastically-scattered laser light. Commercial instruments utilize two or three gratings for thorough filtering. Finally, real-time measurements are facilitated by the use of spectrographs equipped with multichannel detectors, allowing the accumulation of spectra in the order of fractions of seconds to minutes (4). Accumulation times are limited usually by the practical concern of avoiding excessive temperature rises or damaging the sample. The "acceptable" power density is system specific, varying greatly with the material being analyzed and its environment, type of laser (pulsed vs CW), exciting wavelength, and so on.

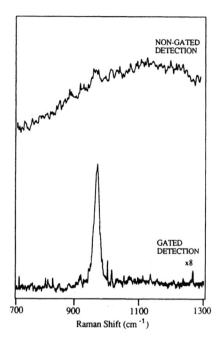

Fig. 2.   Illustration of the use of pulsed laser and gated detection to obtain Raman spectra in highly luminescent environments. Spectra obtained from a $Na_2SO_4$ deposit (peak at 985 cm$^{-1}$) growing on a substrate immersed in the exhaust gases of a $Na_2SO_4$-seeded, $CH_4$/air flame. In the upper spectrum, the detector was continuously on, whereas in the lower spectrum, the detector was active *only during the duration of the approximately 20-ns laser pulses.

## EXAMPLES OF THE ANALYSIS OF MATERIALS AT HIGH TEMPERATURES USING RAMAN SPECTROSCOPY

### Flame Deposition

The combustion of fossil fuels frequently results in the formation of inorganic deposits on the surfaces of combustion equipment. By impeding heat transfer processes and causing corrosion (often catastrophic), deposits result in material failure and costly loss of efficiency. Sodium sulfate ($Na_2SO_4$), a common deposit, is produced from organically and inorganically bound sodium and sulfur of both petroleum and coal-derived fuels (5). $Na_2SO_4$ deposits can react with contaminants containing iron, vanadium, potassium, and chromium to form low-melting-point eutectic mixtures. The formation of these mixtures can transform what would otherwise be a benign solid into a highly corrosive liquid deposit.

Figure 3 illustrates the Raman spectra obtained during the exposure of a typical high-temperature alloy (Incoloy 800 at 550°C) to the exhaust gases of a $Na_2SO_4$-seeded flame (2) The first spectrum, taken before the seed was turned on and labeled 0 min, indicates only the presence of

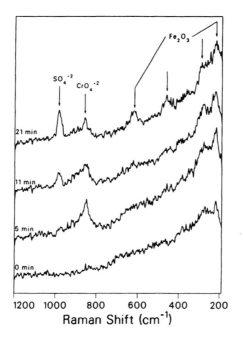

Fe₂O₃

SO₄⁻²  CrO₄⁻²

21 min

11 min

5 min

0 min

1200  1000  800  600  400  200

Raman Shift (cm⁻¹)

Fig. 3. Raman spectra obtained during exposure of Incoloy 800 substrate (550°C) to Na₂SO₄-seeded flame. The deposition of Na₂SO₄ on the substrate destroys the protective chromium oxide layer of the Incoloy 800 by forming chromate salts.

metal oxidation products, most notably, the 220 cm$^{-1}$ band of $Fe_2O_3$. After exposure to the seeded flow for only 5 min, a new broad Raman band appears in addition to the 985 cm$^{-1}$ band of sulfate. This new band, centered at 850 cm$^{-1}$, results from chromate anions ($CrO_4^{-2}$) and is initially stronger than the sulfate band. The deposit grows with time and after 21 min, the sulfate band is more intense than the chromate band. The interaction of the sulfate deposit and the chromium-containing alloy is readily explained by the basic fluxing mechanism of hot corrosion (6). The decomposition of sulfate anions results in the formation of $SO_3$ and oxide anions

$$SO_4^{-2} \rightarrow SO_3 + O^{-2}$$

The oxide anions, being strongly basic, react with the chromium oxides of the oxidized alloy to form chromate anions

$$\text{Chromium oxides} + O^{-2} \rightarrow CrO_4^{-2}$$

This basic fluxing prevents the alloy from forming a protective chromia ($Cr_2O_3$) layer and can lead to severe erosion of the substrate. Here, the formation of chromate has been observed to be rapid even at temperatures far below the melting point of $Na_2SO_4$ (884°C).

In addition to $Na_2SO_4$, deposition from flames seeded with iron pyrite ($FeS_2$) (7) and boron (via $BCl_3$) has been studied (8,9). In the latter case, the rate of deposition was governed by the rate of reaction of gaseous boron precursors at the surface of the $B_2O_3$ deposit.

### Oxidation of Fe–Cr Alloys:
### In Situ and Raman Microprobe Studies

The initial stages of metal oxidation have been identified as being critical in determining the long-term oxidation resistance of high-temperature alloys (10). Raman spectroscopy is a powerful technique for determining the composition of oxide films formed during the oxidation of alloys, such as Fe–Cr alloys. The principal oxides of iron and chromium are the spinel solid solution ($Fe_{3-x}Cr_xO_4$ with $0 \leq x \leq 2$) and the corundum solid solution ($Fe_{2-x}Cr_xO_3$ with $0 \leq x \leq 2$). Since each of these phases is uniquely characterized by its phonon spectrum (11), Raman spectroscopy can be utilized to identify rapidly the phases formed during the oxidation of Fe–Cr alloys.

Oxidation studies of Fe–13Cr and Fe–24Cr alloys were performed at atmospheric pressure in a flowing-gas furnace equipped with optical access for an incident laser beam and Raman-scattered radiation (12). Well-polished alloy coupons were oxidized in either a $10^{-24}$ atm oxygen environment (generated by saturating a hydrogen flow with water vapor) or a $10^{-1}$ atm oxygen environment (generated by a flow of 7% $O_2$ in Ar). The furnace was heated to 700°C over the course of about 8 min and held at 700°C for 80 min. After this exposure, the furnace was shut off, and the composition of the oxide film was reexamined after the sample had cooled overnight. The Raman-scattered radiation, excited by 514.5-nm light from an Ar-ion laser, was dispersed by a Spex Triplemate spectrometer onto a multichannel diode-array detector. Spectra were obtained using 30-s integration times. The thickness of the oxide film was determined by optical interferometry using a photodiode to monitor the intensity of the reflected laser beam.

Figure 4 illustrates the evolution of oxide phases for an Fe–13Cr alloy oxidized in $10^{-24}$ atm partial pressure of oxygen. After about 1 min at 700°C (9 min of heating, including the temperature ramp), bands at 550 and 665 $cm^{-1}$ were present in the Raman spectrum, corresponding to chromia ($Cr_2O_3$) and spinel ($Fe_{3-x}Cr_xO_4$), respectively (11). Thus, both chromia and spinel are formed initially. With time, the spinel began to transform to chromia and after about 90 min, the film contained only chromia. Dramatic changes occurred in the oxide upon cooling from 700°C to room temperature (RT). The Raman band at 665 $cm^{-1}$ reappeared, indicating that spinel reformed in the oxide film as it cooled to RT. The identification of the transient spinel phase during the film growth and its reappearance upon cooling highlights the necessity of conducting in situ, real-time measurements in order to properly understand corrosion mechanisms.

The role of alloy grain boundaries in high-temperature oxidation was addressed by a Raman microprobe study (12). Results in the literature suggest that the grain boundaries of Fe–Cr alloys are paths along which chromium atoms rapidly diffuse from the bulk of the alloy to the growing oxide film (13). The rapid diffusion of chromium is critical to the formation of adherent and protective chromia ($Cr_2O_3$) films on the alloys. If an

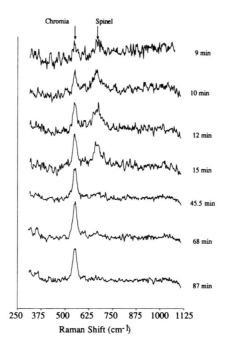

Fig. 4. Evolution of oxide phases during the oxidation of an Fe–13Cr coupon in $10^{-24}$ atm of oxygen at 700°C. Transformation of an oxide film initially containing both chromia ($Cr_2O_3$) and spinel ($Fe_{3-x}Cr_xO_4$) to a film containing only chromia is illustrated.

insufficient amount of chromium is available at the oxide interface, iron-containing, nonprotective oxides, such as spinel ($Fe_{3-x}Cr_xO_4$), form. Fe–13Cr alloys were oxidized in environments with oxygen partial pressures ranging from $10^{-30}$ to $10^{-20}$ atm at 700°C. For alloys oxidized in the middle of the range of oxygen partial pressure, dramatic differences were observed between the bulk and grain-boundary regions. Figure 5 shows the Raman spectra obtained by scanning an approximately 1-μm diameter laser beam from the interior of one alloy grain across the grain-boundary region and into the interior of the adjacent grain. Two different Raman peaks are observed. The 545-cm$^{-1}$ peak arises from chromia, whereas the 675-cm$^{-1}$ peak arises from spinel. Figure 6 illustrates the ratio of chromia to spinel as a function of distance from the grain boundary. It is clearly seen that chromia dominates the oxide film over the grain boundary, whereas spinel dominates the oxide over the bulk of the grain. Raman microprobe measurements have also been made at elevated temperatures in a controlled-atmosphere furnace. This capability allows the evolution of the oxide phases to be determined as functions of time and proximity to grain boundaries.

## High-Temperature Processing
## of Oxide Superconductors

The electrical properties of the new oxide superconductors, particularly $YBa_2Cu_3O_{7-x}$, depend critically upon the oxygen content and

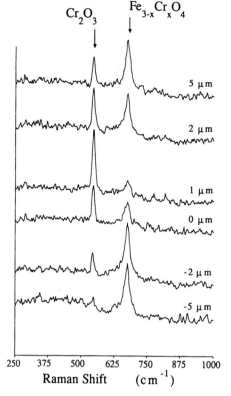

Fig. 5. Raman spectra obtained as a function of distance from the alloy grain boundary for an Fe–13Cr alloy oxidized for 105 min in $10^{-24}$ atm partial pressure of oxygen at 700°C. The peak at 545 cm$^{-1}$ results from chromia ($Cr_2O_3$), whereas the peak at 675 cm$^{-1}$ results from spinel ($Fe_{3-x}Cr_xO_4$).

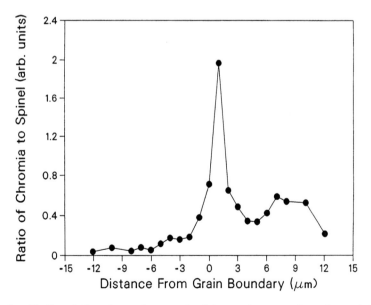

Fig. 6. Ratio of the chromia to spinel intensity as a function of distance from the boundary of a grain of an oxidized alloy of Fe–13Cr (*see* Fig. 5). The formation of chromia is greatly enhanced over the grain boundary, whereas spinel forms over the bulk of the grain.

26

thermal history of the material (14). The processing of the high-temperature superconductor, $YBa_2Cu_3O_{7-x}$, at elevated temperatures was investigated using *in situ* Raman spectroscopy (15). The Raman-active lattice vibrations of $YBa_2Cu_3O_{7-x}$ are highly sensitive to oxygen content. In fact, the frequency of a particular phonon is a sensitive indicator of oxygen content and, thus provides a contactless measure of the superconducting transition temperature ($T_c$) (16). The observed Raman peaks are dominated by Cu–O vibrations and, therefore, probe the structure responsible for superconductivity.

In order to determine directly the relationship between temperature, oxygen content, and phonon structure, Raman spectra were obtained from single-crystal $YBa_2Cu_3O_{7-x}$ at temperatures from 20 to 770°C in oxygen and from $YBa_2Cu_3O_{6.2}$ at temperatures from 20 to 545°C in argon. Figure 7 illustrates spectra obtained as $YBa_2Cu_3O_{7-x}$ is cooled in oxygen. A peak at $\sim570$ cm$^{-1}$ appears above $\sim500$°C, the temperature at which oxygen begins to leave the structure upon heating. The $\sim570$ cm$^{-1}$ peak is normally only IR active, but develops Raman intensity as oxygen vacancies occur in the structure. As oxygen is lost, the frequency of the vibration of the "bridging" oxygen decreases strongly from the RT value of 500 cm$^{-1}$, and the peak loses intensity. These changes result from the formation of oxygen vacancies at both the bridging oxygen sites and at the "chain" oxygen sites (17). There is a third, distinct type of oxygen in the structure, the oxygen of the Cu–O planes. Since few vacancies form at this site, the Raman peaks associated with vibration of this oxygen (the peaks at 330 and 435 cm$^{-1}$ at RT) change little in intensity and frequency upon heating.

High-temperature Raman measurements have also been made of Tl–Ba–Ca–Cu–O superconducting phases (18). Although the $T_c$ of these phases is influenced by variations in processing conditions (19), the variations are less dramatic than those of $YBa_2Cu_3O_{7-x}$. These smaller variations are largely the result of the relative lack of labile oxygen in the Tl-based structures compared to $YBa_2Cu_3O_{7-x}$. Figure 8 displays Raman spectra of one Tl-based phase, $TlCaBa_2Cu_2O_7$, as the crystal was cooled from 600°C to RT in oxygen. $TlCaBa_2Cu_2O_7$ is nearly isostructural with $YBa_2Cu_3O_{7-x}$ (20). In $YBa_2Cu_3O_{7-x}$, Cu replaces Tl and the locations of the O atoms in the planes of these cations are different; this results in the Cu–O "chain" sites in $YBa_2Cu_3O_{7-x}$, whereas the O atoms are centered between four Tl atoms in the Tl–O planes of $TlBa_2Cu_2O_7$.

In contrast with $YBa_2Cu_3O_{7-x}$ (*see* Fig. 7), there are no significant changes in the Raman spectra when $TlCaBa_2Cu_2O_7$ is heated in either oxygen or argon. With increasing temperature, however, the peaks do shift downward in frequency, an expected consequence arising from the anharmonic nature of the lattice. $TlCaBa_2Cu_2O_7$ could be reversibly cycled between RT and about 600°C in either $O_2$ or argon. However, at temperatures above about 650°C, irreversible changes occurred that are believed to result from degradation of the material resulting from Tl loss. Similar results were obtained from another Tl-based superconductor, $Tl_2CaBa_2Cu_2O_8$ (18). That is $Tl_2CaBa_2Cu_2O_8$ could be reversibly cycled

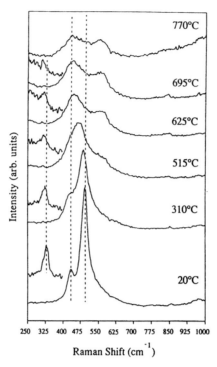

Fig. 7. Raman spectra of single-crystal $YBa_2Cu_3O_{7-x}$ during cooling from 770°C to RT in $O_2$. The estimated oxygen content is ~6.5 at 770°C and rises to ~7 at 450°C. Below 450°C, a constant oxygen content of ~7 exists. Absolute scaling of spectra of the same polarization is the same, with linear offsets for clarity.

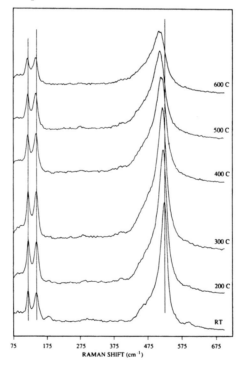

Fig. 8. Raman spectra of a $TlCaBa_2Cu_2O_7$ crystal during cooling from 600°C to RT in $O_2$. Unlike $YBa_2Cu_3O_{7-x}$ (*see* Fig. 7), $TlCaBa_2Cu_2O_7$ does not lose appreciable amounts of oxygen upon heating, and there are no significant changes in the phonon spectra.

28

between RT and ~600°C, and no significant changes in the Raman spectra were found upon heating in this temperature range. As for $TlCaBa_2Cu_2O_7$, however, irreversible changes occurred above about 650°C.

## SUMMARY

In many cases, it is desirable to measure the properties of materials at high temperatures rather than rely on postexposure analysis of cooled samples. Raman spectroscopy can be applied readily at high temperatures. The technique can determine the chemical phase of growing or evolving films, monitor phase changes remotely from hostile environments, and function as an online monitor for process control. In addition, materials at high temperatures can be examined by Raman-microprobe techniques, allowing the determination of phase with high spatial resolution.

## ACKNOWLEDGMENT

This work was supported by the US Department of Energy, Office of Basic Energy Sciences, Division of Materials Sciences, under contract DEAC04-76DP00789.

## REFERENCES

1. Long, D. A., *Raman Spectroscopy*, McGraw-Hill, New York, 1977.
2. McCarty, K. F. and Anderson, R. J., *Comb. Sci. Tech.* **54**, 51 (1987).
3. Lapp, M. and Penney, C. M., *Advances in Infrared and Raman Spectroscopy*, vol. 3, chapter 6, Clark, R. J. H. and Hester, R. E., eds., Heyden, London, 1977.
4. Chang, R. K. and Long, M. B., *Topics in Applied Physics*, vol. 50 Cardona, M. and Güntherodt, G., eds., Springer-Verlag, New York, (1982), p. 179.
5. Reid, W. T., *Combustion Technology*, chapter 2, Academic, New York, 1974.
6. Birks, N. and Meier, G. H., *Introduction to High Temperature Oxidation of Metals*, Edward Arnold, London, 1983, p. 149.
7. McCarty, K. F., Hamilton, J. C., Boehme, D. R., and Nagelberg, A. S., *J. Electrochem. Soc.* **136**, 1223 (1989).
8. McCarty K. F., Anderson, R. J., Nagelberg, A. S., and Lapp, M., *High Temp. Sci.* **23**, 75 (1987).
9. Seshadri, K. and Rosner, D. E., *AICHE J.* **30**, 187 (1984).
10. Whittle, D. P. and Stringer, J., *Phil. Trans. R. Soc. Lond.* A **295**, 309 (1980).
11. McCarty, K. F. and Boehme, D. R., *J. Solid State Chem.* **79**, 19 (1989).
12. McCarty, K. F., Hamilton, J. C., and King, W. E. in press.
13. Jensen, C. P., Mitchell, D. F., and Graham, M. J., *Corrosion Sci.* **22**, 1125 (1982).
14. Cava, R. J., Batlogg, B., Chen, C. H., Rietman, E. A., Zahurak, S. M., and Werder, D., *Nature* **329**, 423 (1987).
15. McCarty, K. F., Hamilton, J. C., Shelton, R. N., and Ginley, D. S., *Phys. Rev.* B **38**, 2914 (1988).
16. Hangyo, M., Nakashima, S., Mizoguchi, K., Fujii, A., and Mitsuishi, A., *Solid State Commun.* **65**, 835 (1988).

17. Jorgensen, J. D., Shaked, H., Hinks, D. G., Dabrowski, B., Veal, B. W., Paulikas, A. P., Nowicki, L. J., Crabtree, G. W., Kwok, W. K., Nunez, L. H., and Claus, H., *Proceedings of the International Conference on High-Temperature Superconductors and Materials and Mechanisms of Superconductivity*, Interlaken, Switzerland, Feb. 29–March 4, 1988.
18. McCarty, K. F., Morosin, B., Ginley, D. S., and Boehme, D. R., to be published.
19. Shimakawa, Y., Kubo, Y., Manako, T., Satoh, T., Iijima, S., Ichihashi, T., and Igarashi, H., *Physica C* **157**, 279 (1989).
20. Morosin, B., Ginley, D. S., Hlava, P. F., Carr, M. J., Baughman, R. J., Schirber, J. E., Venturini, E. L., and Kwak, J. F., *Physica C* **152**, 413 (1988).

# Raman Characterization of High Temperature Materials Using an Imaging Detector

GERD M. ROSENBLATT* AND D. KIRK VEIRS

*Lawrence Berkeley Laboratory, Materials and Chemical Sciences Division, 1 Cyclotron Road, Berkeley, CA 94720*

## ABSTRACT

The characterization of materials by Raman spectroscopy has been advanced by recent technological developments in light detectors. Imaging photomultiplier-tube detectors are now available that impart position information in two dimensions while retaining photon-counting sensitivity, effectively greatly reducing noise. The combination of sensitivity and reduced noise allows smaller amounts of material to be analyzed. The ability to observe small amounts of material when coupled with position information makes possible Raman characterization in which many spatial elements are analyzed simultaneously. Raman spectroscopy making use of these capabilities has been used, for instance, to analyze the phases present in carbon films and fibers and to map phase-transformed zones accompanying crack propagation in toughened zirconia ceramics.

**Index Entries:** Raman spectroscopy; imaging; carbon; zirconia, toughened.

## INTRODUCTION

Understanding the chemistry and molecular architecture that underly the physical properties of many materials requires chemical characterization of the material, i.e., characterization of the chemical composition, bonding, and thermodynamic phase. With films, fibers, and many high-tech materials, sample volumes may be small and a sensitive chemical characterization technique is required. For instance, thin films of pure carbon exhibit a wide variety of electrical, mechanical, and physical properties that depend on the microscopic morphology that may encom-

*Author to whom all correspondence and reprint orders should be addressed.

pass diamond, graphite, and a wide range of amorphous or intermediate structures. Although the two common crystalline forms of pure carbon, graphite and diamond, are easily distinguished in bulk quantities by X-ray diffraction, thin films of pure carbon are extremely difficult to characterize using X-ray diffraction because of small signals and, in the case of amorphous films, the absence of long-range order.

In some cases, macroscopic physical properties depend on the chemical nature of the material changing across small distances. In these cases, spatially resolved chemical characterization is needed. For instance, ceramics that are strong but also unusually resistant to brittle fracture are produced by the controlled use of martensitic transformations, i.e., at least partly irreversible transformations between two solid phases having the same chemical composition (1). The anticipated increase in toughness has been demonstrated for cubic zirconia ceramics containing precipitates of tetragonal and monoclinic $ZrO_2$. When a crack propagates in this material, the metastable tetragonal zirconia particles martensitically transform to the monoclinic phase in the high stress field associated with the crack tip, forming a transformed zone surrounding the crack and crack tip. This transformation is thought to be a significant factor in the toughening of these materials. Theoretical approaches to understanding the fracture resistance of these materials indicate that toughening depends on the zone width, the shape of the transformed zone around the crack tip, the extent of transformation within the zone, and the degree of irreversibility of the martensitic transformation. Understanding and optimizing the fracture resistance of such materials requires determining the phase composition in the transformed zone surrounding the crack with a spatial resolution of better than 50 $\mu M$.

For both of these examples, the distinguishing feature between the various forms of the materials is the difference in bonding (bond lengths, bond angles, and bond strengths) and not the chemical composition. This difference alters the fundamental atom–atom vibrations in the microcrystallites composing the material and can be detected using vibrational spectroscopy. Vibrational Raman spectroscopy is an optical technique in which scattered light is shifted in frequency by an energy equal to the energy difference between vibrational energy levels in the scattering material. The vibrational spectrum of a material is obtained by measuring the shift in energy of light scattered from a material illuminated with a monochromatic source. All materials have Raman-active vibrations with the one exception of a purely cubic phase.

The difficulty with Raman spectroscopy is that only a very small fraction of the incident light is Raman scattered. This means that it is not always easy to achieve reasonable signal:noise levels because of interference from other optical processes such as fluorescence or because of small signals in the case of small volumes. Recent advances in optical detectors have produced two-dimensional "imaging" detectors that have low noise and single-photon sensitivity. Such detectors, when carefully coupled with other recent advances in instrumentation, make it possible to apply Raman spectroscopy to materials problems that were formerly inaccessible to study because of the small sample volumes involved or the spatial resolution required.

In this paper, we summarize some of the characteristics of a Raman apparatus built in our laboratory that utilizes an imaging photomultiplier tube for materials characterization. The salient capabilities are low noise, high sensitivity, and the capability to obtain one-dimensional profiles of chemical phase from a single illumination without moving the sample or scanning the spectrometer. Moving the sample allows subsequent profiles to be combined to produce a two-dimensional map of chemical phases. The characteristics and capabilities of the approach are illustrated by summaries of recent experiments in our laboratory on carbon films and fibers and on the phase-transformed zone surrounding cracks in phase-stabilized zirconia.

## EXPERIMENTAL METHOD

The major problem with the use of Raman scattering for the characterization of materials is that it is a weak phenomenon; the Raman cross section for most materials is $10^{-31}$ cm$^2$ sr$^{-1}$. This means that a 100 mW laser beam that illuminates a typical solid sample with $2 \times 10^{17}$ photons s$^{-1}$ produces about $10^{10}$ Raman scattered photons s$^{-1}$ sr$^{-1}$; these must be collected and spectrally dispersed to yield useful information. Even with bright light sources (lasers) and sensitive detectors (photon-counting photomultiplier tubes (PMTs)), long integration times are necessary to obtain good signal-to-noise ratios (S/N). Advances in multichannel detectors, in particular the advent of the intensified photodiode array (IPDA), have increased the sensitivity of Raman spectroscopy. Using an IPDA coupled to a microscope, commercial Raman microprobe instruments have demonstrated utility in investigating a number of materials problems that require a single point analysis of a small volume (2). At the present time, a new generation of sensitive detectors is available that produces data resolved in two position dimensions (X and Y) in contrast to the one-dimensional resolution (X) of the IPDA. These new detectors, which include the imaging PMT and charge coupled devices (CCD), also have less background and hence lower noise characteristics than the IPDA. These sensitive and low noise detectors can be used to extend the capabilities of Raman spectroscopy to investigate small amounts of material and to obtain multiplexed, spatially resolved Raman spectra with a few microns spatial resolution.

The results presented here were obtained with a Raman spectroscopy apparatus incorporating an imaging PMT detector. This detector has several intrinsic features that allow a Raman spectrum to be obtained with both high sensitivity and multiplexed spatial resolution. First, the 25 mm diameter active region is partitioned into a $1024 \times 1024 \times \pi/4$ data array (the photoresponsive region is a disk that inscribes a $1024 \times 1024$ square). The detector is oriented so that one dimension (X) corresponds to the wavelength dispersion of the spectrometer and the second dimension (Y) corresponds to distance along the sample. Second, the detector dark count which totals 30 counts per second (cps) at $-30°C$ is spread out over the $8 \times 10^5$ active pixels yielding a dark count rate per pixel of just $4 \times 10^{-5}$ cps. Thus, detector dark count is virtually eliminated as a noise source.

The Raman system has been described previously (3) and only the main features and major changes will be emphasized. The apparatus consists of a cw $Ar^+$ laser usually operated at 488.0 nm, focusing and collection optics, single spectrometer, imaging PMT detector, and microcomputer data acquisition. The Raman shifted light is isolated by a six-cavity interference filter that rejects the unshifted Rayleigh line by about six orders of magnitude while passing the region of interest with about 70% transmission. The interference filter used for Rayleigh-scattering rejection adds an additional (rapidly varying) wavelength response to the (normally slowly varying) response of the spectrometer and detector. If accurate determinations of intensity ratios using an apparatus of the type described here are needed, the instrument response should be measured and used to correct the raw data.

Data are collected in two distinct modes depending on whether or not a spatially-resolved profile is desired. When spatial resolution is not being sought, the laser is focused to a spot of about 50 $\mu M$ diameter on the sample using a single lens. The scattered light is collected and focused onto the entrance slit of the spectrometer. The vertical dimension data are summed over a defined active region (typically ca. 100 rows) while wavelength dispersed information is retained. In this case, the detector dark count per wavelength channel is given by the number of rows which are active ($\sim$100) times the dark count per pixel, approximately $3 \times 10^{-3}$ cps per wavelength channel. The S/N of solid samples is always dominated by noise owing to background signals from the sample. Many materials fluoresce strongly and special techniques not practical in our apparatus such as Fourier Transform Raman are needed to obtain their Raman spectra (4).

When spatial resolution is required the laser beam is expanded and then focused on the sample using a cylindrical lens. This produces a slit like illumination of 30 $\mu M \times 2.5$ mM, which is imaged onto the entrance slit of the spectrometer. The cylindrical lens is rotated to align the image with the entrance slit. Each row in the detector array, corresponding to one spatial element along the illumination line at the sample, contains a unique Raman spectrum that can be analyzed to determine the chemical composition within that spatial element of the sample. Computer memory restrictions in our apparatus dictate that the data be acquired in a 256k position array. When desired, this can be accumulated as a 256 wide $\times$ 1024 high matrix where the spatial dimension retains the high resolution. Because approximately 800 rows of the detector data array are illuminated, this configuration simultaneously yields 800 Raman spectra from 800 consecutive spatial elements along the sample; from these a 800-position chemical profile can be derived. The integration times are typically less than 400 seconds with a maximum detected total count rate of about $10^4$ cps which yields $5 \times 10^3$ counts over one row (256 wavelength channels) or an average of 20 counts per wavelength channel. The low noise characteristics of the detector allow such small signals to be obtained and analyzed; the detector contributes 0.01 total counts of background to the detected signal in one row over 400 s.

## RESULTS AND DISCUSSION

### Carbon Thin Films and Fibers

Graphite and diamond have quite different Raman spectra. Spear (5) notes that Raman spectroscopy is the most reliable technique for distinguishing between crystalline diamond and other various forms of carbon and suggests that a working definition for "crystalline diamond material" in vapor deposited thin films include having a Raman spectrum typical of crystalline diamond (Fig. 1). Detailed analysis of the Raman spectrum of amorphous carbon, e.g., relative peak intensities, linewidths and absolute peak positions, can be used to estimate physical and chemical properties such as the average microcrystalline domain size and the $sp^2/sp^3$ bonding ratio (6).

The Raman spectrum of carbon in the 1000 cm$^{-1}$ to 2000 cm$^{-1}$ range has been used to study the microscopic morphology of various forms of carbon. The data are corrected for the wavelength dependent intensity response that was calibrated using a tungsten intensity calibration lamp traceable to NBS. Single crystal graphite has a single, narrow Raman peak (the G-band) at 1575 cm$^{-1}$ and a disorder induced peak (D-band) at 1355 cm$^{-1}$ (7). Diamond's single, sharp Raman peak at 1332 cm$^{-1}$ is distinct from the D-band of graphite, Fig. 1. The ratio of the integrated intensities of the D-band to the G-band has been correlated with average microcrystalline domain size, $L_\alpha$, as determined by X-ray data (7) and transmission electron microscopy (8). The observed increase in the intensity of the D (disorder) band relative to the G (graphite) band with decreasing microcrystallite size may reflect that the D-band intensity arises primarily from carbon atoms near the crystallite surface while the G-band arises from bulk-like carbon atoms in the interior volume. Linewidths of the D-band have also been correlated with $L_\alpha$ (8,9). Shifts of the G-band to lower frequencies have been related to the $sp^3$ bonding fraction (10,11).

Theoretical models of amorphous carbon with differing percentages of threefold ($sp^2$) and fourfold ($sp^3$) coordinated atoms indicate that the addition of fourfold coordinated atoms produces a gradual transition in the vibrational spectra to lower frequencies rather than a mixture of distinct features typical of graphite and diamond (10). Richter et al. (11) modeled the frequency of the G-band of graphite using force constants for both $sp^2$ and $sp^3$ bonded carbon atoms and found the shift in the frequency to lower energies to be essentially linear with increasing numbers of $sp^3$ bonded carbon atoms.

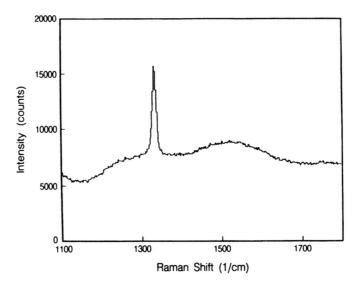

Fig. 1.   A single sharp Raman peak at 1332 cm$^{-1}$ is a distinguishing feature of diamond thin films. The weak, broad peaks extending from 1100–1600 cm$^{-1}$ are due to amorphous carbon, which has a much larger Raman cross-section.

Thin, sputtered carbon films ~20–40 nm thick used as protective overcoats on commercial 5½-inch computer hard disks have been investigated. The results are reported elsewhere (6), and the salient features will be summarized here. Samples from two manufacturers designated as K and L were obtained; sample K2 (450°C for 10 h) and K5 (350°C for 0.5 h) were annealed in vacuum. Raman spectra between 1000 and 2000 cm$^{-1}$ were collected for four samples using 6.5 mW of laser power and a 1000 s integration time, Fig. 2. The spectra were fit to the sum of a linear background term and two damped oscillator line shapes. Each plot in Fig. 2 contains the corrected experimental data and three curves calculated from the fitted parameters. Two of the curves are calculated from the parameters representing the G and D bands, individually, and the third curve is the sum of the calculated background, D, and G bands. The standard error of the fit, defined as the RMS average of the differences between the fit and the data divided by the data, ranged from 0.027 to 0.022 for all samples. Wavelength calibration was achieved by fitting 17 Ne emission lines resulting in a 0.33 cm$^{-1}$ standard error (less than one pixel).

From the fitting parameters one can calculate the peak position and linewidth of both the D and G bands, as well as the integrated intensity ratio, $I_d/I_g$. The results are shown in Table 1. The observed ratios $I_d/I_g$ are considerably larger than one. Available literature data (7) on the correlation of this ratio with domain size indicate that values close to or greater than one are always associated with graphite microcrystallite domains less than 5 nm across, but those data are limited to bulk samples having $I_d/I_g \leq 1$. Ratios considerably larger than one are routinely observed in thin films (8,11). The correlation with domain size in fine-grained films

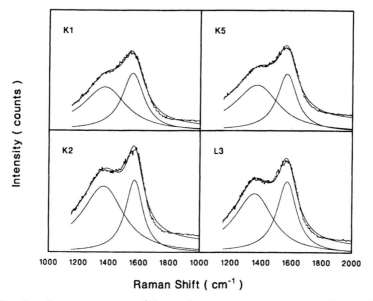

Fig. 2.   Raman spectra of the carbon overcoats on magnetic media disks from two different manufacturers. Samples K1, K5, and K2 were sectioned from the same disk and annealed at 30, 350, and 450°C, respectively. The data and fit, comprised of two peaks and a background (not shown), are shown for the four samples.

Table 1
Positions, Linewidths, and Intensity Ratios of the Graphite
and Disorder Induced Raman Bands of Carbon Thin Films Obtained
from Fitting the Observed Data to a Theoretical Model[a]

| | | G-Band | | D-Band | | | |
|---|---|---|---|---|---|---|---|
| Sample | Anneal T °C | Position cm$^{-1}$ | Linewidth cm$^{-1}$ | Position cm$^{-1}$ | Linewidth cm$^{-1}$ | $I_d/I_g$ | % sp$^3$ |
| K1 | — | 1562 | 176 | 1393 | 347 | 1.47 | 5 |
| K5 | 350 | 1571 | 160 | 1391 | 349 | 1.72 | 2 |
| K2 | 450 | 1575 | 140 | 1386 | 335 | 2.16 | 0 |
| L3 | — | 1575 | 159 | 1372 | 300 | 1.54 | 0 |

[a]The percentage of sp$^3$ bonding is obtained from the shift of the G-band.

has not been measured, but measurements (8) of $I_d/I_g$ as a function of annealing temperature suggest that for very small grains (<2 nm) $I_d/I_g$ increases with grain size (rather than decreases) to a maximum of 2.5 for 2 nm grains, followed by a decrease in $I_d/I_g$ (similar to that observed with coarser carbon (7)) as the grains grow further. It appears reasonable to postulate that the large $I_d/I_g$ ratios in Table 1 correlate with grains smaller than 5 nm and that the grain size increases with increased annealing temperature (and also with increasing $I_d/I_g$) in these thin, fine-grained films. From comparison of the peak positions and widths, the micro-crystallites in L3 are larger (>2 nm) than those in the K series samples (<2 nm) even though the $I_d/I_g$ ratios for L3 and K1 are nearly equal.

For the K series carbon films, the Raman data suggest that these are composed of microcrystallites of less than 2 nm with 5% $sp^3$ bonds. Upon annealing, the $sp^3$ bond percentage drops essentially to 0% as the grains grow. In the L3 sample, there are very few $sp^3$ bonds and the microcrystals still are much smaller than 5 nm, but probably larger than in the K series. Diamond spectral peaks were not observed in any sample. However, the small grains and extensive $sp^3$ crosslinking of the K series films correlate with these having the "diamond-like" hardness required for the intended application.

Carbon fibers used to increase the strength of composite materials have much better separated D- and G-bands. Mechanical properties of polymer fibers have been related to their crystal quality as measured by X-ray diffraction (12). Graphite crystal quality is directly related to observables in the Raman spectra, namely the peak positions, linewidths, and intensity ratios of the 1355 (D) and 1575 (G) $cm^{-1}$ peaks.

Fibers annealed at 2400, 2600, and 2800°C have been examined and the $I_d/I_g$ ratio, line positions, and linewidths have been determined. The wavelength was calibrated using Ne and Ar discharge lamps and the Th lines from a hollow cathode lamp to a standard error of 0.4 $cm^{-1}$. The data were taken with 5 mW of 488.0 nm laser light incident on a fiber with an integration time of 500 s and a slit width of 121 $\mu M$. Representative Raman spectra are shown in Fig. 3. At laser powers above 200 mW the samples (a single carbon fiber) are destroyed by the focused laser beam.

The results are shown in Table 2. The fibers have narrower D and G bands, smaller $I_d/I_g$ intensity ratios, a D-band shifted by ca. 40 $cm^{-1}$ to lower frequency, and an unchanged G (graphite) band position relative to the carbon films. These changes are consistent with much smaller surface:volume ratios, and larger microcrystallites, ca. 7–13 nm, than the films. Heat treatment to 2600°C shifts both the D- and G-bands to higher frequencies by about 5 $cm^{-1}$ and decreases their linewidths by about 20% as compared to the 2400°C samples. Further heat treatment to 2800°C does not significantly change the line positions or linewidths. In contrast, heat treatment to 2800°C continues to change the intensity ratio. As discussed above, intensity ratios less than one have been correlated with the inverse of the domain size (7) yielding a linear relation. Using this relation, the microcrystallite domain size of the carbon fiber samples appears to nearly double on increasing the temperature from 2400 to 2800°C.

### Transformed Zones in Partially Stabilized Zirconia

As mentioned in the introduction, elucidation of the toughening mechanism in high tech phase-stabilized zirconia ceramics requires mapping the phase transformed zone surrounding cracks with a resolution of better than 50 $\mu M$. Under static loading a crack initiated in PSZ grows until a zone is formed around the crack tip where the tetragonal phase has transformed martensitically into the monoclinic phase. For the crack

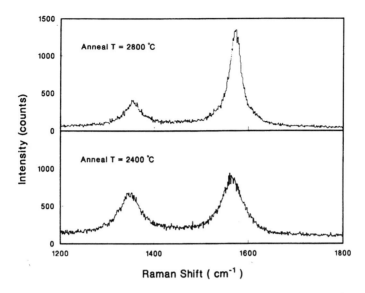

Fig. 3. The effect of annealing on single carbon fibers is seen in the Raman spectra. At higher annealing temperatures the G-line at 1575 cm$^{-1}$ grows at the expense of the D-line at 1350 cm$^{-1}$ and the linewidths decrease. This is an indication of graphite microcrystallite growth from 7 to 13 nm.

Table 2
Intensity Ratios, Domain Sizes, Line Positions and Linewidths
of Raman Peaks from Carbon Fibers Annealed at 2400°, 2600°, and 2800°C

| | $I_d/I_g$ | $L_a$ | G-Band | | D-Band | |
|---|---|---|---|---|---|---|
| | | | Position | Linewidth | Position | Linewidth |
| | | nm | cm$^{-1}$ | cm$^{-1}$ | cm$^{-1}$ | cm$^{-1}$ |
| CF2400 | 0.63 | 6.7 | 1565.6 | 38.6 | 1348.8 | 38.2 |
| CF2600 | 0.48 | 8.3 | 1572.8 | 32.0 | 1353.1 | 29.0 |
| CF2800 | 0.28 | 12.5 | 1570.3 | 29.4 | 1354.3 | 31.6 |

to grow further an increased static applied stress in necessary. The increased applied stress in turn produces a larger transformed zone and the crack again stops. Thus, under static loading, continuously increasing applied loads are necessary for continued crack growth. However, under cyclic loading, cracks in PSZ grow without cessation (13). Thus failure is somewhat analogous to fatigue-crack growth in metals. The mechanism of crack growth under cyclic load is not fully understood and is being investigated.

Raman spectroscopy utilizing the two-dimensional detector and techniques described above can be used to map the transformed zone (14). A semi-automated experimental setup yields the fraction of monoclinic phase relative to the sum of the monoclinic and tetragonal phases, determined from the Raman spectrum, as a function of distance across the surface of the specimen. First, a one-dimensional chemical profile of

800 elements is determined by analysis of the data resulting from a single 400 s slit-like illumination of a line across the sample. Second, the sample is translated a known amount perpendicular to the slit-like illumination and another one-dimensional chemical profile is determined. This process is repeated automatically and successive profiles are then combined to produce a map of the phase distribution.

The partially stabilized zirconia used for this study is composed of 50 $\mu M$ grains of cubic-phase zirconia with lenticular precipitates of tetragonal and monoclinic phases with a maximum dimension of 300 nm. The tetragonal-phase precipitates, which otherwise would transform martensitically to monoclinic at room temperature, are stabilized by the addition of 9 mol% MgO. The precipitates form 35–40% by volume of the cubic grains.

A Raman spectrum typical of PSZ used in these studies is shown in Fig. 4. To calculate a phase profile, the fraction of monoclinic is computed using the integrated intensities (minus the background) of the 181 and 192 cm$^{-1}$ peaks for the monoclinic phase and the integrated intensity (minus the background) of the 264 cm$^{-1}$ peak for the tetragonal phase. The monoclinic and tetragonal phases are assumed to produce the same relative Raman intensities; this assumption is consistent with literature relative Raman intensities (*15*) and yields results that are not inconsistent (within the large, perhaps a factor of two, uncertainty in the X-ray determination) with the composition of samples determined by X-ray diffraction. The relative Raman response of the two phases affects the scale, but not the topography of the compositional maps that are determined.

The application of spatially-resolved Raman spectroscopy to cracks grown in peak-toughened PSZ is described in detail in (*14*). That report describes results on a crack grown using a constant cyclic load. In subsequent work we have investigated a crack grown with a varying amplitude cyclic load, during fatigue-crack growth in a peak-toughness MgO-PSZ sample 3 mm thick (*16*). The crack-growth rate exhibits sharp changes which depend on changes in the magnitude of the load. Because the crack-growth rate depends on the transformed zone, detailed measurements of the transformed zone are required in order to understand the behavior of the rate of crack growth and the implications for failure mechanisms under varying load conditions.

The specimen was cyclically loaded at a load ratio of 0.1 ($K_{min}$ = 0.1 $K_{max}$; where $K_{min}$ and $K_{max}$ are, respectively, the minimum and maximum stress intensities in a given fatigue cycle; $K_{max}$ and $K_{min}$ are changed together in varying-load experiments) and frequency of 50 Hz in a high-resolution, computer-controlled electro-servohydraulic testing machine (*16*). Crack-growth rates, dA/dN, were determined during computer controlled applied stresses over a range of $\Delta K$ ($\Delta K = K_{max} - K_{min}$). Figure 5 shows the applied stress and the crack-growth rate as a function of crack extension. A constant high cyclic load is first applied ($\Delta K$ = 9.5 MPa$\sqrt{m}$ ) and the crack-growth rate increases toward a steady state value. On reducing the cyclic loads somewhat (to $\Delta K$ = 8.5 MPa$\sqrt{m}$ ), a

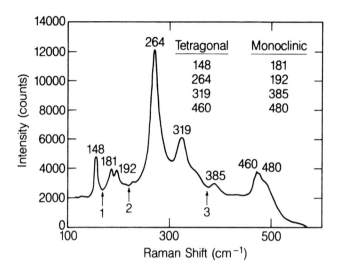

Fig. 4. The tetragonal and monoclinic phases of zirconia have Raman peaks at different frequencies that can be used to determine the relative amounts of each phase. In this work the 181, 192 cm$^{-1}$ pair is used to measure the monoclinic phase and the 264 cm$^{-1}$ peak is used to measure the amount of tetragonal phase.

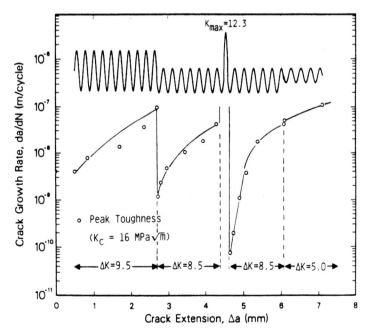

Fig. 5. The crack growth rate and applied stress (schematic) in toughened PSZ are shown as a function of crack extension for a fatigue-crack grown under varying amplitude cyclic loads.

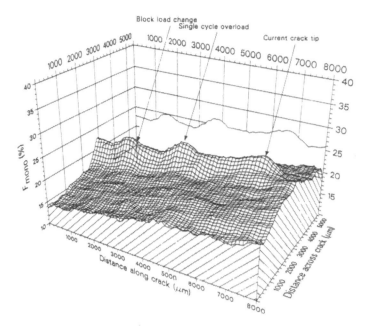

Fig. 6.    A map of the monoclinic phase fraction of the zirconia specimen whose applied stress and crack growth rate are shown in Fig. 5 indicates that the stress history of the material is revealed in the extent and degree of transformation of the transformed zone. Large stresses induce a larger transformed zone around the crack tip that remains after the crack tip moves forward.

marked transient retardation in crack velocity is seen followed by a gradual increase in crack-growth rate toward a (new) steady-state velocity. A significant retardation is seen following a single tensile overload to a $K_{max}$ of 12.3 MPa$\sqrt{m}$ ) (the fracture toughness of this material is $K_c = $ 16.0 MPa$\sqrt{m}$ )).

The Raman map of the phase distribution of this sample, Fig. 6, is constructed from 65 chemical profiles taken perpendicular to the crack every 125 $\mu$M. A laser power of 150 mW and an integration time of 200 s per profile were used. Calculation of the chemical profile from each illumination takes 132 s so the total time for data collection and analysis was 6 hours ( (200 s + 132 s) × 65). Further development of the data collection and computation algorithm now allow this same profile to be collected in less than 4 h (200 s × 65). The count rate was 32000 cps and the average count per pixel was 24. The transformation zone shows considerable changes in the zone width and extent of transformation which are related to the applied stress.

The crack-growth rate can be understood qualitatively by considering the size of the transformed zone and the magnitude of the applied stress. If the transformed zone is large the rate of crack growth (under a given cyclic applied stress) will be relatively small. Conversely, if the transformed zone is small, the crack-growth rate will be relatively large. The size of the transformed zone surrounding the crack tip is dependent upon the stress that has been applied. When the applied stress is lowered from $\Delta K$ = 9.5 MPa$\sqrt{m}$ to $\Delta K$ = 8.5 MPa$\sqrt{m}$ the crack tip is surrounded by a larger transformed zone associated with the larger applied stress, and the crack-growth rate is observed to decrease sharply. As the crack grows out of this large transformed zone to the smaller transformed zone associated with an applied stress of 8.5 MPa$\sqrt{m}$ , the rate of crack growth is observed to increase. A large zone associated with the application of a single overload of 12.3 MPa$\sqrt{m}$ caused a rapid decrease in the crack-growth rate followed by a gradual increase in growth rate as the crack tip moved out of the large zone. Changing $K_{min}$ without changing $K_{max}$ does not appear to change the crack-growth rate or the size of the transformed zone.

## SUMMARY

The applicability of Raman spectroscopy to materials—and to the characterization questions raised by high-technology, high-temperature and energy-related materials—has been extended by recent advances in photon detectors. The utilization of two-dimensional array detectors for Raman spectroscopy has increased the sensitivity of the technique by virtually eliminating the detector dark count as a source of noise and has resulted in the development of a new method for obtaining multiplexed, spatially-resolved Raman spectra. By using the spatial resolution of the detector to achieve spatial resolution at the sample, spectra from many spatially resolved elements at the sample can be collected and analyzed simultaneously. This simultaneous multiplexing of both wavelength and position makes use of the inherent low dark count per pixel of the detector and results in the rapid determination of a chemical composition profile. By translating the sample (in one direction) across the illuminating slit-like beam a map of the chemical composition of the sample surface can be determined. This technique for chemical mapping can be adapted to most forms of optical spectroscopy when an appropriate two-dimensional detector is used. This paper has illustrated some of these capabilities by describing results from on-going work on carbon films and fibers and on phase-stabilized zirconia ceramics.

## ACKNOWLEDGMENTS

This work was supported by the Director, Office of Energy Research, US Department of Energy, at Lawrence Berkeley Laboratory under contract DE-AC03-76SF00098. The authors wish to thank the following people and institutions: Prof. K. E. Spear of the Pennsylvania State University for supplying the diamond thin film, Prof. Fitzer and Dr. Kunkele of the Institute fur Chemische Technik der Universitat Karlsruhe for supplying the carbon fibers, Dr. D. Marshall of Rockwell International for supplying the partially stabilized zirconia, and Komag, Inc., and Lin-data Co. for supplying the rigid carbon-covered disks. The studies of carbon overcoats on rigid disks were in collaboration with Prof. D. Bogy of the University of California, Berkeley and Dr. H.-c Tsai, Dr. M. K.

Kundmann, Dr. M. R. Hilton, and Dr. S. T. Mayer. The studies of transformed zones in partially stabilized zirconia were in collaboration with Prof. R. O. Ritchie and Dr. R. H. Dauskardt of the Lawrence Berkeley Laboratory and the University of California, Berkeley.

## REFERENCES

1. Evans, A. G. and Heuer, A. H., *J. Am. Ceram. Soc.* **63**, 241 (1980).
2. Soto, L. and Adar, F., *Microbeam Analysis-1984*, Romig, Jr., A. D. and Goldstein, J. I., eds., San Francisco Press, 1984, p. 121; Huong, P. V., Boutinaud, P., Kasaoui, S., and Leycuras, A., *Microbeam Analysis-1988, Newbury, D. E., ed., San Francisco Press, 1988, p. 167, and others in that volume.*
3. Veirs, D. K., Chia, V. K. F., and Rosenblatt, G. M., *App. Opt.* **26**, 3530 (1987).
4. Chase, D. B., *J. Am. Chem. Soc.* **108**, 7485 (1986); Zimba, C. G., Hallmark, V., Swalen, J. D., and Rabolt, J. F., *Appl. Spec.* **41**, 761 (1987).
5. Spear, K. E., *J. Am. Ceram. Soc.* **72**, 171 (1989); Messier, R., Badzian, A., Badzian, T., Spear, K. E., Bachmann, P., and Roy, R., *Thin Solid Films* **153**, 1 (1987).
6. Tsai, H.-c., Bogy, D. B., Kundmann, M. K., Veirs, D. K., Hilton, M. R., and Mayer, S. T., *J. Vac. Sci. Technol.* **A6**, 2307 (1988).
7. Tuinstra, F. and Koenig, J. L., *J. Chem. Phys.* **53**, 1126 (1970).
8. Beny-Bassez, C. and Rouzaud, J. N., *Scan. Elec. Micros.* **1985**, 119 (1985).
9. Johnson, C. A., Patrick, J. W., and Thomas, K. M., *Fuel* **65**, 1284 (1986).
10. Beeman, D., Silverman, J., Lynds, R., and Anderson, M. R., *Phys. Rev. B* **30**, 870 (1984); Di Domenico, Jr., M., Wemple, S. H., Porto, S. P. S., and Bauman, R. P., *Phys. Rev.* **174**, 522 (1968).
11. Richter, A., Scherbe, H.-J., Pombe, W., Brzezinka, D.-W., and Muhling, I., *J. Non-Crystal Sol.* **88**, 131 (1986).
12. Day, R. J., Robinson, I. M., Zakikhani, M., and Young, R. J., *Polymer* **28**, 1833 (1987).
13. Dauskardt, R. H., Yu, W., and Ritchie, R. O., *J. Am. Ceram. Soc.* **70**, C248 (1987).
14. Dauskardt, R. H., Veirs, D. K., and Ritchie, R. O., *J. Am. Ceram. Soc.*, in press.
15. Clarke, D. R. and Adar, F., *Advances in Materials Characterization*, Rossington, D. R., Condrate, R. A., and Synder, R. L., eds., Plenum, p. 199 (1983).
16. Dauskardt, R. H., Marshall, D. B., and Ritchie, R. O., *Proceedings of the Materials Research Society International Meeting*, 1988.

# Spontaneous Raman Spectroscopy in Flames Containing High Concentrations of Silica Particles

M. D. ALLENDORF* AND R. E. PALMER

*Combustion Research Facility, Sandia National Laboratories, Livermore, CA 94551–0969*

## ABSTRACT

Spontaneous Raman spectra of nitrogen have been measured in hydrogen/methane/oxygen/nitrogen flames containing high concentrations (up to $10^9/cm^3$) of small ($<0.2$ μm diameter) silica particles. In contrast with carbon soot particles, which emit large quantities of broadband visible radiation at high temperatures, the addition of silica particles at these concentrations has little or no effect on the SRS measurement at temperatures up to 2100 K. The spatially-resolved, time-averaged SRS measurements reported here are thus the first obtained in the submicron, high number density regime, similar to that found in flame synthesis processes. They indicate that spontaneous Raman spectroscopy should be an effective technique for the measurement of temperatures and species concentrations in flames produced by burners used in the manufacturing of optical fiber.

**Index Entries:** Raman spectroscopy; flame synthesis; silica particles; optical fiber; light scattering.

## INTRODUCTION

Combustion processes are currently used by industry to produce large quantities of refractory oxides such as fumed silica, titanium dioxide, and uranium oxide (1). In addition to these bulk materials, communications-grade optical fiber is manufactured by the vapor axial deposition process (VAD), which uses a flame to decompose silicon and germanium chlorides and codeposit the resulting oxides on a preform (2). Combustion and other high-temperature gas-phase synthesis tech-

---

*Author to whom all correspondence and reprint orders should be addressed.

niques are also receiving increased attention as potential manufacturing processes for high-purity materials such as ceramic powders (3,4). These gas-phase synthesis techniques offer high purity and control over powder particle size, level of agglomeration, and other powder characteristics. Flame formation of oxides has also been used as a model experimental system for the study of the nucleation and growth of aerosols (5–7).

Increasingly, optimization and control of complex manufacturing processes such as flame synthesis are being approached through the development of computational models that predict production rates and materials properties as a function of key process variables. Verification of these models, as well as insight into the physical and chemical phenomena controlling synthesis, requires detailed knowledge of temperature, gas-phase species concentrations, particle size, and particle number density. Temperature is a particularly critical variable; in addition to being a monitor of the combustion process, gas temperatures exert a strong influence on the properties of synthesized materials through temperature-dependent materials properties such as viscosity. Diagnostic techniques developed to monitor the size of submicron particles in flames also require knowledge of the gas temperature. Accurate, spatially resolved temperature measurements in high-temperature, particle-laden flows can only be obtained by optical techniques, since the presence of large numbers of particles and high temperatures (>2200 K) preclude the use of thermocouples.

One technique that has been used extensively to obtain temperatures and concentrations in clean flames is spontaneous Raman spectroscopy (SRS) (8–14). SRS has several important advantages for flame measurements (13,15): It is applicable to a wide variety of species present in flames; is relatively straightforward to implement, requiring only a single fixed-frequency laser; and, unlike laser-induced fluorescence, it is not affected by collisional quenching. SRS also provides good spatial resolution, which is critical for some complex burner geometries used in optical fiber synthesis.

The most serious limitation of SRS, the low cross section for Raman scattering, has largely prevented the extension of the technique to sooting hydrocarbon flames. At flame temperatures, carbon soot particles produce considerable quantities of broadband background radiation (Mie scattering, fluorescence, and thermal radiation) that can prevent detection of the Raman signal. Although this problem can, in principle, be surmounted by either signal averaging or the use of pulsed lasers and gated detection techniques, the additional complication of laser-modulated incandescence (13,15), in which soot particles radiate at a temperature above that of the flame owing to absorption of the laser beam, prevents the use of SRS in flames containing high concentrations of carbon soot particles (>$10^9$/cm$^3$) (10,16). Most metal oxides, however, do not absorb significantly, if at all, in the visible portion of the spectrum. Consequently, these particles are expected to be poor black-body radiators and should generate little if any visible light, even at flame temperatures (nor will they be heated further by absorption of the laser beam).

Experiments in counterflow hydrogen/oxygen flames producing silica particles by oxidation of silane have demonstrated that this is indeed the case; the darkest part of the flame corresponded to the region of highest particle number density (5).

SRS has been successfully applied to arc-heated flows containing low densities ($\sim$10/cm$^3$) of 50- or 100-$\mu$m-diameter glass beads (16), indicating that it could be a useful technique for the measurement of temperatures in flames containing much higher numbers of transparent, submicron particles. In this article we report SRS measurements of temperature in the post-flame gases of lean hydrogen/methane/oxygen/ nitrogen flames containing up to $10^9$/cm$^3$ small ($<$150 nm diameter) flame-generated silica particles, conditions approaching those observed in flames used in the manufacture of optical fiber. To our knowledge, the data shown here are the first spatially-resolved SRS measurements obtained in the submicron particle size, high particle number-density regime typical of flame synthesis processes.

## METHODS

All measurements were made in the post-flame gases of $CH_4$/$H_2$/$O_2$/$N_2$ flames stabilized by a stainless steel honeycomb burner with dimensions 25 mm $\times$ 100 mm (Fig. 1). Nitrogen and oxygen were fed through the honeycomb, whereas fuel gases (with nitrogen added to match fuel/ oxidizer cold-gas velocities) flowed through tubes inserted into the honeycomb. Flowrates were maintained by calibrated mass flow controllers and were as follows (in liters/minute at STP): 0.25 $CH_4$, 5.14 $H_2$, 0.75 $N_2$ through the fuel tubes and 4.36 $O_2$, 11.8 $N_2$ through the honeycomb. These flowrates were chosen so that fuel and oxidizer cold-flow velocities were matched; this yields the flattest temperature profiles across the length and width of this type of burner. Although a lean gas mixture was used, resulting in excess oxygen (mole fraction 0.06) and substantial nitrogen (mole fraction 0.63) in the post flame gases, the exact amounts present were not known since the absence of a chimney (which would rapidly become coated with white silica particles) allowed room air to be entrained by the flame. Silica particles were formed by the decomposition of hexamethyldisiloxane ($\{(CH_3)_3Si\}_2O$; HMDS) vapor, fed to the flame by bubbling nitrogen through liquid HMDS at room temperature. The HMDS-saturated nitrogen gas was premixed with the flame oxygen and entered the flame through the burner honeycomb.

Figure 1 shows the experimental layout for SRS measurements. A CW argon-ion laser (Coherent Innova 100) producing 8.25 W at 514.5 nm was used to generate the Raman signal. Following rotation of the plane of polarization to vertical using a $\lambda$/2 plate, the beam entered a retro-reflecting multi-pass cell of the type described by Hill (17). The beam diameter was reduced to 1.0 mm using a Galilean telescope prior to entering the cell. Typically, the laser beam traversed the cell 10 to 13 times; after optical losses, this increased the average power in the sample volume to 60–75 W. Spatial resolution obtained with this cell was excel-

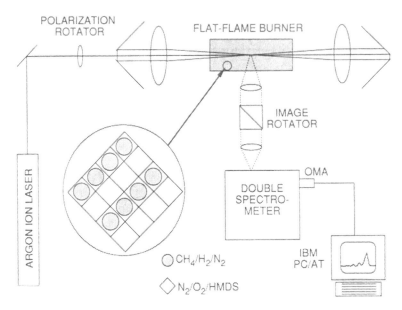

Fig. 1.   Schematic of the experimental configuration used for spontaneous
Raman spectroscopy (SRS) measurements. A detailed view of the burner surface
is shown in the inset.

lent; the illuminated sample volume in the direction of the flame gas flow
(determined by decreasing the width of the monochromator slits until
the intensity of the room temperature nitrogen signal began to decrease)
was estimated to be 75 μm. Resolution along the beam path is determined by the 1.0-cm slit height; since the magnification of the collection
optics was 3.3, the length of sample volume actually imaged was 3.0 mm.

Scattered Raman light was collected by a 50-mm diameter f/2.4
achromatic lens at 90° to the laser beams and focused onto the monochromator slits (after rotating the image parallel to the slits) using a
second 50-mm-diameter f/8 achromatic lens. A SPEX 0.85-m double
monochromator dispersed the spectrum; the entrance slit width was 300
μm. The signal was detected by a 1024-element, intensified diode array
(Princeton Instruments), controlled by an IBM PC/AT. Raman spectra
were corrected for background flame radiation by subtracting the spectrum recorded with the laser off from that obtained with the laser on. The
typical integration time required to record both spectrum and baseline
was about two minutes. After background subtraction, all spectra were
corrected for the diode-to-diode response of the OMA by dividing by the
white light OMA response function.

SRS has been used for many years to determine flame temperatures
(*14*). In our experiments, we chose to use the "quick-fit" approach of Hall
and Boedeker (*18*), modified for SRS. Theoretical spectra were calculated
using a code developed at Sandia for CARS (*19,20*) and modified to
generate the imaginary part of the susceptibility for SRS. These spectra
were convoluted with the experimental slit function to generate library
spectra, spaced 50 K apart. A second program (*20*) was used to interpo-

late between library spectra to obtain a best fit to the experimental data. In addition to adjusting the temperature by interpolation between library spectra, the code adjusts the horizontal shift, wave number expansion, and intensity expansion of the data, as well as a vertical shift, to determine the best fit. The vertical shift is added to the theoretical spectrum to correct for experimental offsets and noise in the data. For the data shown here, only the (0-1) and (1-2) vibrational bands were fit.

Particle sizes were determined by the dynamic light scattering technique (DLS), as described in reference (21). Briefly, light scattered by particles in Brownian motion produces photocurrent fluctuations when imaged onto a photomultiplier. The Fourier transform of this time-dependent signal yields a Lorentzian line centered at zero frequency whose half-width is related to the particle diffusion constant. The nature of the fluid flow determines the details of the relationship between the particle diameter and its diffusion constant. Particles in these flames are in the transition regime between free-molecular and continuum flow ($1 < \lambda/d < 10$, where $\lambda$ is the mean free path of the gas molecules and d is the particle diameter). The relationship between the diffusion constant $D$ and the particle diameter $d$ is (22)

$$D = (kT/3\mu\pi d)[1 + (2A\lambda/d)(1 + B\exp(-Cd/2\lambda))] \qquad (1)$$

where k is Boltzmann's constant, $T$ is the temperature, $\mu$ is the viscosity, and $A$, $B$, and $C$ are constants equal to 1.257, 0.318, and 1.10, respectively. Equation (1) was solved iteratively to obtain $d$; temperatures determined from SRS measurements (*see below*) were used in the calculation. The gas flowrates used yield maximum hot-gas velocities that are sufficiently small that the particle residence time in the scattering volume is long compared to the diffusion time, allowing the transform of the scattering signal to be fit by a Lorentzian line shape. DLS experiments were performed using the argon-ion laser used for the SRS measurements (without the multipass cell) and collecting scattered light at angles to the incident beam between 10 and 17°. The scattered light was focussed onto a photomultiplier tube and the resulting signal analyzed by a Scientific-Atlanta FFT spectrum analyzer (spectral range 0–300 kHz). Although the spatial resolution obtained by this near-forward scattering geometry was poor (10–17 mm of the beam was imaged onto the detector), flame conditions were sufficiently uniform in the measurement volume to permit direct correlation of particle size profiles with the SRS temperature profiles, which were obtained with much higher spatial resolution.

Particle number densities were obtained from measurements of the particle scattering intensity at a fixed angle $\theta$, which is proportional to the number of particles and the particle scattering cross section (21):

$$V = CI_o\sigma(d,\theta)\Omega N \qquad (2)$$

$I_o$ is the intensity of the laser beam and $\Omega$ is the collection solid angle. The single particle scattering cross section, $\sigma(d,\theta)$, is calculated (23) from the particle size determined by the DLS measurements. The particle number density $N$ was thus determined by measuring the absolute scattering from particles at 90° to the incident beam, using the voltage $V$ produced by a photomultiplier detecting the scattered light. A proportionality constant, $C$, which is a function of the collection optics and detection system, is determined by measuring the scattering from atmospheric nitrogen (for which the concentration of scattering centers is known). Calculations of $d$ and $N$ using Eqs. (1) and (2) assumed that a mono-disperse distribution of spherical particles was present in the flame.

## RESULTS AND DISCUSSION

The flame produces small amounts of background radiation in the absence of particles; the post-flame gases are red-orange up to ~2.5 cm above the burner surface. As expected, there is no visible increase in flame radiation upon introduction of HMDS to the flame, although it was necessary to record the Raman baseline in the presence of particles to achieve an accurate subtraction. In the region closest to the burner surface (5–7 mm above the burner), an increase in the background luminosity of ~20% was observed when HMDS was introduced into the flame. Little or no change was observed above this point in the flame. The laser beam becomes clearly visible, however, when particles are present.

For purposes of comparison with other experiments performed in flames containing silica particles, profiles of particle size and number density as a function of height above the burner were measured, using DLS to determine particle diameter and absolute scattering measurements to obtain particle number densities. The increase in particle diameter with height above the burner surface (Fig. 2a) agrees with that observed by Flower and Hurd (24) in a similar system. Comparison of the number densities (Fig. 2b) with other silica-producing systems reported (25,26) indicates that conditions produced by this burner are similar to counterflow flames that have been studied (25) but contain ~100 times fewer particles per unit volume than have been found in burners used in the optical fiber manufacturing industry (based on $SiCl_4$ flowrates given in (26) and assuming 100% conversion to $SiO_2$). The decrease in particle number density with increasing residence time in the flame is consistent with coalescence and agglomeration of particles; silica particles sampled in regions where the flame temperature is below the melting point of silica are found to be long-chain, highly branched clusters (6). Thus, the particle sizes and number densities quoted here are quantitative only to the extent that the particles are spherical and the size distribution is monodisperse. Since the Rayleigh cross section, upon which these two optical techniques rely, is proportional to $d^6$, the observed particle size is heavily weighted toward large particles. It is, therefore, possible that the true mean diameter may be as much as a factor of two smaller than the DLS measurements indicate (21).

A typical background-corrected nitrogen Raman spectrum is shown in Fig. 3. The (0-1), (1-2), and (2-3) nitrogen Q-branches are clearly

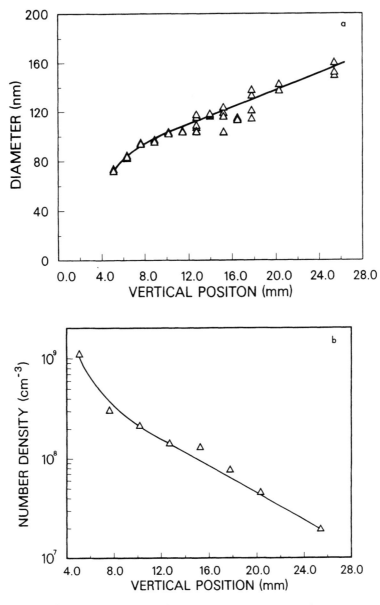

Fig. 2.   Profiles of silica particle diameter (a) and number density (b).

visible. Also shown is the instrument slit function (dashed line), determined by recording the spectrum of Rayleigh-scattered light with the flame on, in the absence of particles; the full-width at half-maximum of the instrument slit function is 4.0 cm$^{-1}$. It is clear from the data in Fig. 3 that the presence of as many as 10$^9$ silica particles/cm$^3$ does not seriously affect our ability to obtain high-quality Raman spectra. Data obtained in the presence of particles are only slightly noisier than data obtained in the clean flame. No broad-band radiation from fluorescence or laser heating of the particles was evident. The large signal owing to Rayleigh scattering of the laser beam by the particles appears to have been completely rejected by the spectrometer. Since the Rayleigh scattering signal

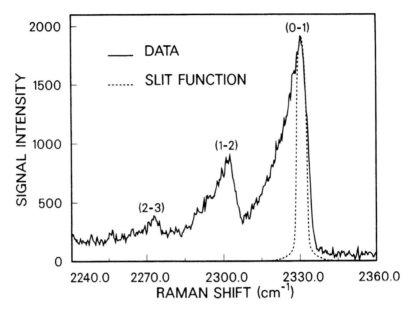

Fig. 3. A typical nitrogen SRS spectrum obtained 12.5 mm above the burner in the presence of silica particles. Numbers in parenthesis above the peaks indicate the vibrational transition. The dashed line is the corresponding instrument slit function (monochromator entrance slit width of 300 μm).

at 514.5 nm is $\sim 10^{10} - 10^{12}$ times larger than the Raman signal, the superior stray light rejection of a double monochromator $(10^{-14})$ is required to make this rejection possible without prefiltering. (Typical rejection ratios for single monochromators are only $10^{-4} - 10^{-7}$.) These results are in contrast to SRS measurements in hydrocarbon flames containing similar concentrations of submicron soot particles (16), in which Raman data could not be obtained owing to intense background radiation from particles and laser-modulated incandescence.

It is interesting to compare the amount of thermal background radiation produced by carbon soot particles with the radiation increase observed in these experiments upon introduction of silica to the flame. The radiation energy density produced by soot particles has been calculated as a function of temperature (13,15); for $10^8/cm^3$ 40-nm diameter carbon soot particles at 2000 K, ~250 nW/cm³/nm/sr is emitted at 580 nm (13) (the location of the nitrogen Raman spectrum for 514.5-nm excitation). To estimate the amount of background owing to silica particles in the flames under study here, we apply the following expression for the Raman signal/interference ratio (13):

$$S/I = P(\partial\sigma/\partial\Omega)N\delta f/R(\Delta\lambda)V]  \qquad (3)$$

Using the measured value of the S/I ratio, the background radiation energy density $R$ can be calculated from Eq. (3). Under typical experimental conditions, the power $P$ of the scattering laser in the illuminated sample volume was 65 W. The detection bandwidth $\Delta\lambda$ was 0.14 nm (corresponding to the 4-cm$^{-1}$ full-width half-maximum of the slit function), yielding a bandwidth factor $f$ of 0.05 (*13*). The length of the imaged sample volume, $\delta$, was 0.30 cm. The collection volume $V$ was assumed to be a rectangular slab with dimensions $\delta w(2\delta f_\#)$, in which $w$ is the magnification-corrected slit width (90 $\mu$m) and the $f$-number $f_\#$ of the collection lens is 2.4. The Raman scattering cross section $\partial\sigma/\partial\Omega$ of nitrogen for 514.5-nm excitation is $4.3 \times 10^{-31}$ cm$^2$/sr (*13*), and the nitrogen concentration $N$ at 2000 K for a mole fraction of 0.63 is $2.31 \times 10^{18}$/cm$^3$. At 13 mm above the burner (temperature = 2050 $\pm$ 75 K), the number of particles observed is $\sim 10^8$/cm$^3$ with equivalent sphere diameter on the order of 100 nm. Using the S/I ratio measured at this location of $\sim 0.5$, a value of R of 3.6 nW/cm$^3$/nm/sr is calculated for the flame without particles. An increase in the background radiation of 5% or less over the clean flame is observed at this point when particles are present, yielding a value of $R$ for the particles of 0.2 nW/cm$^3$/nm/sr. Carbon soot particles in the submicron size range thus produce approximately 1300 times more visible thermal radiation at 2000 K than silica particles of comparable size. It is likely that this factor represents a lower limit, since we could not distinguish between radiation produced by particles and any chemiluminesence that could result from the combustion of HMDS (*see below*). This demonstrates that the injection of up to $10^9$/cm$^3$ of transparent particles into a flame or other hot environment should not prevent the use of time-averaged SRS to measure flame temperatures or concentrations.

We have applied SRS to determine flame temperatures in the presence of silica particles. Spectra obtained in these experiments were not shot-noise limited; the primary source of noise is fluctuations in flame radiation. For this reason, a rigorous statistical treatment of fitting errors was not applied. Uncertainty in the fit, as determined by the amount the temperature can be changed without significantly affecting the fit (based on visual inspection and the value of the variance), was typically $\pm 75$ K, with slightly higher uncertainty at points below 8 mm ($\pm 100$ K). A typical theoretical fit, using the data and slit function shown in Fig. 3, is shown in Fig. 4. The agreement between theory and experiment is good.

A temperature profile as a function of vertical height above the burner for the flame with the silica particles is shown in Fig. 5. Measurements were not made below 5 mm owing to vignetting of the signal by the burner. The adiabatic flame temperature of 2224 K is indicated by the horizontal dashed line. Error bars indicate the range of fitted temperatures for five sets of data; measurement precision was poorest at points low in the flame (below 8 mm), owing to the somewhat higher flame emission observed in this region. The profiles show that the flame temperature is fairly constant up to $\sim 15$ mm above the burner, with temperatures 100–150 K below the adiabatic flame temperature, consistent with some heat loss to the burner surface. Above 15 mm, the temperature drops as room air mixes with the flame. Owing to the absence of a chimney, room air currents introduced substantial flicker into the flame at heights above 18 mm. Spectra measured in the presence of flame

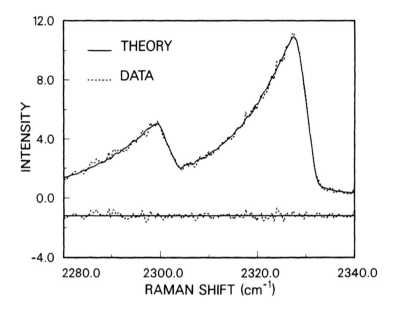

Fig. 4.    A theoretical fit to the data shown in Fig. 3, using the interpolation computer code described in the text. The difference between theory and data is shown on the horizontal line below the spectra. The temperature obtained from this fit was 1933 K.

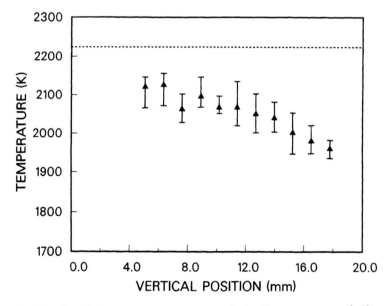

Fig. 5.    Profile of flame temperature made in the presence of silica particles. The dashed horizontal line indicates the adiabatic flame temperature. Error bars show the range of fitted temperatures obtained from five sets of data.

flicker are a combination of high and low temperatures and thus could not be fit reliably. Comparison of these profiles with temperatures found in the absence of silica particles (not shown) suggests that the temperature increases by approximately 50 K when HMDS is added to the flame. Although some increase would be expected since the addition of HMDS moves the fuel/oxidizer ratio closer to the stoichiometric value, this change is smaller than the uncertainty limits of our measurement.

Particle loadings in methane/oxygen flames employed in optical fiber manufacturing are estimated to be as much as 100 times higher than those obtained here; since particle thermal radiation should scale with the number of particles, thermal background radiation from particles will be on the order of 20 nW/cm$^3$/nm/sr. Additional broadband radiation has been observed when silica precursors such as silicon tetrachloride are added to VAD flames (27) and hydrogen/oxygen counterflow flames doped with silane (5). In the latter case, its location in the flame does not correspond to the maximum of the particle number density. A chemiluminescent reaction of a gas-phase silicon species with oxygen prior to the formation of particles may be partly responsible (28). This luminosity is not evident in the flames studied here, perhaps because the reaction zone is virtually on the burner surface. These luminous conditions will preclude measurements using CW lasers and detection techniques, requiring instead the use of pulsed lasers and gated signal detection. As discussed earlier, silica particles do not absorb visible light to any significant extent, so that laser-modulated incandescence should not prevent detection of the Raman signal as it does when pulsed lasers are used in sooting hydrocarbon flames (10,16).

Since the problems that preclude the measurement of Raman spectra in highly luminous environments, such as sooting hydrocarbon flames, can be overcome by the application of coherent anti-Stokes Raman spectroscopy (CARS) (13,15,27), it is of interest to compare the two techniques regarding the application to flames containing high concentrations of silica particles. The coherent nature of the CARS signal combined with its greater peak intensity permits measurements to be made in flames containing high concentrations of carbon soot particles, even though the very high peak powers required cause absorbing soot particles in the measurement volume to be vaporized (29). Combined with gated detection, CARS should thus be considered for measurements in the most luminous VAD flames. We have performed a preliminary evaluation of broadband CARS as a diagnostic for temperature in a VAD burner consisting of four concentric quartz tubes surrounding a central tube (outer diameter = 25 mm) (30). Silicon tetrachloride in nitrogen carrier gas (0.2 L/min) flowed through the central tube, resulting in the deposition of ~0.7 g/min of silica in the flame. The surrounding rings contained (proceeding outward) 0.5 L/min nitrogen, 3.0 L/min hydrogen, 3.0 L/min nitrogen, and 3.5 L/min oxygen. Since the particles were largely confined to a stream along the central axis of the flame, particle number densities were as high as $10^{11}$/cm$^3$. The CARS apparatus has been previously described (31). The energy of the pump beams was 10–

20 mJ/pulse (10 Hz 2 × Nd:YAG) and that of the Stokes beam was 2–5 mJ/pulse. The beams were focused to a common probe volume using a single-element, 239-mm-focal-length lens.

The quantities of nitrogen present along the centerline were sufficient to obtain CARS spectra with a signal/noise ratio > 10 for 100 laser shots. The introduction of silica particles into the flame caused dielectric breakdown in a significant percentage of laser shots, however, requiring the pump laser energies to be decreased to a point at which the CARS signal could no longer be detected. This behavior is somewhat surprising, since electron micrographs of particles sampled in the region of the CARS measurement show particles with diameters on the order of 100 nm or less (32). Particles smaller than several microns in diameter are thought to have no effect on the breakdown threshold owing to their high electron diffusion losses (13). With such large number densities, however, it is likely that there will be significant numbers of particles with diameters 10–100 times larger than the mean, owing to agglomeration of smaller particles. Studies of breakdown thresholds using suspended particles have shown that a single 50-μm carbon particle can reduce the breakdown threshold for 1.06-μm radiation by a factor of 50 (33). Since this effect is expected to become more serious at shorter wavelengths, large reductions (up to several orders of magnitude) could occur in flames containing such large concentrations of particles. Although these results do not represent a complete study, they do indicate that considerable difficulties may be encountered when attempting to apply optical techniques employing focussed laser beams (such as those produced by frequency-doubled Nd:YAG lasers) with high peak powers to silica-laden flames. This further indicates that, even though laser-modulated incandescence should not interfere with pulsed-laser SRS measurements in flames laden with transparent particles, the decrease in the breakdown threshold could be the limiting factor in the application of pulsed lasers. Cavity-dumped lasers with moderate peak powers and high repetition rates represent a possible solution to this problem (15,16), so that gated SRS measurements could be made in even the most highly particle-laden VAD flames without optical breakdown.

## SUMMARY

Spontaneous Raman spectroscopy (SRS) has been applied to hydrogen/methane/oxygen/nitrogen flames containing high concentrations (>$10^9$/cm$^3$) of small (<0.2 μm diameter) silica particles. In contrast with carbon soot particles, which emit large quantities of broadband visible radiation at high temperatures, the addition of large numbers of silica particles has little or no effect upon the SRS measurements. The data reported here are thus the first spatially-resolved, time-averaged SRS measurements obtained in the submicron, high-number-density regime, conditions that approach those typical of flame synthesis processes. SRS thus appears to be a promising technique for measurement of flame temperatures and concentrations in VAD flames containing higher densities of silica particles than observed in these experiments. Preliminary CARS measurements in a hydrogen/oxygen VAD flame containing silica particles indicate that measurement techniques employing high-power pulsed lasers (including SRS) may be limited by the breakdown threshold in the flame, which decreases when large numbers of submicron particles are present.

## ACKNOWLEDGMENTS

The authors would like to thank R. A. Hill for the loan of the multipass cell, and R. W. Dibble, W. L. Flower, D. R. Hardesty, and L. A. Rahn for helpful technical discussions.

This work was supported by the US Department of Energy.

## REFERENCES

1. Ulrich, G. D., *Chem. Eng. News* **62**, 23 (1984).
2. Li, Tingye, Ed., *Optical Fiber Communication*, Academic Press, New York, 1985.
3. Suyama, Y., Marra, R. A., Haggerty, J. S., and Bowen, H. K., *J. Amer. Ceram. Soc. Bull.* **64**, 1356 (1985).
4. Alam, M. K. and Flagan, R. C., *Aerosol Sci. Tech.* **5**, 237 (1986).
5. Chung, S. and Katz, J. L., *Comb. Flame* **61**, 271 (1985).
6. Hurd, A. J. and Flower, W. L., *J. Colloid Interface Sci.* **122**, 178 (1988).
7. Ulrich, G. D. and Riehl, J. W., *J. Colloid Interface Sci.* **87**, 257 (1982).
8. Lapp, M., Goldman, L. M., and Penney, C. M., *Science* **175**, 1112 (1972).
9. Blint, R. J., Bechtel, J. H., and Stephenson, D. A., *J. Quant. Spect. Rad. Transfer* **23**, 89 (1980).
10. Drake, M. C., Lapp, M., Penney, C. M., Warshaw, S., and Gerhold, B. W., (1981), *Eighteenth Symposium (International) on Combustion*, The Combustion Institute, Pittsburgh, 1980, p. 1521.
11. Dibble, R. W., Kollmann, W., and Schefer, R. W., *Comb. Flame* **55**, 307 (1984).
12. Vanderhoff, J. A., Anderson, W. R., Kotlar, A. J., and Beyer, R. A., *Twentieth Symposium (International) on Combustion*, The Combustion Institute, Pittsburgh, 1984, p. 1299.
13. Eckbreth, A. C., *Laser Diagnostics for Combustion Temperature and Species*, Abacus Press, Cambridge, 1988.
14. For example, *see* Bechtel, J. H., *App. Opt.* **18**, 2100 (1979); Hill, R. A., Mulac, A. J., and Aeschliman, D. P., *J. Quant. Spect. Rad. Transfer* **21**, 213 (1979); Schoenung, S. M. and Mitchell, R. E., *Comb. Flame* **35**, 207 (1979); Stricker, W., *Combust. Flame*, **27**, 133 (1976).
15. Rahn, L. A., Mattern, P. L., and Farrow, R. L., *Eighteenth Symposium (International) on Combustion*, The Combustion Institute, Pittsburgh, 1981, p. 1533.
16. Flower, W. L., *Sandia National Laboratories Report* SAND81-8608, 1981.
17. Hill, R. A., Mulac, A. J., and Hackett, C. E., *App. Opt.* **16**, 2004 (1977).
18. Hall, R. J. and Boedeker, L. R., *App. Opt.* **23**, 1340 (1984).
19. Farrow, R. L., Lucht, R. P., Clark, G. L., and Palmer, R. E., *App. Opt.* **24**, 2241 (1985).

20. Palmer, R. E., *Sandia National Laboratories Report*, SAND89-8206, 1989.
21. Flower, W. L., *Comb. Sci. Tech.* **33,** 17 (1983).
22. Flower, W. L., *Phys. Rev. Lett.* **51,** 2287 (1983).
23. Bohren, C. F. and Huffman, D. R., *Absorption and Scattering of Light by Small Particles*, Wiley-Interscience, New York, 1983.
24. Flower, W. L. and Hurd, A. J., *App. Opt.* **26,** 2236 (1987).
25. Zachariah, M. R., Chin, D., Semerjian, H. G., and Katz, J. L., accepted for publication in *Comb. Flame*, 1988.
26. Bautista, J. R., Potkay, E., and Scatton, D. L., *Process Diagnostics: Materials, Combustion, Fusion* **117,** 151 (1988).
27. Potkay, E., private communication, 1988.
28. Van de Weijer, P., Zwerrer, B. H., and Suijker, J. L. G., *Chem. Phys. Lett.* **153,** 33 (1988).
29. Farrow, R. L., Lucht, R. W., Flower, W. L., and Palmer, R. E., *Twentieth Symposium (International) on Combustion*, The Combustion Institute, Pittsburgh, 1984, p. 1307.
30. Allendorf, M. D. and Potkay, E., unpublished results.
31. Lucht, R. W., submitted to *Comb. Flame*, 1988.
32. Clark, H. R., Stawicki, R. P., Smyth, I. P., and Potkay, E., submitted to *J. Amer. Ceram. Soc.* 1988.
33. Lencioni, D. E., *App. Phys. Lett.* **25,** 15 (1974).

# Composition Determination at High Temperatures by Laser Produced Plasma Analysis

Yong W. Kim

*Department of Physics, Lehigh University,
Bethlehem, PA 18015*

## ABSTRACT

Materials in a high temperature condensed phase require new means of characterization. The method of analysis by laser produced plasmas (LPP) combines interaction of a high power laser pulse with the surface in nanosecond time scales and time- and space-resolved detection of the emission spectrum from the plasma plume. In order to meet the requirement that the plasma plume has the same elemental composition as in the condensed phase, a rule of thumb for operation of the laser has been first established in terms of the relationship between the thermal diffusion speed and the rate at which the surface layer is removed by evaporation. This paper describes the physical processes of the laser–matter interaction and the evolution of the plasma plume in time and space, as revealed by experiment and numerical simulation. An account of a new application of the LPP method to compositional analysis of molten metals is presented.

**Index Entries:** High temperature; laser produced plasmas (LPP); plasma plume; condensed phase; elemental composition analysis.

## INTRODUCTION

Materials in a high temperature condensed phase are of increasing interest both from the standpoints of exotic states of matter and processing of high performance hybrid materials. Examples are the formation of molecular clusters of refractory elements, exotic compounds, thin film growth of high $T_c$ superconductors, high performance coatings, and production of precision alloys, just to name a few. Such a high temperature state presents extreme demands on the means of characterization of materials. The objects of characterization may be the transport properties, mechanical properties or chemical or elemental composition.

The nature of challenge from the standpoint of the material characterization has fundamentally to do with the following two issues: the characterization of the interface region of the materials that come into contact with a given diagnostic means as it relates to the properties of the bulk; and the integrity of the diagnostic means in terms of its predictable, i.e., calibrated, response to the thermodynamic state of the material under question. Specifically, the high temperature keeps the material volatile and vulnerable to chemical reactions and also much of the front-end sensing elements exposed to thermal drift, deformation, and the threat of destruction. The exact definition of the representative sample of the bulk material becomes subject to interpretation and dependent on the intricate details of the environmental conditions.

In this paper, we will describe a full set of physics involved in the new method for compositional analysis of metallic alloys in a high temperature condensed phase utilizing laser produced plasmas (LPP). The investigation has been carried out in a two-pronged approach: laboratory experiments and numerical simulation by computer codes. The experiments rely heavily on time- and space-resolved spectroscopy and non-linear optics. We have also developed two computer simulation codes, one for laser heating and another for radiative transport within the plasma. The experiments and simulation are carried out in an interactive fashion, so the experimental results are needed to verify the code predictions, and the code outcome helps strategize the measurements until the two sets of results become consistently reconciled to each other. This then results in a satisfactory understanding of the physical processes.

Metallic alloys, such as steel, in a molten state represent the most severe thermal burden on any analysis technique and its associated instrumentation. In addition, there are wide variations in the temperature, pressure, and composition of the ambient gas atmosphere, the depth, and composition of the oxide or slag layer on the molten metal pool and in the composition of the molten alloy itself. The successful analysis technique must be able to function consistently and with great precision and accuracy over such wide dynamic ranges of variables.

The LPP technique meets all of the above requirements. The basic concept is to generate a robust plasma plume off a representative surface of a condensed phase alloy by delivering a giant laser pulse and carry out spectroscopic analysis of the plasma emission spectra for determination of elemental composition. Preparation and identification of the representative surface of the molten metal, prior to laser firing, requires penetration of the slag layer and precise distance ranging. The spectroscopic measurement of the LPP plume is subject to temporal and spatial coordinates of the plasma. The elemental composition of the molten metal follows a detailed interpretation of the measured spectral data according to the calibration data base.

The above sequence of control, measurement, and analysis has been established through a rigorous understanding of the physics of laser–matter interaction. Implementation of the LPP methodology into a survivable and functional sensor probe has, on the other hand, been achieved by detailed characterization of the pyrometallurgical environment and assembly of special purpose controllers, detectors, and instruments into an integrated system.

In the following sections, we will describe the physical processes of the laser–matter interaction and the evolution of the plasma plume in time and space, as revealed by experiments and numerical simulation. The fundamental requirement that the LPP plume have the same elemental composition as the condensed phase will also be elucidated in the form of a rule of thumb for operation of the laser.

## NUMERICAL SIMULATION

The numerical simulation of the generation of the LPP plume and its subsequent spatio-temporal evolution has been carried out by means of two separate numerical codes developed for this work: a laser heating code and radiation transport code. Each code is one-dimensional, but together they give rise to a two-dimensional description of the LPP plume.

In the laser heating code, two different stacks of cells are constructed, one on the condensed phase side and another on the gas phase side. The cells are 1 mm in diameter on both sides. In the early stage of the laser heating, the time step is set at $10^{-13}$ s and the cell thicknesses are 200 Å and 2500 Å for the condensed and gas phases, respectively. As the heating progresses, the rate of change of the state variables decreases, and the time step is increased eventually to $10^{-10}$ s in several stages. The cell thickness has been similarly increased to 1 and 50 μm for the condensed and gas phases, respectively.

The total number of cells has been allowed to increase in time in order to keep up with the growing regions of a high temperature gas/plasma and a condensed phase on the opposite sides of the interface.

For a given laser pulse of known time profile and power density, the computation commences with the first installment of the laser energy absorbed at the surface, heating the first condensed-phase cell. The state of the cell is determined by requiring that the first gas-phase cell comes to equilibrium with it though heating of both cells and evaporation of the condensed-phase matter. Diffusion of gas-phase species, thermal conduction, work done by expansion, and internal excitations including ionization and radiative transfer, are then accounted for successively for the adjoining cells in two directions away from the gas-condensed matter interface, all during the single time interval. At the end of such calculations, the two interface cells end up with different temperatures. During the second time interval, the entire procedure is repeated, except that the gas-phase cells may begin to absorb some part of the laser energy.

The surface of the target heats up rapidly in the first few hundred picoseconds, resulting in a large loss of the condensed-phase matter. The number of gas-phase cells multiply quickly while undergoing rapid heating. The heating brings about significant ionization, as well as excitation of the internal states of the atomic and ionic species. This category of excitation and ionization is handled by invoking a local thermodynamic equilibrium. A full set of the Saha equations for multicomponent gas with multiple stages of ionization for each species are solved self-consistently with the energy balance equation. A predict-and-check numerical algorithm is used for all such calculations.

The first condensed-phase cell loses a significant portion of its total mass owing to evaporation, resulting in a movement of the interface. In order to deal with this development, we have adopted a procedure by which both the gas- and condensed-phase cells are renormalized to their respective regular sizes after each time step. The temperature, pressure, density, elemental composition, and the degrees of ionization are reassigned to each cell by interpolating the respective profiles, as determined at the end of the preceding time interval.

The requirement that the two interface cells attain the same temperature in the beginning of each new time step is a very important physical statement of the nature of the process under question. We have used this method successfully, first to obtain the heat transfer coefficient of a number of different metal–gas interfaces and then to treat the problem of sputtering of a copper electrode into a discharge-generated plasma (1–4).

Once the gas-phase cells acquire free electrons, the heating by the inverse bremsstrahlung process is activated (5–8) and the heating of the gas-phase cells is accelerated. The early stage production of electrons is primarily owing to the process of thermal ionization, but multiphoton ionization also contributes (although only weakly) when the laser power density is low (9). Multiphoton inverse bremsstrahlung has been ignored, however, for the power densities considered here (10).

It is important to note that the rate of evaporation of a given condensed-phase species not only is a strong function of temperature, but also varies greatly from a species to another. The evaporation rates are obtained from experimental data available in the literature (11). What this means is that for a multielement target, the evaporation process can take place preferentially in favor of those elements with larger rates of evaporation under conditions of a modest laser power density (11, 12). If, on the other hand, the laser power density is set above a certain threshold value, the total mass loss rate by evaporation can keep pace with the thermal diffusion front into the bulk of the condensed-phase target, thus eradicating the elemental dependence of the evaporation rate. The plasma formed under such a condition becomes truly representative of the multielement target. We will return to this point later.

The radiation transport code is used to predict the emission spectrum of the plasma in the direction normal to the laser beam axis. The results of the laser heating code are used to define the properties of the plasma core, and the temperature and density profiles of the outer layer of the plasma are fashioned after the computed axial profile of the plasma. Known atomic properties, such as energy levels, degeneracies, transition probabilities, and line broadening parameters, are compiled into the code (13–15), and a step by step radiation transport computation is carried out starting from the plasma core. In view of the fact that the typical LPP plume at the core is optically dense, the core radiates from its surface as a blackbody at the core temperature. The net effect of the cooler outer layers is to develop a line reversal spectrum. At the later stages of the plasma afterglow, beginning with the expiration of the laser pulse, the emission spectrum exhibits the line spectral characteristics with attendant cooling in both the radial and axial directions.

Figure 1 shows a sample calculation of the laser heating code in the very early stage of a laser heating of a molten iron target. The temporal profile of the laser pulse used here is a triangular one with base width of 20 ns and maximum power density of $6.4 \times 10^{10}$ W/cm$^2$. The temperature profiles in the gas phase, as well as in the liquid phase, are shown at 10, 50, 100, and 200 picoseconds (ps). Already in this short period, the precursor of the shock wave develops in the form of a hump in the temperature profile growing away from the gas–liquid interface. It also shows an increasing temperature profile within the bulk of the molten metal. During the first 200 ps, the temperature of the liquid iron exceeds the critical temperature of iron. This is because the laser heating outpaces the cooling by evaporation. It also means that the latent heat of evaporation needs to be established as a function of temperature. The critical temperature and the latent heat data for iron are not available in the literature and, consequently, we have calculated such data by means of the law of corresponding states and statistical mechanics. A description of the calculation is given in an accompanying paper (16,17). Another interesting consequence of the rapid heating by laser is that the supercritical liquid metal contributes elemental particles at thermal velocities to the gas-phase cells well beyond the expiration of the laser pulse.

The rate at which the surface layer of the liquid is removed is comparable to the thermal diffusion velocity in this example. This, of course, necessitates repartitioning of the cells on both sides of the liquid–gas interface, as stated earlier.

We note that execution of the laser heating code entails a great deal of subsidiary computations to establish the thermodynamic properties of the condensed- and gas-phase matter, such as the equation of state, thermal conductivity, and diffusion constants. These vary a great deal as

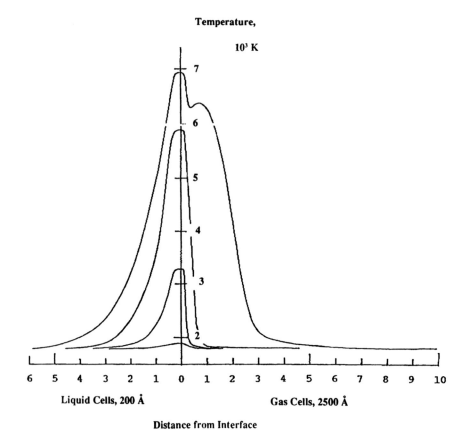

Temperature,

10³ K

Liquid Cells, 200 Å          Gas Cells, 2500 Å

Distance from Interface

Fig. 1.   Calculated temperature profiles on both sides of the gas–liquid interface of a molten iron target. A laser pulse of a triangular profile is incident on the surface at $t = 0$. The pulse is 20 ns long at the base and the peak power density is set at $6.4 \times 10^{10}$ W/cm$^2$. The profiles correspond to $t = 10, 50, 100,$ and 200 ps in the order of increasing temperature. Note that the basic cells are 2500 Å thick for the gas side and 200 Å for the liquid. Each cell has been further broken down to four and two zones in the computation, respectively, in order to obtain smoother profiles. The time step used for this calculation is 0.1 ps.

a result of the extremely large variations in temperature and pressure encountered during the laser heating.

In the case of solid targets, the latent heat of melting plays a role, and the melting transition contributes to a sharp change in the optical reflectance of the target surface with significant consequences in the heating scenario in the early stages (*18*).

Continuation of the laser heating code reveals a sharp increase of the plasma density into a supercritical regime, where the plasma frequency exceeds the frequency of the incident laser beam. The onset of the critical plasma density is detectable experimentally because the interaction of the laser beam with the critical density plasma results in a generation of (integer) harmonics (*5,6*). The time of the first detection of the second harmonic emission is one of the important verification points of the laser heating code. The onset of the shock formation is another verification point.

The laser heating code computations under varying conditions of the laser pulse and target properties have defined the window of applicability of the LPP plume for compositional analysis of the target. An excessively high laser power density and short risetime leads to a premature termination of the laser heating of the target, because the plasma becomes supercritical, preventing the laser beam from reaching the target. A basic rule of thumb, thus, has been obtained. Productive LPP plumes are generated when the laser heating can sustain an evaporation rate that keeps the movement of the interface in pace with the thermal diffusion velocity within the bulk of the condensed-phase target. Under such a condition, the elemental dependence of the evaporation rate can be overcome and a spectroscopically productive plasma, which is truly representative of the target composition, may be generated.

Another lesson of the numerical simulation is that the state of the LPP plume changes rapidly in time and over space, and its attendant spectroscopic properties undergo similar, but even more dramatic changes. This accurately bears out the experimental observations. Consequently, the quantitative spectroscopic measurements must be made in a time- and space-resolved manner. For example, the early part of the emission spectrum from the core region of the LPP plume is dominated by the plasma effects, which bears little resemblance to the elemental atomic line emission characteristics, and, thus, has no role to play in the composition analysis.

## EXPERIMENT

The experimental arrangement consists of three target chambers, lasers, several spectrographs, timing electronics, a streak camera, and a large assortment of high speed detectors. Three different setups of the metallic targets have been used: a solid target in atmospheric air, a solid target in a vacuum chamber constructed of a fused quartz body, and a molten metal target in a controlled-atmosphere furnace. The primary laser used for the experiment is a Q-switched Nd:glass laser rated for a maximum energy of 20 J with a pulse width of 20 ns at FWHM. In most of the experimental runs, we have operated it at 2–7 J with 65 ns FWHM. High-dispersion spectrographs are used in conjunction with optical multichannel analyzers (OMA). The streak camera is equipped with a channel-plate intensifier and is coupled to a spectrograph or imaging optics through a ribbon-shaped linear array of optical fibers. The OMA's are equipped with gated, channel-plate intensifiers.

Each laser pulse has been monitored for its energy by a calorimeter and its pulse shape by a fast-response photodiode. By means of a strict protocol of operation, the laser has been made to perform in a highly reproducible manner. Figure 2a shows the laser pulse shape. Five succes-

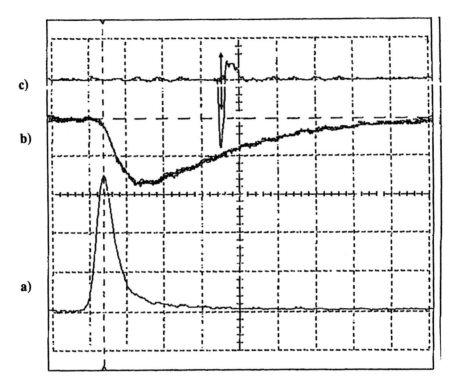

**Time, 125 nsec/div**

Fig. 2. A digital oscilloscope display of a typical LPP run for the spectral
data shown in Figs. 3 and 4: a) the laser power density profile, b) five successive
photomultiplier outputs of the white light emitted from five separately produced
LPP plumes, and c) a 20 ns gate pulse generated by a delay generator for gated
acquisition of the emission spectrum by the optical multichannel analyzer. A
very high degree of reproducibility has been achieved for LPP plume production.

sive readings of the white light output from the LPP plume off a steel
target are shown in Fig. 2b. It indicates the exceptionally good reproducibility of the laser output and the laser–target interaction. Also shown
in Fig. 2c is a 20 ns gating pulse used to enable the OMA's channel plate
intensifier for a fixed time exposure of the detector array.

As indicated in the preceding section, the LPP plume evolves
through several plasma regimes in the course of the laser heating and
afterglow cooling. The time-resolved spectrum reveals such changes. In
Fig. 3, we show 50 successively acquired spectra that are separated by 20
ns. Each spectrum consists of 1024 independent, intensified photodiode
readings with exposure time of 20 ns, covering the spectral range of 1850
to 6200 Å. The target used here is a specialty steel alloy in an ambient
atmosphere of 0.015 torr of argon. The entrance slit of the spectrograph is
focused onto a region of the LPP plume 3 mm away from the target
surface and 2 mm off the laser beam axis.

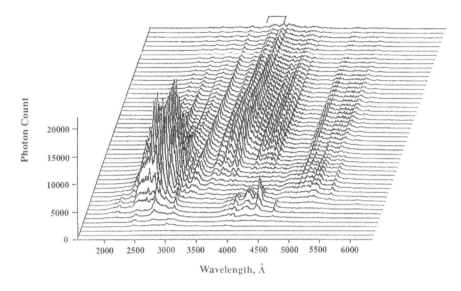

Fig. 3.   Time-resolved emission spectra from an LPP plume generated off a specialty steel alloy target. Each trace represents a 20 ns exposure spectrum covering the spectral range of 1850 to 6200 Å. Each successive trace is delayed by 20 ns and the 50 traces shown cover the first 1 μs of the LPP plume. The laser energy is 3.38 J, and the ambient gas is argon at 0.015 torr at room temperature.

It is quite apparent that the spectral character undergoes a dramatic change from one of a continuum, with prominent broad features at early times, to another dominated by narrow emission lines in the later stages of the plasma evolution. In fact, the spectrum taken at about the time when the laser pulse peaks out in its power density reveals a definite second harmonic at 5290 Å of the incident laser. This indicates that the particular region of the LPP plume has attained the critical electron density at which the plasma frequency equals the incident laser frequency. This is in excellent agreement with the prediction of the laser heating code. The second harmonic peak has also been found to have satellite structures with characteristic spacing of 11.8 Å, and this corresponds to the predicted ion acoustic wave frequency corresponding to the ion number density of $10^{21}/cm^3$ (*16,17*).

The streak photographs show that there emerges a shock wave driven by the growth of the plasma core, consistent with the code prediction (*16,19*). In addition, we have observed, for the first time, the generation of fractional harmonics of the 1.06 μm laser. It represents the first observation of such a phenomenon in plasma physics; we are now reasonably certain that it results from the chaotic nonlinear interaction of the plasma and the laser beam (*20*).

The above agreement between the numerical simulation and measurements demonstrates a satisfactory understanding of the physics of the laser–matter interaction. The richness of the observed phenomena, such as the nonlinear laser–plasma interaction and shock wave formation, helps define the necessary and sufficient conditions for successful analytical spectroscopy of the LPP plume.

The essential step to high-precision compositional analysis of condensed-phase metallic alloys is to obtain the line emission spectrum in a reproducible manner. Figure 4 gives a definitive illustration of the reproducibility with the LPP technique. There are 10 successive spectra displayed in their raw data form without any rescaling, which are obtained with an integration time of 100 ns. The individual line intensities can be linked to the elemental abundances through calibration by certified alloy standards. The result for a single LPP shot, such as that shown in Fig. 4, gives rise to determination of elemental concentrations within ± 2% of the mean values for all elements in a given alloy. We estimate that by further fine tuning of parameters at our disposal, the uncertainty figure can be reduced to ±0.5%.

## CONCLUSIONS

We have shown by numerical simulation and experimental measurements that the LPP plume can provide a quantitative measure of the elemental composition of metallic alloys in a high temperature state, given that a set of strict conditions is satisfied. The primary requirements are that (1) the laser pulse shape and power density be such that the evaporation front advances into the bulk in pace with the thermal diffusion front, (2) it be recognized that the LPP plume has three characteristic regimes of the fast ion plasma, plasma core, and the shock wave, each with distinct spectroscopic properties, and (3) the measurements be made by means of space- and time-resolved quantitative spectroscopy. It is important to stress that the state of the LPP plume evolves as a result of a competition among the laser heating process, material response, and the process of plasma expansion into the ambient atmosphere. Consequently, one needs to clarify the basic question of whether the LPP plume is in a state of local thermodynamic equilibrium and also optically thin. Considerations regarding this question are treated in detail elsewhere (21).

The LPP methodology for composition analysis of molten metal alloys has been implemented into a full sensor-probe system for initial applications in primary metals production as a process control device. Robust plasmas have been produced under both laboratory and small-scale production conditions in a reproducible manner, setting the stage for quantitative calibrations. In the case of molten metals, there are no reliable alloy standards because in a molten state, the elemental composition of an alloy changes in time owing to element-dependent rates of evaporation. A new protocol is currently being developed for calibration purposes.

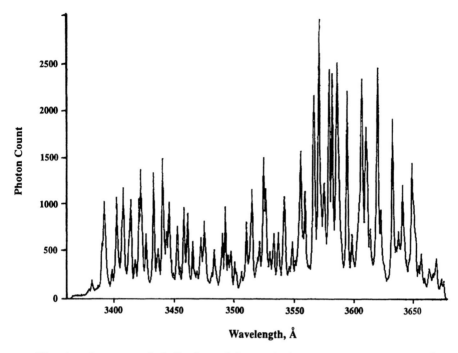

Fig. 4.   An expanded display of the emission spectrum corresponding to the 300 Å segment, indicated by a bracket at the top of Fig. 3 at 450 ns after the start of the laser pulse. In fact, there are 10 spectra that are superimposed in their raw data form without any rescaling. This clearly demonstrates the high degree of reproducibility that is essential for precision compositional analysis of metallic alloys.

It is quite clear that production of the LPP plume off a molten metal requires preparation of the surface because the top surface of any molten metal is covered with a layer of oxides of the constituent species. Uncovering the layer and presenting the representative surface to the laser pulse are an exacting process of real time control. This aspect has been incorporated into the sensor-probe system. The entire operation entails penetration of the oxide layer by the leading end of the sensor probe, firing of the laser pulse at the appropriate moment when the representative surface of the molten metal is found at a predetermined distance from the laser, and subsequent acquisition of the spectroscopic data in a time- and space-resolved manner. Data analysis follows to determine the elemental composition. The leading end of the sensor probe is then withdrawn from the molten metal bath, completing the full measurement cycle of less than a minute duration.

The LPP methodology is applicable to many other analysis activities. In different implementations, the methodology can provide information on the thermodynamic, transport, and mechanical properties of matter, as well as the chemical composition under extremely challenging conditions.

## ACKNOWLEDGMENT

The author acknowledges numerous valuable contributions to this work by K. S. Lyu, J. Kralik, H. Y. Chang and S. H. Kim.

## REFERENCES

1. Lee, H. S., *Ph.D. dissertation*, Lehigh University, Bethlehem, PA (1980).
2. Sincerny, P., *Ph.D. dissertation*, Lehigh University, Bethlehem, PA (1980).
3. Kim, Y. W., Lee, H. S., and Sincerny, P., *Proceedings of XVI International Conference on Phenomena in Ionized Gases*, Düsseldorf, FRG, 29 August–2 September, 1983, pp. 2–256.
4. Kim, Y. W. and Sincerny, P., *Proceedings of XVII International Conference on Phenomena in Ionized Gases*. Budapest, Hungary, 8–12 July, 1985, p. 1001.
5. Hora, H., *Physics of Laser Driven Plasma*, Wiley, New York, NY (1981).
6. Bobin, J. L., *Physics Reports* **122**, 173 (1985).
7. Duston, D., Clark, R. W., Davis, J., and Apruzese, J. P., *Phys. Rev.* **A27**, 1441 (1983).
8. Johnston, T. W. and Dawson, J. W., *Phys. Fluids* **16**, 722 (1973).
9. Eberly, J. H. and Lambropoulous, P., eds., *Multiphoton Processes*, Wiley, New York, NY (1978).
10. Brysk, H., *J. Phys. A.* **8**, 1260 (1975).
11. *American Institute of Physics Handbook*, 3rd edition, McGraw-Hill, New York, NY (1972).
12. Porter, J. R., Goldstein, J. I., and Kim, Y. W., *Physics in Steel Industry: AIP Conference Proceedings No. 84*, Schwerer, F., ed., American Institute of Physics, 1982, p. 377.
13. Moore, C., *Atomic Energy Levels, vol. I, II, and III*, National Bureau of Standards, Washington, DC (1958).
14. Wiese, W. L., et al., *Atomic Transition Probabilities*, vol. 1 and 2, National Bureau of Standards, Washington, DC (1969).
15. Griem, H. R., *Spectral Line Broadening by Plasmas*, Academic, New York, NY (1974).
16. Lyu, K. S., Kralik, J., and Kim, Y. W., *Proceedings of the Sixth International Conference on High Temperatures, Gaithersburg, MD 3–7 April, 1989* (in press).
17. Lyu, K. S., *Ph.D. dissertation*, Lehigh University, Bethlehem, PA (1988).
18. Walters, C. T. and Clauer, A. H., *Appl. Phys. Lett.* **33**, 718 (1978).
19. Kim, Y. W., Kralik, J., and Lyu, K. S., *Proceedings of the 17th International Symposium on Shock Waves and Shock Tubes*, Bethlehem, PA 17–21 July, 1989 (in press).
20. Kim, Y. W., Lyu, K. S., and Kralik, J., *Bull. Am. Phys. Soc.* **33**, 2070 (1988).
21. Kim, Y. W., *Laser-Induced Plasmas and Applications*, Radziemski, L. J. and Cremers, D. A., eds., Dekker, New York, NY (1989).

# EXAFS, Matrix Isolation,
# and High Temperature Chemistry

I. R. BEATTIE,* N. BINSTED, W. LEVASON, J. S. OGDEN,
M. D. SPICER, AND N. A. YOUNG

*Department of Chemistry, University of Southampton,
Southampton, UK*

## ABSTRACT

The advantages resulting from the combination of the EXAFS (extended X-ray absorption fine structure) and matrix isolation techniques to the study of high temperature molecules are outlined. Equipment allowing the collection of EXAFS data from matrix isolated molecules at *ca* 9 K, combined with *in situ* monitoring of the isolated materials by vibrational spectroscopy is described.

Metal K-edge EXAFS data of *cis*-[Fe(CO)$_4$I$_2$] isolated in CH$_4$ and of FeCl$_2$ and CoCl$_2$ isolated in the potentially reactive matrix CO are presented and described. The FeCl$_2$ data show a mixture of *trans*-[Fe(CO)$_4$Cl$_2$] and "free" FeCl$_2$ are present, whereas for CoCl$_2$, only van der Waals interaction between linear CoCl$_2$ and the CO matrix was observed. EXAFS can also provide information about the short range order (<4th shell) in the matrix material and this is illustrated by data for solid Kr and for Kr in Ar.

**Index Entries:** EXAFS; matrix isolation; metal dihalides; carbonyl halides.

## INTRODUCTION

The difficulties encountered in the direct study of high temperature molecules are well documented (1). Apart from being experimentally challenging, such systems often present problems of interpretation, and the principal vapor phase techniques currently in use may be summarized as follows.

### Mass Spectrometry

This technique is highly sensitive and, provided fragmentation problems are not serious, readily leads to identification of parent ions and

*Author to whom all correspondence and reprint orders should be addressed.

hence, by inference, neutral species. However, the spectra in general do not provide structural data.

## High Resolution Rotational Analyses

Unless the molecule has a very high rotational constant, it is usually necessary to cool the species (e.g., in a supersonic jet) in order to be able to analyze the spectra of polyatomic species. In particular, laser induced fluorescence rivals mass spectrometry in sensitivity, but is restricted to species with the requisite transitions. Microwave spectroscopy requires the presence of a permanent electric dipole.

## Electron Diffraction

This technique requires a pressure of the order of 1 torr, and the observed pattern is averaged over the thermal motions of all the molecular species in the beam. Interpretation becomes considerably more difficult as the temperature is raised, and as the number of atoms increases, and some reassessment of early high temperature work has been necessary.

Matrix isolation (2) offers an alternative approach to the characterization of high temperature molecules whereby many of the experimental difficulties may be circumvented, at the same time preserving the integrity of the information obtained. Thus, the ir spectra of matrix isolated molecules yield vibrational transitions that are very similar to those obtained from the vapor phase, and the vast majority of molecules studied show no substantial differences in geometry between the gas phase and isolation in an inert matrix (3). As a result, there have been numerous studies on the characterization of high temperature molecules using matrix isolation in combination with one or more types of spectroscopy (4).

However, although this approach has established the basic shape of many high temperature molecules, and has in favorable cases also led to estimates of bond angles, there has been no comparable strategy for the determination of internuclear distances. In principle, such information could still be obtained for matrix isolated species by traditional diffraction techniques, but sensitivity would be severely affected by the presence of the matrix material.

Two alternative techniques that could be used, however, are nmr spectroscopy and EXAFS (extended X-ray absorption fine structure). Both these methods have the advantage that they are element (or nucleus) specific, and so effectively probe only the molecule of interest. Both are routinely used for exploring nuclear environments in solids, and EXAFS in particular has been applied to dilute systems (5). In common with electron diffraction, however, the technique may not readily differentiate between, for example, "directly bonded" distances in monomers and dimers. With the present state of the art, EXAFS does not easily yield reliable angular information except in certain favorable cases, although this deficiency may eventually be overcome using near-edge techniques *viz.* XANES (X-ray Absorption Near Edge Structure).

With these caveats in mind, and noting the success of others in this field (6), we have started to use EXAFS in order to study interatomic distances in matrix isolated molecules, and this paper reports some preliminary experiments carried out using the Synchrotron Radiation Source (SRS) at Daresbury, UK. Our principal aim was to attempt to establish internuclear distances in a range of high temperature molecules isolated in inert gas matrices, but it was realized at an early stage that in addition to yielding "directly bonded" internuclear distances, the EXAFS experiment would in principle also provide data on van der Waals-type interactions between isolated molecules and the matrix environment, and so perhaps establish a firm basis for the weak perturbations routinely observed in spectroscopic studies on matrix isolated molecules. With this aim in view, we therefore include a brief account of our supporting EXAFS studies on solid krypton and Kr/Ar mixtures, where the krypton atom is used as a probe of matrix environment.

Complementary matrix isolation ir studies were also carried out in order to establish appropriate experimental parameters such as furnace temperatures and deposition rates.

## EXPERIMENTAL

The apparatus used for EXAFS studies incorporates an Air Products "Displex" unit which operates at a base temperature of ca. 9 K. This unit is mounted, via a rotatable vacuum seal, in the vacuum shroud shown in Fig. 1. This housing was designed primarily to allow both transmission and fluorescence EXAFS studies on material condensed on the cold central window (W) from two possible deposition ports (C), and it also affords the possibility of obtaining reflectance or transmission ir spectra on the same sample.

The outer windows (A), which allow passage of the incident and transmitted X-rays, are made of mylar (thickness 75 micron), whereas the larger diameter fluorescence window (B) is made of beryllium. Two types of material were used for the cold central window depending upon the wavelength of X-rays used. EXAFS data for the lighter elements in this study (e.g., Fe, Co) were obtained using beryllium (thickness 0.5 mm), but for the heavier elements (e.g. Br) aluminum (0.5 mm) was found to be a very satisfactory alternative. As in the majority of matrix rigs, thermal contact between this central window and the cryogenic reservoir was improved by the use of indium gaskets, and the design also incorporates temperature monitoring and control via a thermocouple/tip heater.

Sample deposition typically takes place from one of the ports (C), and in the majority of experiments, a resistively heated silica tube served as the high temperature furnace. During deposition, the central window is typically oriented at 45° to the beam path (A---A). Not only is this orientation desirable for deposition, but it is the optimum orientation for

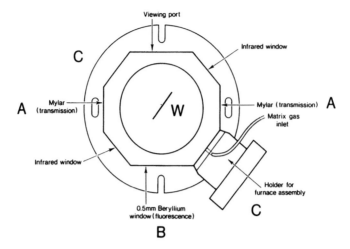

Fig. 1.   Experimental Arrangement for matrix isolation EXAFS data collection.

fluorescence studies, and allows the simultaneous collection of transmission data. In this orientation it is therefore possible to monitor the deposition directly by tuning the EXAFS monochromator to the X-ray edge position of the element under examination, and noting the change in transmission or fluorescence counts with time.

The nature of the EXAFS experiment imposes an additional constraint on the range of materials that can be used as matrices. Not only must the matrix be chemically inert, but as it is present in a large excess, it is essential that it has a low absorption cross-section to the X-ray energy range of interest. Krypton would, therefore, be quite inappropriate for studies on elements in the first transition series, and even argon would pose problems for the lighter elements of this group (e.g. Ti or V). The ideal, totally inert matrix for our experiments would be neon, but at the base temperature accessible to us (9 K), this might not result in effective isolation. However, the conventional constraint of optical transparency is relaxed, and we have explored the possibility of using methane as an "inert" matrix in addition to potentially reactive hosts such as carbon monoxide and nitrogen.

The complementary matrix ir studies were carried out using an identical "Displex" refrigerator mounted in a standard housing equipped with CsI optics. A Perkin-Elmer 983G spectrometer interfaced with a 3600 Data Station was used to obtain spectra. As in our earlier ir studies (7), matrix ratios were estimated to be 1,000 :1 and once a suitable furnace temperature had been established, sample deposition typically extended over the time period 60–90 min. The same deposition rates were used in the EXAFS experiments, but times were generally increased to ca 120–150 min.

The EXAFS spectra were collected at the Daresbury SRS with a typical beam current of 200 mA at 2 GeV. The K-edge spectra of iron and cobalt were recorded on Station 8.1 using a double crystal silicon (111) monochromator. The krypton K-edge spectra were measured on Station

9.2, using a double crystal silicon (220) monochromator. In both of these monochromators the second crystals were detuned to achieve rejection of higher harmonics. The detectors were either standard ion chambers containing rare gases for transmission studies, or in the case of fluorescence experiments, solid state scintillation counters (Tl/NaI) were used to measure the fluorescence. Data were typically collected for 400 eV before the edge and 800–1000 eV above the edge and each scan would take 1–2 h. To improve the statistics at high k-values (where k is the wavevector $\text{Å}^{-1}$) two or three spectra were averaged. Details of the data analysis used in the treatment of our EXAFS results and definitions of all symbols and abbreviations are described in the Appendix.

## MODEL SYSTEMS

### Krypton and Krypton in Argon

The Kr K-edge EXAFS of solid Kr at 10 ($\pm 2$) K shows a great deal of structure in spite of the relatively high Debye-Waller factors. The spectrum of the matrix annealed at 30 K and recooled to 10 K shows no significant differences from that of the unannealed matrix. At 30 K the amplitude is reduced by about 50% and the resolution of features owing to distant peaks is poorer. The results of a preliminary data analysis of Fourier filtered spectra are shown in Table 1 and Fig. 2. Structure resulting from the first four shells is included. The quality of fit and the extent of the data used was limited by difficulties in background subtraction. The interatomic distances for the first four shells give excellent agreement with crystallographic distances for FCC Kr at 10 K (8). In particular, agreement for the first shell is within .01 Å, confirming the validity of the theoretical phaseshifts and the method of analysis. The largest discrepancy is in the fourth shell for which the refined value of the shell radius is too high, and the Debye-Waller term ($\alpha = 2\sigma^2$) is much lower than would be expected. Both these features can be explained as a result of enhancement of the shell by multiple scattering effects that are particularly strong when two or more atoms are colinear with the absorbing atom (9,10). These results are therefore consistent with an ordered FCC lattice rather than a less ordered structure with similar shell radii. We have not yet included multiple scattering in the analysis as the treatment of the Debye-Waller factor in our refinement programs is inappropriate for systems with high thermal disorder except where relatively rigid units occur. The values of AFAC (.65) and VPI ($-2.25$) obtained from the 10 K data were used in the analysis of other Kr K-edge spectra to permit meaningful comparison of the Debye-Waller factors in different spectra. Refinement of the 30 K data gave almost exactly the same distances as the 10 K data, although with a lower degree of precision, but $\alpha$ was about 50% higher for all shells. Although the quality of fit ($R = 21.8$) is not great and the higher disorder reduces the accuracy of the refinement, we might have expected to have detected the thermal expansion ($\sim.02$ Å) observed in crystallographic refinements (8). In fact we observed a very

Table 1
Optimum Parameters for FCC Kr and 0.7% Kr in Ar'

| Atoms | Krypton (10 K) R = 14.1 | | | Krypton (30 K) R = 21.8 | | | Krypton in argon (10 K) R = 17.1 | | |
|---|---|---|---|---|---|---|---|---|---|
|  | $r_{X\text{-ray}}$ | $r_{exafs}$ | $2\sigma^2_{exafs}$ | $r_{X\text{-ray}}$ | $r_{exafs}$ | $2\sigma^2_{exafs}$ | $r_{Ar\text{-}Ar}$ | $r_{exafs}$ | $2\sigma^2_{exafs}$ |
| 12 | 3.99 | 3.98(0) | .027(0) | 4.01 | 3.98(0) | .039(0) | 3.76 | 3.78(0) | .030(0) |
| 6 | 5.64 | 5.62(3) | .049(6) | 5.65 | 5.55(6) | .074(10) | 5.31 | 5.34(2) | .056(5) |
| 24 | 6.91 | 6.92(1) | .039(2) | 6.94 | 6.99(2) | .061(3) | 6.50 | 6.49(1) | .048(2) |
| 12 | 7.98 | 8.09(2) | .025(2) | 8.01 | 8.07(2) | .041(3) | 7.51 | 7.62(1) | .021(2) |

'R is a measure of goodness of fit in text.
$r_{X\text{-ray}}$ (Å) are crystallographic interatomic distances for solid Kr at 10 K and 30 K.
$r_{exafs}$ (Å) are interatomic distances from EXAFS analysis.
$r_{Ar\text{-}Ar}$ (Å) are crystallographic distance for argon at 10 K.
$2\sigma^2$ (Å$^2$) is twice the mean square variation in interatomic distance.

slight decrease in the first shell distance. This phenomenon has been recognized by several authors (*11, 12*) and is a consequence of the inadequacy of treatment of thermal disorder in EXAFS for high values of $\alpha$, which gives rise to systematically short distances compared to other techniques. Although the first shell distances are still within the precision expected of the technique, we suggest that systematic errors owing to this effect give rise to distances which are too short by about .01 Å in Kr at 10 K and .03 Å in Kr at 30 K. As a further test of the validity of our analysis, we also attempted to refine the spectrum assuming a HCP structure. This was unsuccessful in predicting either the correct distances or reasonable values of $\alpha$ beyond the second shell.

The Kr K-edge spectrum of the Ar matrix containing 0.7 mole % Kr, measured at 10 K, had a poorer signal to noise ratio than the spectra of solid Kr, as is to be expected given the much lower concentration of Kr. This restricted the range of data to be analyzed at high k but the greater ease of background subtraction allowed lower k data to be included (Fig. 2). Analysis again showed clearly that the argon matrix possessed an ordered FCC structure. Values of $\alpha$ are between those of solid Kr at 10 K and those of solid Ar (Malzfeldt et al. (*13*) reported $\alpha = .040$ for the first shell of Ar at 5K). They are, however, only slightly higher than in Krypton at 10 K, a result that indicates that substitution of a Kr atom does not give rise to a significant component of static disorder (Table 1). The interatomic distances are also of interest as, even if a systematic error of .01–.02 Å is occurring, the distances, including the first shell, are substantially less than those calculated for a hard sphere model. In this case, the first shell distances would be given by the mean of the Kr and Ar values (3.873). Malzfeldt et al. (*14*) found a Kr-Ar distance of 3.81 Å in a Kr-Ar matrix at 23 K, and a Kr-Kr distance in solid Kr of 4.03 Å at the same temperature. As their value for solid Kr was rather higher than the 23 K crystallographic value of 4.0 Å (suggesting +ve rather than −ve systematic errors in this analysis) there is good agreement with our results for the nearest neighbour distance. Our 2nd and 3rd shell distances are however, shorter than those of Malzfeldt et al. and our 4th shell distance suggests that after taking multiple scattering into account a value only .01 to .02 Å higher than solid Ar would be observed.

We conclude from this analysis that although lattice dislocations may be relatively numerous, the structure of rapidly deposited Kr and Ar matrices is at least locally an ordered FCC lattice. The substitution of Kr in an Ar matrix at 10 K appears to result in minimal local disruption of the FCC Ar lattice, with the Kr accommodating itself to an argon site far more readily than a hard sphere model would suggest. The study illustrates that EXAFS can provide limited information on 3-dimensional structure and thermal disorder as well as giving accurate near-shell atomic distances, although care must be taken in interpreting results on systems with high thermal or static disorder.

## cis-[Fe(CO)$_4$I$_2$]

The eighteen electron compound *cis*-[Fe(CO)$_4$I$_2$] is readily prepared (*15*) from iron pentacarbonyl and iodine. Although it is both thermally and photolytically unstable, ir studies of the vapor isolated in argon matrices were found to give infrared spectra with peaks in the $v_{co}$ region which correspond closely with those observed for solutions in cyclohexane (*16*).

For the EXAFS experiments, the sample was sublimed at 2–4°C and the vapour cocondensed with a large excess of methane on to a cold beryllium plate at ca 10 K. The deposition period was of the order of 2 h. At the end of this period the beryllium window exhibited a canary yellow color. Fluorescence EXAFS spectra were recorded for three periods each of one and a half hours. The absorbance and EXAFS for averaged data are shown in Fig. 3 and 4 respectively. Figure 5 shows the Fourier filtered spectrum (obtained using a window of 1 to 5 Å) together with the best fit (R = 25.9) using the parameters below and a Fe-C-O bond angle of 180°.

|       | r/Å            | $\alpha(= 2\sigma^2)$  |
|-------|----------------|------------------------|
| Fe-C  | 1.81 (±0.004)  | 0.005 (±0.0008)        |
| Fe-O  | 2.97 (±0.006)  | 0.015 (±0.0012)        |
| Fe-I  | 2.60 (±0.005)  | 0.010 (±0.0009)        |

The Fourier transform in Fig. 5 is phaseshifted with respect to the first shell and therefore the distances for Fe-C at 1.81 Å and Fe-O at 2.97 Å match the peaks in the Fourier transform very well. (The oxygen phaseshifts are very similar to those for carbon.) The second peak does not occur at the calculated Fe-I distance because of significant contributions from the first and third shells and because the iodine back scattering phaseshift is much larger than carbon.

The high intensity of the oxygen shell is owing to enhancement by the very strong second and third order multiple scattering contributions which occur in terminal metal carbonyls (*10,17*). The importance of the multiple scattering is owing not only to the linear (or near linear) atomic arrangement, as previously discussed, but also to the short bond lengths involved. It can be seen that the correspondence of the phases between

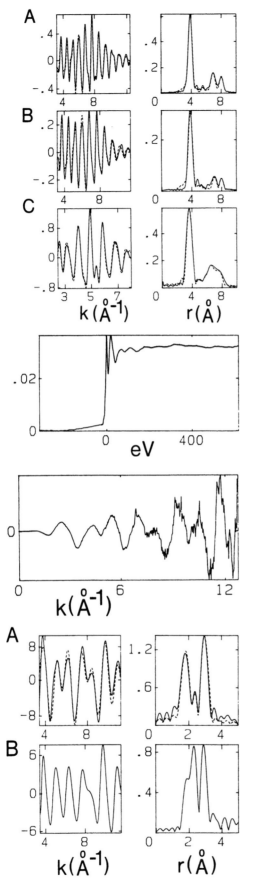

Fig. 2. EXAFS x $k^2$ (left) and Fourier transforms of EXAFS x $k^3$, corrected for phase shift (right) for: A Krypton at 10 K; B Krypton at 30 K; C 0.7% Krypton in argon at 10 K. (———) Fourier filtered experimental data. (----) Best fit theory using parameters in Table 1.

Fig. 3. Fluorescence K-edge X-ray Absorption Spectrum for cis-[FeI$_2$ (CO)$_4$] isolated in methane (0 = 7119 eV).

Fig. 4. EXAFS x $k^3$ of cis-[FeI$_2$(CO)$_4$] isolated in methane.

Fig. 5. Fourier filtered EXAFS x $k^3$ (left) and Fourier transform of EXAFS corrected for phaseshift (right). A cis-[FeI$_2$(CO)$_4$] isolated in methane (———) experimental. (----) theory. B FeCl$_2$ isolated in carbon monoxide.

78

experiment and theory is very good, although there is some discrepancy in amplitudes, probably associated with difficulties in background subtraction.

The resultant bond lengths (Fe-C 1.81 Å; Fe-O 2.97 Å) are in good agreement with those obtained by Binsted et al. for $Fe_2(CO)_9$ and $Fe_3(CO)_{12}$ (17). The Fe-I distance of 2.60 Å is comparable to that of the related compound $[Fe(CO)_2(PH_3)_2I_2]$ (18). Although the amplitude due to the iodine shell (Fig. 5) is relatively small, it can be shown, by using the significance tests of Joyner et al. (19) that the probability of it not being significant is less than 1%.

## TRANSITION METAL CHLORIDES

The vibrational and electronic spectra of first row transition element dihalides have been extensively studied in matrices. However, the most commonly used matrix—argon—is far from ideal for EXAFS experiments due to its high mass absorption coefficient. We therefore report data for $FeCl_2$ or $CoCl_2$ isolated in carbon monoxide, which has the additional interest that it is potentially a reactive matrix gas.

### FeCl₂ Cocondensed with Carbon Monoxide

Recently we have shown (20) that when $FeCl_2$ vapor is cocondensed with argon doped with carbon monoxide a new species *trans-*$[Fe(CO)_4Cl_2]$ is obtained. The infrared spectrum shows one intense $\nu_{CO}$ band and two $\delta_{MCO}$ modes. The higher frequency of the two $\delta_{MCO}$ modes on partial (1:1) isotopic substitution gives a quintet identifying the presence of four equivalent carbonyl groups, and there are close analogies to the previously reported *trans-*$[Fe(CO)_4I_2]$ obtained from the action of light or heat on the *cis-*form (21).

In our EXAFS experiment, the vapors from $FeCl_2$ heated at 450–500°C were cocondensed with a large excess of carbon monoxide on to a cold beryllium window. The deposition period was of the order of 2–3 h. The Fourier filtered spectrum (1–5 Å) and Fourier transform are shown in Fig. 5. Although a theoretical spectrum was computed, it was not possible to obtain a good fit on the basis of $[Fe(CO)_4Cl_2]$ alone. Indeed visual inspection of Fig. 5 identifies the problem. The carbon shell is not well resolved from the chlorine shell; and the chlorine shell itself is too intense.

It is clear from the intense oxygen shell at ca 3Å that linear Fe-C-O groups are indeed present, in agreement with the infrared data. We then attribute the enhanced amplitude of the Fe-Cl peak to the presence of uncomplexed $FeCl_2$. In principle the matrix infrared studies could resolve the problem by identifying "free" $FeCl_2$ molecules. However, it is well known that M-Cl modes of transition metal chlorides can shift substantially when they are isolated in carbon monoxide matrices. Further, the bands are frequently broad and featureless.

The Fe-Cl distance obtained from the EXAFS experiment is ca 2.25 Å, which compares favorably with the value of 2.15 Å found from electron diffraction of $FeCl_2$ vapor (22), allowing for the expected increase in bond length on complex formation.

### Crystalline $CoCl_2$ and $CoCl_2$ in CO at 10 K

The K-edge spectrum of $CoCl_2$ in CO, measured at 10 ($\pm$ 2) K is dominated by two shells of atoms at about 2 Å and 2.8 Å (Fig. 6). The first shell can be assigned to two Cl atoms, although small contributions owing to additional species cannot be ruled out. The refined value of the Co-Cl distance is 2.19 Å with $\alpha$ equal to 0.009 Å$^2$ (Table 2). The second shell may be identified as consisting of light elements, with an occupation number of 5 or greater. Models that included additional Cl atoms in the second shell or bonded terminal carbonyl groups were quickly ruled out as the heavier Cl atoms give a very different phase and amplitude envelope at high k. Linear terminal carbonyl groups are readily eliminated as they would show a short metal-carbon bond and a very significant oxygen shell resulting from strong multiple scattering contributions as discussed above.

A reasonable fit was obtained for a model that involved 4 carbon and 4 oxygen atoms at 2.77 Å, which are taken to be a nonbonded distance. During refinement, the distances and values of $\alpha$ were constrained so that the values were the same for C and O, as the information content of the spectrum did not merit the introduction of additional variables. It is apparent from the Fourier transform (Fig. 6) that the peak is asymmetric with a maximum at a rather longer distance than 2.77 Å. This nonGaussian disorder, and the higher value of $\alpha$ will both contribute to significantly under-estimating the mean interatomic distance.

A structural model compatible with the EXAFS data involve substituting two CO molecules in cubic CO (23) by Cl so that the Co lies at the centre of the unit cell as suggested in Fig. 7. Rotation of the four nearest neighbor CO molecules toward the cube faces is also required. Although the real situation is undoubtedly more complex it explains the EXAFS data and generates reasonable distances for van der Waals bonds. Solid $CoCl_2$ measured at room temperature, was also analyzed in order to validate the phaseshifts and also to obtain a value of AFAC. A reasonable fit (Fig. 6) was obtained that showed good agreement with crystallographic parameters for the first two shells (24). The slightly low first shell distance can again be partly explained by the relatively high disorder. The more remote shells are less precise because of strong correlation effects between variables of different shells, but show a clear correspondence to crystallographic data.

## CONCLUSION

In this report of our preliminary work we have been able to show that the combination of matrix isolation and EXAFS can yield significant data that is not obtainable by other techniques. The bond lengths of high

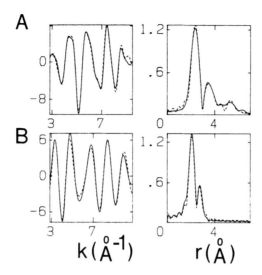

Fig. 6. EXAFS x $k^3$ (left) and Fourier transform EXAFS x $k^3$, corrected for phaseshift (right) for: A Crystalline $CoCl_2$ at 298 K; B $CoCl_2$ in CO matrix at 10 K; (———) Fourier filtered experimental data. (----) Best fit theory using parameters in Table 2.

temperature species can be obtained from gas phase electron diffraction or matrix EXAFS. The latter can also provide information about host-matrix interaction and, in particular, the interaction of coordinatively unsaturated species with reactive hosts.

## ACKNOWLEDGMENTS

We thank J. Evans for help; the Director of Daresbury Laboratory for access to the Synchtrotron Radiation Source; the Science and Engineering Research Council for financial support.

## APPENDIX

### Data Reduction and Analysis

The oscillatory part of the absorption spectrum, $\chi(E)$, is given by

$$\chi(E) = [\mu(E) - \mu_a(E)]/\mu_a(E)$$

where $\mu_a(E)$ is the atomic contribution that varies relatively smoothly away from an absorption edge. The total absorbance, $\mu(E)$, is obtained from experimental data by subtraction of a pre-edge background function. The modulus of the Fourier transform

$$F(r) = \int k^n \chi(k) e^{-i2kr} dk$$

gives a radial distribution function with peaks corresponding approximately to shells of scattering atoms. Distances are shorter than the actual

Table 2
Optimum Parameters for Crystalline $CoCl_2$ and $CoCl_2$ in CO at 10 K[a]

| $CoCl_2$ (crystalline, 298 K) R = 13.1 | | | | $CoCl_2$ in CO (10 K) R = 14.3 | | |
|---|---|---|---|---|---|---|
| Atoms | $r_{X-ray}$ | $r_{exafs}$ | $2\sigma^2_{exafs}$ | Atoms | $r_{exafs}$ | $2\sigma^2_{exafs}$ |
| 6 Cl | 2.45 | 2.43(0) | .023(0) | 2 Cl | 2.19(0) | .009(0) |
| 6 Co | 3.55 | 3.53(0) | .028(1) | 4 C | 2.77(3) | .029(1) |
| 6 Cl | 4.32 | 4.24(1) | .035(3) | 4 O | 2.77(3) | .029(1) |
| 2 Cl | 4.44 | 3.99(2) | .015(4) | | | |
| 12 Cl | 5.59 | 5.49(2) | .028(6) | | | |
| 12 Cl | 5.69 | 5.79(2) | .045(2) | | | |

[a]*See* footnotes to Table 1.

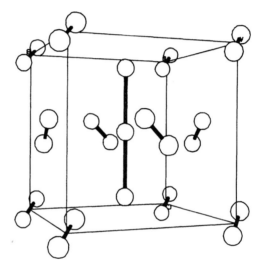

Fig. 7.   EXAFS compatible model of $CoCl_2$ in cubic carbon monoxide.

interatomic distances by 0.1 to 0.6 Å, depending on the phaseshift experiended by the photoelectron. Peaks cannot usually be resolved beyond about 5 to 10 Å as the amplitude of the contribution of a shell to $\chi(k)$ decays as $1/r^2$. the wavevector k is given by

$$k^2 = (8\pi^2 m_e/k^2)(E - E_{zero})$$

where the threshold energy $E_{zero}$ should strictly be measured from the lowest continuum state rather than the experimentally observed absorption edge. $\chi(E)$ was extracted by approximating $\mu_a$ by polynomials or spline functions. The optimum atomic background function was selected using criteria such as the minimisation of low frequency contributions in $|F(r)|$ and maximising the correspondence between $\chi(k)$ and a Fourier filtered spectrum obtained by back-transforming F(r). The window used in Fourier filtering at this stage was selected so as to exclude contributions of too low or too high frequency to represent significant shells of scattering atoms. Background subtraction is often the most difficult stage of data analysis and frequently determines the useful data range at low k. At high k, where the EXAFS is progressively damped, signal to noise ratio is the major limitation. The background subtraction process is illustrated in Fig. 8, in which the inadequacy of the background function is apparent in the low frequency contribution to the Fourier transform and in the large difference between $\chi(k)$ and the back-transformed spectrum below $k = 6$ Å$^{-1}$.

The spectra were analyzed by fitting $k^n\chi(k)$ with a theoretical spectrum. The weighting term of $k^2$ or $k^3$ is used to generate a function whose amplitude is approximately equal over all k. The theory was calculated using the spherical wave formulation of Lee and Pendry (9). The data required for the theory calculation are:

1.  Atomic phaseshifts for the central (excited) atom and scattering atoms. Ab-initio phaseshifts were calculated within single element clusters using the prescription of Mattheis (25), with a muffin-tin radius equal to the approximate covalent or van der Waals radius as appropriate. Where this resulted in substantially different values of the inter-sphere potential for different atoms, the term $E_o$ which determines the origin of the wavevector was allowed to vary for each shell to ensure that contributions from all shells were in phase. Relativistic Hartree-Fock atomic potentials were used in the calculation. The excited atom was modeled by assuming that the core hole was screened by an additional valence electron and that atomic wave functions had fully relaxed after creation of the core hole. 26 phaseshifts ($l = 0$ to 25) were used in theory calculations. The suitability of the phaseshifts is best tested by analysis of suitable model compounds for which interatomic distances are known by other means. EXAFS is rather insensitive to atom type, although differences in atomic number as large as one row of the periodic table can be detected. It is usually possible, for example, to obtain fits to a carbon shell using oxygen phase shifts, obtaining different distances and amplitude dependent factors, and it is usually necessary to rely on additional chemical information when devising theoretical models to be tested.

2.  **The interatomic distances (r)**

Distances for significant shells were refined from initial values taken from X-ray structure determinations or calculated for the structural model being tested. The precision of the distances is typically $\pm .01$ to .03 Å for the first shell, depending on quality of data, degree of thermal or structural disorder and other factors. For more remote shells the precision is usually poorer, particularly where several shells of low occupation number are closely spaced.

3.  **The Debye-Waller factors** $e^{-\alpha k^2}$**, where** $\alpha = 2\sigma^2$ **is twice the mean square variation in interatomic distance for the shell**

These terms include both thermal and static disorder, which are assumed to be Gaussian. Correlation in the thermal motion of neighboring atoms causes $\alpha$ to rise significantly with interatomic distance. During refinement $\alpha$ is strongly correlated with other

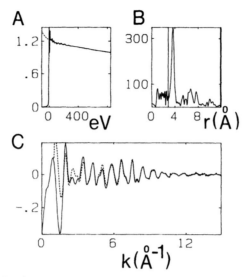

Fig. 8.    Background subtraction for solid Kr (10 K): A Absorbance (——) and atomic background function (----); B Fourier transform of EXAFS x $k^3$ with window; C EXAFS x k (——) and backtransform (----).

amplitude dependent variables and also depends on the pre-edge background function and on experimental factors. The precision of $\alpha$ is, therefore, probably no better than $\pm 30\%$.

### 4. Occupation number N

Occupation numbers cannot be obtained with great precision owing to correlation with other variables as discussed above. When analyzing a known compound the correct value was assumed. When testing a model for an unknown system the only constraint is that the Debye-Waller factors and other amplitude dependent variables are physically realistic, leading again to a precision of $\pm 30\%$.

### 5. AFAC

This is a constant term accounting for amplitude reduction due to processes which do not contribute towards EXAFS, such as multielectron excitation. It is largely dependent on the central atom, but its refined value may reflect inadequacies in experimental technique and background subtraction. Here we used values obtained from known compounds and transferred them to unknowns, thus reducing errors in the precision of $\alpha$ and N.

### 6. VPI

Is the constant imaginary potential that describes the lifetime of the final state. It mainly affects the amplitude of the EXAFS at low k. As it depends on both central and scattering atoms it can be transferred from a known compound only if the compounds are similar. In general values decrease (become more negative) with atomic number and are higher in insulators than conductors.

In most cases only single scattering events were considered, but multiple scattering to third order was included when fitting $Fe(CO)_4I_2$ as discussed above. Multiple scattering contributions were calculated using the small atom approximation (26).

Fourier filtered spectra were fitted using all shells that were considered to be statistically significant (10). The r-space window was selected so as to exclude any additional contributions. The goodness of fit was measured using a factor R (10) defined as

$$R = [\int k^n |\chi_t(k) - \chi_{exp}(k))| \, dk / \int k^n |(\chi_{exp}(k))|] \, dk \times 100$$

although R is rather sensitive to the choice of the minimum value of k, a value of less than 10 can always be taken as an indication of a convincing fit. Errors quoted in tables are the statistical uncertainties obtained during least-squares refinement. These ignore correlations between variables and systematic errors. The errors thus indicate the relative accuracy of variables in different shells or different compounds rather than their absolute precision.

# REFERENCES

1. Margrave, J. L., *The Characterization of High-Temperature Vapors*, John Wiley, New York, 1967.
2. Moskovits, M., and Ozin, G. A. S., *Cryochemistry*, John Wiley, New York, 1976.
3. Perutz, R. N., *Chem. Rev.* **85**, 77 (1985).
4. Weltner, W., *Ber. Bunsenges Phys. Chem.* **82**, 80, (1978); Hastie, J. W., Hauge, R. H., and Margrave, J. L., *Ann. Rev. Phys. Chem.* **21**, 475 (1970).
5. Sham, T. K., *Acc. Chem. Res.* **19**, 99 (1986).
6. Purdum, H., Montano, P. A., Shenoy, G. K., and Morrison, T., *Phys. Rev. B*. **25**, 4412, (1982); Niemann, W., Malzfeldt, W., Rabe, P., and Haensel, R., *Phys. Rev. B* **35**, 1099 (1987).
7. Beattie, I. R., Blayden, H. E., Hall, S. M., Jenny, S. N., and Ogden, J. S., *J. Chem. Soc. Dalton* 666 (1976).
8. Pollack, G. L., *Rev. Mod. Phys.* **36**, 748 (1964).
9. Lee, P. A. and Pendry, J. B., *Phys. Rev. B* **11**, 2795 (1975).
10. Binsted, N., Cook, S. L., Evans, J., Greaves, G. N., and Price, R. J., *J. Amer. Chem. Soc.* **109**, 3669 (1987).
11. Marques, E. C., Sandstrom, D. R., Lytle, F. W., and Greegor, R. B., *J. Chem. Phys.* **77**, 1027 (1982).
12. Eisenberger, P. and Brown, G. S., *Solid State Comm.* **29**, 481 (1979).
13. Malzfeldt, W., Niemann, W., Rabe, P., and Haensel, R., *Springer Proc. Phys.* **2**, 445 (1984).
14. Maltzfeldt, W., Niemann, W., Rabe, P., and Schwentner, N., *Springer Ser. Chem. Phys.* **27**, 203 (1983).
15. Hieber, W. and Bader, G., *Ber. Chem. Ges.* **61**, 1717 (1928).
16. Butler, I. S. and Spendjian, H. K., *J. Organomet. Chem.* **18**, 145 (1969); Pankowski, M. and Bigorgne, M., *J. Organomet. Chem.* **19**, 393 (1969); Bigorgne, M., Poilblanc, R. and Pankowski, M., *Spec. Chim. Acta* **26A**, 1217 (1970).
17. Binsted, N., Evans, J., Greaves, G. N., and Price, R. J., *J. Chem. Soc. Chem. Comm.* 1330 (1987).

18. Birck, J. L., Le Cars, Y., Baffier, N., Legendre, J. J., and Huber, M., *Comp. Rend. Acad. Sci. Paris* **C273**, 880 (1971).
19. Joyner, R. W., Martin, K. J., and Meehan, P., *J. Phys.* **C20**, 4005 (1987).
20. Beattie, I. R., McDermott, S. D., Mathews, E. A., Millington, K. R., and Willson, A. D., *Angew. Chem. Int. Ed. Eng.* **27**, 1161 (1988).
21. Noack, K., *Helv. Chim. Acta* **45**, 1847 (1962).
22. Hargittai, M., *Coord. Chem. Rev.* **91**, 35 (1988).
23. Krupskii, I. N., Prokhvatilov, A. I., Erenburg, A. I., and Yantsevich, L. D., *Phys. Stat. Solid* **19A**, 519 (1973).
24. Wilkinson, M. K., Cable, J. W., Wollan, E. O., and Koehler, W. C., *Phys. Rev.* **113**, 497 (1959).
25. Mattheiss, L. F., *Phys. Rev.* **A133**, 1399 (1964).
26. Gurman, S. J., *J. Phys.* **C21**, 3699 (1988).

# High Temperature X-Ray Diffraction and Landau Theory Investigations of Thermal Symmetry-Breaking Transitions

## The W Point of Fm3m and the Structure of $NbN_{1-x}$

HUGO F. FRANZEN* AND SUNG-JIN KIM

*Ames Laboratory-DOE and Iowa State University, Ames, IA 50011*

## ABSTRACT

The Landau theory of symmetry and phase transitions is applied to the thermal symmetry-breaking transitions corresponding to the special point W in the case of the NaCl-type structure. The results are shown to include the ordered structure of defect niobium mononitride.

**Index Entries:** Structure of $NbN_{1-x}$; Landau theory; phase transitions; distorted NaCl-type structure.

## INTRODUCTION

Changes in symmetry in solid materials can occur via nucleation and growth mechanisms or via continuous changes in structure punctuated by points of symmetry breaking. Changes of the first kind are first-order phase transitions while those of the second kind are second-order. Landau's (1–3) theory of symmetry and phase transitions provides four conditions that a phase transition must meet in order that it can occur continuously, i.e., as a second-order process.

1. The symmetries of the two solids must be in a group–subgroup relationship;

*Author to whom all correspondence and reprint requests should be addressed.

High Temperature Science, Vol. 26     © 1990 by the Humana Press Inc.

2. The difference between the particle density functions of the two solids must project onto the basis functions of a single irreducible representation of the high-symmetry group;

3. Those basis functions must form no third-order invariant combinations; and

4. There must be no wave vector that is invariant to all of the operations of the point group of the wave vector to which the symmetry change corresponds.

## Definition of Symmetry-Breaking Transitions

Symmetry changes can meet all of the criteria of the theory and still be first order. Transitions that are first-order in spite of meeting all of the Landau criteria, together with those that are second-order and those that are first-order because they fail to meet the third or fourth criteria, form a class of transitions that have in common the characteristic that they are characterized by a single irreducible representation (irr. rep.). This means that the average particle density function of the low symmetry phase, $<\rho>$, can be related to that of the high symmetry phase, $<\rho^o>$, by a single order parameter, $\eta$, multiplying a combination of particle density functions, $\phi_i$, which are basis functions for a single irreducible representation

$$<\rho> = <\rho^o> + \eta\Sigma\gamma_i\phi_i \qquad (1)$$

where $\gamma_i$ is a normalized coefficient ($\Sigma\gamma_i^2 = 1$).

It follows that it is possible, in principle, to consider the Gibbs free energy of the system as a function of $\eta$ such that $G \longrightarrow G^o$ as $\eta \longrightarrow 0$. The structural implication is that as, for example, the temperature of a lower-symmetry phase increases and as a consequence $\eta$ decreases toward zero, the structure of the lower-symmetry phase changes continuously in the direction of (but not necessarily all the way to) that of the higher-symmetry phase. Such behavior is different from that of a transition, such as one between phases with no meaningful symmetry relationship, for which no continuous path can be conceived. Therefore, those transitions with a single order parameter corresponding to a single irr. rep., whether they are first- or second-order, are meaningfully characterized as "symmetry-breaking" transitions. The symmetry-breaking transitions considered here are those that occur with changing temperature (although the changes can presumably be brought about through the change of other variables such as composition or pressure) and are thus identified as "thermal symmetry-breaking" transitions. The examples considered are those corresponding to the W point of Fm3m, a heretofore incompletely analyzed case important to the understanding of symmetry breaking in $NbN_{1-x}$.

## Classification of Symmetry Breaking Transitions

Symmetry breaking transitions can be categorized according to whether they are pure displacive (positions continuously change away from the high-symmetry positions without any changes in site occupation, e.g., the NiAs → MnP distortion (4), pure order-disorder, e.g., the β → β' brass distortion (5) or a mixture of both (e.g., the $NbN_{1-x}$ distortion discussed below). They can also be categorized according to whether the broken symmetry does not include pure translational operations (a transition that occurs at $\vec{K} = 0$ or the $\Gamma$ point) or does include broken transitional symmetry, resulting in "superstructure."

Thermal symmetry breaking transitions are purposely or inadvertently studied by a variety of techniques, one of which is high-temperature X-ray diffraction. The information obtained by this technique can be difficult to interpret for two reasons: the superstructure reflections, when they occur, are typically weak (because their intensity vanishes with $\eta$); and the line splittings, when they occur, typically require high resolution in order that they be correctly characterized. Thus, when these sometimes subtle effects occur, both high resolution and high sensitivity are required for their study, requirements that are somewhat at odds with high-temperature diffractometry. Thus, any theoretical help in limiting choices is more than welcome and the Landau theory plays a crucial role.

The group-subgroup condition is, in many cases, not sufficiently restrictive to be useful. For example, if one considers subgroup types (in a number of cases there is more than one subgroup of the same type, e.g., Fm3m $(0_h^5)$ has a Pm3m $(0_h^1)$ subgroup that, in turn, has a Fm3m subgroup, ad infinitum) then all space groups except some of those that correspond to primitive hexagonal cells are subgroups of Fm3m. Thus, in order to consider possible symmetry breaking transitions in the NaCl-type structure, it is necessary to consider the second, third, and fourth conditions of Landau. Application of the standard test for vector invariance, i.e., the test for intersection of the antisymmetrized square representation and the vector representation (2) has shown that all of the irr. reps. of Fm3m at W meet the fourth condition.

The thermal behavior of nitrogen deficient niobium mononitride has been investigated by neutron diffraction (6) and the c/a variation with temperature indicates a thermal symmetry breaking transition, probably of the second order. The reported structure does not fit this observation. A more recent neutron diffraction result (7) yielded a tetragonal cell with c = $2a_{NaCl}$ and I4/mmm symmetry, in agreement with the conclusion reached below. The high-temperature X-ray diffraction patterns of $NbN_{1-x}$ obtained in our laboratory confirmed the apparent symmetry-breaking character of the transition, i.e., c/a approached unity as the temperature approached 1200°C and the intensities of the weak superstructure reflections similarly diminished toward zero at 1200°C. Accordingly, the structures corresponding to the various irreducible representations of Fm3m symmetry were examined. It was clear from the outset that the low temperature diffraction pattern could be indexed on a tetragonal basis, and thus irr. reps. leading to the tetragonal symmetry were sought. It is also clear, because of the existence of superstructure reflections, that the irr. rep. in question is at some reciprocal space point other than $\vec{K} = 0$. The fourth condition of Landau restricts irr. reps., resulting in commensurate structures to the high-symmetry points of the first Brillouin zone (2) and thus a solution at X, L, or W yielding tetragonal symmetry was sought.

The W point corresponds to the set of vectors: 1/2,0,1; 0,1/2,1; 1/2,1,0; 0,1,1/2; 1,1/2,0; 1,0,1/2. There are three equivalent centered-tetragonal supercells implied by these wave vectors, for example both 1/2,0,1 and 1/2,1,0 satisfy a body-centered tetragonal cell with $\vec{c} = 2\vec{a}_{cubic}$ and $\vec{a} = \vec{b}_{cubic}$. Corresponding to this $\vec{K}$ point there are four one-dimensional and one two-dimensional irr. rep. (8). Previous reports of the structures that result from these irr. reps. in the case of the NaCl-type structure (9,10) were incomplete. These cases are considered below in two parts, first the structures that can result from the one-dimensional representations and then those that can result from the two-dimensional irr. rep.

## The One-Dimensional Small Representations

There are four one-dimensional representations of Fm3m at the W point, the basis functions for each of which can be taken as a product of a function totally symmetric to the translations and a function that transforms according to the small representation. The function totally symmetric with respect to the translations (and also all of the operations in the group of the wave vector) for $\vec{K} = \overset{*}{\vec{a}} + \vec{c}^{*}/2$ is

$$e^{\pi i z}\cos 2\pi x + e^{-\pi i z}\cos 2\pi y \qquad (2)$$

which is complex. Since the distortion functions of Eq. (1) must be real this is combined with its complex conjugate to yield

$$\phi_1 = \cos \pi z (\cos 2\pi x + \cos 2\pi y) \qquad (3)$$

which is centrosymmetric in addition to having the $\bar{I}4m2$ subsymmetry of the complex function. In earlier reports, (9,10) the complex nature of the basis function was overlooked, and, thus, it was reported that a transition from Fm3m to $\bar{I}4m2$ could occur continuously at the W point. The symmetry of the real function given above is I4/mmm. The structure of $NbN_{1-x}$ in this space group differs only from that previously reported in the fixed position of the Nb atom at 0,1/2,0 (which was at 0,1/2,0.0034 in the structure refined in $\bar{I}4m2$ (9)). The structure refined equally satisfactorily in I4/mmm and thus this is the preferred description. The result is summarized in Table 1. The space group $\bar{I}4m2$ could result from an irr. rep. of I4/mmm at $\vec{K} = 0$ compatible with the totally symmetric irr. rep. of Fm3m at W, and thus a continuous symmetry breaking to $\bar{I}4m2$ through I4/mmm would be possible, but is not required by the observed data. A second of the four one-dimensional irr. reps. also yields I4/mmm (with a different origin) and the other two yield I4/mcm.

The structure that results from transitions from NaCl-type to I4/mcm has atoms in fixed 4-four positions (0,0,0; 0,1/2,1/4; 0,0,1/4 and 0,1/2,0), and thus results from a pure order-disorder transition. There are super-structure reflections in the $NbN_{1-x}$ pattern that result from displace-

Table 1
Refined Parameters for $NbN_{1-x}$ in Tetragonal Cell

| Atom | x | y | z | Occupancy |
|------|---|---|---|-----------|
| NI   | 0 | 0   | 0       | 0.16 |
| N2   | 0 | 0   | 1/2     | 2.00 |
| N3   | 0 | 1/2 | 1/4     | 4.00 |
| Nb1  | 0 | 0   | 0.24373 | 4.00 |
| Nb2  | 0 | 1/2 | 0       | 4.00 |

$a$ = 4.3860 (2).
$c$ = 8.6606 (5).
R = 13.93.
$R_w$ = 18.90.
Bragg R = 2.97.

ment of Nb atoms from their ideal positions in the NaCl-type structure. Therefore this space-group can be ruled out in this case.

The real, totally-symmetric basis function corresponding to $\vec{K} = \pm(\vec{a}^* + \vec{c}^*/2)$ is $\phi_1 = \phi(x,y,z)$ as given above. The complete irreducible real representation is generated by the basis set $\phi_1 = \phi(x,y,z)$, $\phi_2 = \phi(y,z,x)$, and $\phi_3 = \phi(z,x,y)$ and the generalized particle density function is

$$\langle\rho\rangle = \langle\rho^o\rangle + (\gamma_2\phi_1 + \gamma_2\phi_2 + \gamma_3\phi_3)\eta \qquad (4)$$

to which the Gibbs free energy expansion (to fourth order) is

$$G = G^o + A\eta^2 + [C_1 + C_2(\gamma_1^4 + \gamma_2^4 + \gamma_3^4)]\eta^4 \qquad (5)$$

This function exhibits two possible minima subject to $\gamma_1^2 + \gamma_2^2 + \gamma_3^2 = 1$, namely, $\gamma_1 = \gamma_2 = \gamma_3 = 1/\sqrt{3}$ when $C_2 > 0$, and $\gamma_1 = 1$, $\gamma_2 = \gamma_3 = 0$ (or equivalent) when $C_2 < 0$. The latter solution is the I4/mmm solution discussed above. The former is a solution with the symmetry of $\phi_1 + \phi_2 + \phi_3$. This function has the translational symmetry of a cubic lattice with $\vec{a} = 2\vec{a}_{NaCl}$ and space-group symmetry Pm3m. Similarly a cubic solution with $\vec{a} = 2\vec{a}_o$ and Pn3m symmetry corresponding to the I4/mcm solution results when $\gamma_1 = \gamma_2 = \gamma_3 = 1/\sqrt{3}$ in the second case.

### The Two-Dimensional Small Representation

At W there is a two-dimensional small representation for which $\phi_1 = \phi(z,x) = \cos\pi z \sin 2\pi x$ and $\phi_2 = \phi(z,y)$ form a basis. The complete six dimensional irreducible representation is generated by the basis set $\phi_1 = \phi(z,x)$, $\phi_2 = \phi(z,y)$, $\phi_3 = \phi(x,y)$, $\phi_4 = \phi(x,z)$, $\phi_5 = \phi(y,z)$, $\phi_6(y,x)$. These functions permute under the symmetry operations such that the invariant fourth-order combinations are

$$\phi_1\phi_2 + \phi_3\phi_4 + \phi_5\phi_6$$

$$\phi_1\phi_3 + \phi_3\phi_5 + \phi_1\phi_5 + \phi_2\phi_4 + \phi_4\phi_6 + \phi_2\phi_6$$

$$\phi_1\phi_4 + \phi_3\phi_6 + \phi_2\phi_5$$

$$\phi_1\phi_6 + \phi_2\phi_3 + \phi_4\phi_5$$

Table 2

| Nonzero coefficients | Space group | Lattice parameters |
|---|---|---|
| $\gamma_1 = 1$ | Pmma | $a \cong b \cong a°,\ c \cong 2a°$ |
| $\gamma_1 = \gamma_2 = 1/\sqrt{2}$ | Fmmm | $a \cong b \cong 1/\sqrt{2}a°,\ c = 2a°$ |
| $\gamma_1 = \gamma_4 = 1/\sqrt{2}$ | P4/mbm | $a \cong 2a°,\ c \cong a°$ |
| $\gamma_1 = \gamma_6 = 1/\sqrt{2}$ | P4/nmm | $a \cong 2a°,\ c \cong a°$ |
| $\gamma_1 = \gamma_3 = \gamma_5 = 1/\sqrt{3}$ | Pa3 | $a \cong 2a°$ |
| $\gamma_1,\ \gamma_2,\ \gamma_4$ | P2$_1$/c | $a \cong a°,\ b \cong c \cong 2a°,\ \beta \cong 90°$ |
| $\gamma_1,\ \gamma_2,\ \gamma_6$ | P2$_1$/m | $a \cong b \cong 2a,\ c \cong a°,\ \beta \cong 90°$ |
| $\gamma_1,\ \gamma_4,\ \gamma_6$ | Pnma | $a \cong b \cong c \cong 2a°$ |
| $\gamma_1 = \gamma_2 = \gamma_3 = \gamma_4 = \gamma_5 = \gamma_6 = 1/\sqrt{6}$ | R$\bar{3}$m | $a \cong b \cong c \cong 2a°,\ \alpha \cong 90°$ |

It follows that G, to fourth-order, is of the form

$$G = G° + A\eta^2 + [C_1 + (\gamma_1^2\gamma_2^2 + \gamma_3^2\gamma_4^2 + \gamma_5^2\gamma_6^2)C_2 +$$
$$(\gamma_1^2\gamma_3^2 + \gamma_3^2\gamma_5^2 + \gamma_1^2\gamma_5^2 + \gamma_2^2\gamma_4^2 + \gamma_4^2\gamma_6^2 + \gamma_2^2\gamma_6^2)C_3 + \qquad (6)$$
$$(\gamma_1^2\gamma_4^2 + \gamma_3^2\gamma_6^2 + \gamma_2^2\gamma_5^2)C_4 + (\gamma_1^2\gamma_6^2 + \gamma_2^2\gamma_3^2 + \gamma_4^2\gamma_5^2)C_5]\eta^4$$

This form for the Gibbs free energy, which we have found also to be appropriate to the X points of Fm3m and Im3m, yields the solutions and symmetries given in Table 2. As was previously found to be the case for the X point (10), there result from the fourth-order invariants given above three solutions with variable contributions from several basis functions. These unusual solutions permit consecutive second-order transitions corresponding to the same irreducible representation, for example, an NaCl-type solid could transform via a second-order transition from Fm3m to P4/nmm symmetry at some T, and continue to Pnma or P2$_1$/m (corresponding to nonzero values for $\gamma_1$, $\gamma_4$, and $\gamma_6$ or for $\gamma_1$, $\gamma_2$ and $\gamma_6$, respectively) at some lower T. This kind of possible thermal symmetry breaking has not yet been observed, however, as has been demonstrated here, it is consistent with Landau theory in the cubic cases with two-dimensional small representations and six equivalent wave vectors.

## ACKNOWLEDGMENT

The Ames Laboratory-DOE is operated for the US Department of Energy by Iowa State University under Contract No. W-7405-Eng-82. This research was supported by the Office of Basic Energy Sciences, Materials Sciences Division.

## REFERENCES

1. Landau, L. D. and Lifschitz, E. M., *Statistical Physics*, Pergamon, London, 1962, chapter 14.

2. Tolédano, J.-C. and Tolédano, P., *The Landau Theory of Phase Transitions*, World Scientific, New Jersey (1987).
3. Franzen, H. F., *Physical Chemistry of Inorganic Crystalline Solids*, Springer Verlag, Heidelberg (1986).
4. Franzen, H. F., Haas, C., and Jellinek, F., *Phys. Rev. B* **10**, 1248 (1974).
5. Beck, L. H. and Smith, C. S., *Trans. AIME* **194**, 1079 (1952).
6. Christensen, A. N., *Acta Chem. Scand. Ser. A.* **30**, 219 (1976).
7. Heger, G. and Baumgartner, O., *J. Phys. C.: Solid St. Phys.* **13**, 5833 (1980).
8. Kovalev, O. V., *Irreducible Representations of Space Groups*, translated by A. Murray Gross, Gordon and Breach, NY (1965).
9. Kim, S.-J. and Franzen, H. F., *J. Less-Common Metals* **143**, 339 (1988).
10. Franzen, H. F., *J. Less-Common Metals*, **146**, 229 (1989).

# High Temperature
# X-Ray Diffractometry of Ti-Al Alloys

ROBERT D. SHULL* AND JAMES P. CLINE

*National Institute of Standards and Technology,
Gaithersburg, MD 20899*

## ABSTRACT

High temperature X-ray diffraction, an established technique for high temperature materials characterization, has been applied to the titanium-aluminum system in order to obtain structural information on the material at elevated temperatures. *In situ* X-ray diffraction data for a titanium-45 atomic percent aluminum alloy clearly showed the disappearance of the ordered $Ti_3Al$ structure on heating to 1300°C, but with the fundamental $\alpha$-Ti diffraction peaks remaining. All diffraction peaks are indexed and prove the existence of the previously proposed $Ti_3Al + TiAl \rightarrow \alpha$ Ti eutectoid reaction near 1125°C in this alloy. No BCC $\beta$-Ti phase was detected for this alloy up to 1400°C. In addition, two sets of hexagonal $\alpha$-Ti diffraction peaks in $Ti_{55}Al_{45}$ were detected during both heating and cooling between 1250–1400°C, suggesting the formation of a new high temperature disordered hexagonal $\alpha'$ phase. This conclusion is supported by the discovery of a discontinuity in the volume expansion coefficient for $\alpha$-Ti at the low end of this temperature range. Only slight modifications to the existing Ti-Al phase diagrams are required to account for the present results. High temperature X-ray diffraction measurements on a titanium-52 atomic percent aluminum alloy also showed no $\beta$-Ti phase up to 1350°C. Debye Waller factor analysis of the $\gamma$-TiAl phase diffraction peaks for $Ti_{48}Al_{52}$ also indicated the absence of any phase changes between 850–1250°C. The modified Ti-Al phase diagram presented here includes a shift in the $\gamma$-phase transus lines to higher aluminum contents, the addition of a new $\alpha'$ phase region, and the elimination of the $\beta + \gamma$ phase field.

*Author to whom all correspondence and reprint orders should be addressed.

## INTRODUCTION

The titanium-aluminum system is a system wherein some of the high temperature phases may not be retained upon quenching (*1*). Consequently, room temperature X-ray diffraction on these materials has provided little information on the high temperature structures. This system is of technological interest because it forms the basis for many of the important titanium alloys presently used in industry. These alloys are unique in the respect that they are generally both lightweight and strong, and can be used to higher temperatures ($T \geq 300°C$). However, portions of the phase diagram for this binary system are still uncertain (*1–3*). Part of this uncertainty results from the fast kinetics of the phase transformations that occur on cooling from high temperatures, partially from the effects of short-range atomic ordering, and partially from the effects of inadvertent interstitial (oxygen, nitrogen, and hydrogen) additions (*4*).

In an effort to clarify the Ti-rich end of the Ti-Al phase diagram, a concerted study (via low temperature X-ray and neutron diffraction, differential thermal analysis, and optical and transmission electron microscopy) was initiated several years ago in this laboratory. The results of this study (*5–6*) included the development of procedures for the preparation of low interstitial content (<300 ppm O, <30 ppm H, and <50 ppm N by weight) alloys and the detection of a previously unknown low temperature eutectoid reaction near 45 at% Al at $\approx 1125°C$. Subsequently, Murray (*7*) redrew the phase diagram for this system (Fig. 1) incorporating our eutectoid discovery. In our previous studies, the structure of the high temperature phase involved in the eutectoid reaction at 1125°C in the Ti-45Al alloy (all alloy compositions in this paper are given in atomic percents) could not be determined by low temperature observations. However, from the optical and transmission electron microscopy of high temperature equilibrated samples cooled from this region, and from the magnitude of the thermal arrests in the DTA data for alloys with slightly lower Al content, it was concluded that the unknown phase could not be the body centered cubic, β, phase. It was further concluded that the unknown phase was most likely to be the disordered hexagonal close packed, α, phase of pure titanium. These results have been substantiated by more recent indirect evidence from optical and electron microscopy observations on high temperature equilibrated and both quenched and slow cooled samples (*8*). Because the identification of the above-mentioned unknown elevated temperature phase required an *in situ* structurally-sensitive tool, the present work utilizing the high temperature X-ray diffraction technique was initiated.

Recently, Waterstrat and Prince (*9*) conducted a very systematic neutron diffraction study of several Ti-Al alloys at elevated temperatures,

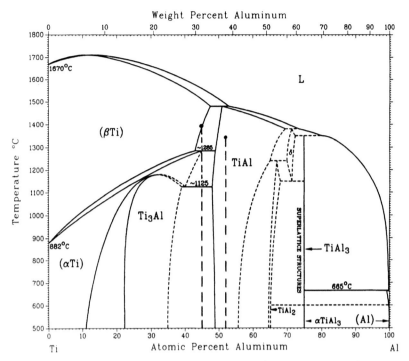

Fig. 1. Titanium-aluminum phase diagram from ref. (7), indicating (bold dashed lines) the temperature range and alloy compositions investigated in the present study.

in which they clarified portions of the binary system as well as the effects of interstitials on the phase boundaries in the Ti-rich end of this system. However, their attempt to identify the unknown high temperature eutectoid phase near $Ti_{55}Al_{45}$ was unsuccessful because of the oppositely signed scattering factors for titanium and aluminum. Their data for $Ti_{60}Al_{40}$ at 1200°C only showed the presence of one diffraction maximum (however, at a position consistent with that expected for the strongest α-Ti line).

Very recently McCullough et al. (10) examined the structure of a $Ti_{50}Al_{50}$ alloy by means of high temperature X-rays up to 1450°C. These investigators found only diffraction maxima owing to the ordered face centered tetragonal, γ-TiAl, structure ($L1_0$-type) and to an unknown hexagonal phase (which they assumed was α-Ti). Since they found no evidence of the expected β + γ phase field for their sample, McCullough et al. modified Murray's phase diagram, as shown in Fig. 2. In drawing this diagram, these investigators assumed that their observed hexagonal phase was the high temperature phase involved in the 1125°C eutectoid reaction (near $Ti_{55}Al_{45}$) and extended it up to the liquidus (thereby eliminating the undetected β + γ phase field). Unfortunately, the 50% Al composition of their alloy did not allow these investigators to study the high temperature α(?) phase to prove its structure or origin at the 1125°C eutectoid.

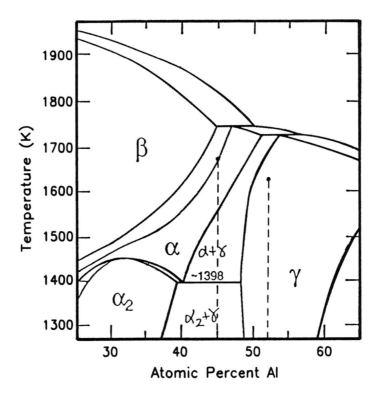

Fig. 2. A portion of the Ti-Al phase diagram (ref. *10*), indicating (bold dashed lines) the temperature range and compositions investigated in the present study.

In the present study, a well characterized (low interstitial content) alloy of lower Al content, $Ti_{55}Al_{45}$, from one of the author's previous studies (*6*) was used for this purpose together with an alloy of larger Al content, $Ti_{48}Al_{52}$, for checking the extent of the $\alpha + \gamma$ field. The high temperature X-ray diffraction technique was chosen for this study because of its high sensitivity to structural modifications in accordance with the well known Bragg diffraction equation, $n\lambda = 2d\sin\Theta$ (where $\lambda$ is the X-ray wavelength, $\Theta$ is the diffraction angle, and $d$ is the atomic spacing between the diffracting crystallographic planes). Measurement at high temperatures is especially important for investigating the present system, since some of the high temperature Ti-Al phases may not be retained on cooling, regardless of the cooling rate. Elevated temperature measurements are also required for ascertaining equilibrium reactions in this system. Such transformations must be reversible when approached from two opposite directions (e.g., via heating and cooling).

## EXPERIMENTAL PROCEDURE

The high temperature X-ray equipment used in this study was a Siemans™ theta-two theta diffractometer (Fig. 3) equipped with a Buhler™ 3000 K hot stage and a Braun™ linear position sensitive propor-

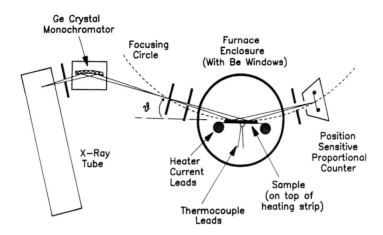

Fig. 3.   A schematic of the high temperature X-ray diffractometer used in the present study.

tional counter (PSPC). An incident beam monochrometer utilizing a curved germanium crystal was installed near the end of this study. The monochrometer reduced the tails of the diffraction peaks, eliminated the Cu $K\alpha_2$ radiation, and provided a higher peak to background ratio. The furnace chamber was a vacuum tight enclosure with Be windows. Tantalum heating elements were used: both as the main specimen heating strip (to which a Pt-Rh thermocouple, Type B, was welded) and as a secondary surrounding heating element. The surround heater is a unique feature of the Buhler™ stage and is effective at reducing temperature gradients within the sample. Tantalum heat shields were also used inside the furnace chamber to reduce heat losses to the chamber walls.

The PSPC used a carbon-coated quartz wire (with P10 methane/argon counting gas) to determine not only the quanta of diffracted X-rays, but also their position with respect to angle two theta. On entering the PSPC chamber, an X-ray quantum ionizes the counting gas, creating a voltage pulse on the quartz wire at the location closest to the ionization event. Position discrimination is achieved by measurement of the differential time of arrival of the pulse at the two ends of the wire. The 5 cm long wire with a 1.5 cm wide PSPC window set the angular range of observation to 4.5 degrees 2Θ. During the measurement of an X-ray spectrum, the whole PSPC enclosure was rotated through angle 2Θ with a multichannel analyzer (MCA) storing the data. Because the PSPC scanned several degrees simultaneously, an increase in the counting rate by over two orders of magnitude was achieved over that of a conventional detector. A typical spectrum with good statistics from 10 to 60 degrees 2Θ required approximately 20 min. One disadvantage to the PSPC was that it required regular maintenance to keep its many adjustable parameters within operational limits.

A vital component of the system was a computer control interface together with sophisticated software, which allowed the programming of any desired time-temperature profile for data collection. Consequently, data collection was reliable and time-temperature profiles were repeatable. Data presentation could also be in the form of a 3-dimensional plot of the X-ray spectra with respect to temperature. Such presentations are especially valuable for visualizing the progression of diffraction line growth or disappearance in the sequence of spectra measured at temperatures leading up to a phase transformation. Common occurrences in materials are phase transformations between similar structures. In such cases, if only a few diffraction lines are observed (because of high texture or a restricted angular range), it may be difficult to determine from two different spectra measured during the course of the suspected transformation whether any line position changes were caused by a change in structure or simply a shift in lattice constant. For these situations, observation on a single graph of a sequence of diffraction spectra measured as the suspected transformation was approached would enable the separation of such phenomena. Such a capability would have been invaluable in an earlier study (11) monitoring the phase evolution in TiCu alloys crystallized from the glassy state, wherein the two phases that formed, TiCu and $Ti_3Cu_4$, possessed very similar structures.

The specimens used in this study, $Ti_{55}Al_{45}$ and $Ti_{48}Al_{52}$ (atomic ratios), were both prepared by arc melting as indicated in (5,6), forming rods of approximately 1 cm in diameter. Following homogenization at 1000°C for 10 d in double encapsulated quartz tubes containing helium gas at one atmosphere, they were quenched into icewater. $Ti_{55}Al_{45}$ was given an additional equilibration treatment at 600°C for 1 mo encapsulated in a quartz tube. Following all these treatments, the 2 samples listed above were analyzed to contain 44.4 and 52.7% aluminum, respectively. Their interstitial contents were both less than 120 ppm oxygen, 40 ppm nitrogen, and 15 ppm hydrogen (by weight). For the X-ray measurements, the polycrystalline specimens were sliced, polished flat, and spot welded onto the tantalum heating strip. Four separate samples of $Ti_{55}Al_{45}$, which provided similar results, and one sample of $Ti_{48}Al_{52}$ were studied. Bulk samples were used in this study rather than powdered materials because our initial attempts using powders resulted in severe sample contamination owing to the large surface area and high trapped air content.

The temperatures quoted herein are those determined by the thermocouple welded to the underside of the heating strip beneath the specimen, and, therefore, serve as upper limits to the actual specimen temperature. Manual optical pyrometer observations (with and without the specimen present) suggested the specimen temperature may be lower than these values by as much as 50–75°C. However, the thermocouple temperatures provided phase transformation values in the present specimens that were consistent with the transition temperatures from previous phase diagrams. Since the optical pyrometer readings were not corrected for the emissivity of the specimen (which is higher than that of Ta by a factor of 2 and which, for example, can cause a 150°C absolute temperature error (*12*) at 1400°C for Pt, an element with an emissivity ½ that of Ta), and, since they contained interference effects from the leaded glass door of the diffractometer chamber, their accuracy was doubted. In view of this uncertainty, the highly precise thermocouple readings were used for consistency and so that subsequent adjustments (if necessary) could be made accurately.

After loading the sample into the furnace, the chamber was evacuated with a roughing pump and backfilled with titanium getter-purified helium several times before heating. Following the last evacuation (usually overnight), the sample was heated to 200°C for 2 h under vacuum to release any adsorbed water vapor on the sample and heating strip. Measurements were then performed in flowing helium gas in order to reduce vaporization of the constituents, minimize scattering of X-rays, and carry away any volatile species adsorbed to the furnace walls. Despite these precautions, however, during the many hours of the experiment, some atmospheric contaminants did manage to enter the furnace chamber and contact the sample, as evidenced by the detection of TiN and $Ti_2AlN$ in some of the spectra. The degree of contamination varied with each experiment. The data shown here for $Ti_{55}Al_{45}$ and for $Ti_{48}Al_{52}$ (Figs. 4–7, 12–13) are for samples that, at the end of the experiment, were still shiny, indicating a minimal amount of interstitial addition.

Following the 200°C treatment under vacuum, helium gas was reintroduced into the system and the sample sequentially measured in order of ascending or decreasing temperature as it was, respectively, heated to the highest temperature and cooled. Each spectrum was measured while the temperature was held constant (starting after a 3-min thermal equilibration period) and represents the data collected after ≈20 min. Between measurement temperatures the sample was heated at a rate of 2 degrees/s.

Data analysis was performed by comparing standard X-ray patterns for the known crystallographic Ti-Al phases (with and without oxygen and nitrogen present) with the measured spectra and observing the appearance and disappearance of sets of diffraction lines as the temperature was varied. This comparison, however, of room temperature standard patterns with the measured elevated temperature X-ray spectra, does require special consideration. As the temperature of the specimen is raised, the atoms become more energetic and, consequently, oscillate further from their equilibrium positions. Because this oscillation is not completely harmonic, the mean positions of the atoms do not remain constant. Thermal expansion of the lattice occurs, and changes in the **d** spacings of the atomic planes will result. However, depending on the crystal structure, not all of these **d** spacings will increase, nor at the same rate. Consequently, assignment of diffraction lines was initially performed on the low temperature patterns (where the differences from the standard patterns were the smallest) and the movement of the diffraction lines was followed as the temperature was sequentially varied (using the line's intensity and shape, in addition to closeness in position, for identification).

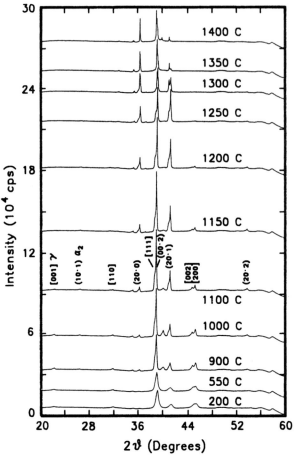

Fig. 4. X-ray intensity vs 2Θ data measured for $Ti_{55}Al_{45}$ (without mono-chromator) at the indicated temperatures during heating. Indices for the γ and $α_2$ lines are indicated by [ ] and ( ) brackets respectively. The zero intensities for the spectra have been shifted; scales are the same.

From some spectra the measured d spacings were used to calculate the lattice parameters, **a** and **c**, of the crystal structure. For precise lattice parameter determinations, high angle diffraction lines (where the error, $\Delta d/d = -cot\Theta * \Delta\Theta$, is the smallest) are preferred (13). Unfortunately, only low angle lines were measured in the present study. Since only a few peaks were observed here, extrapolations of the low angle values using the Nelson-Reilly function (14), $(cos^2\Theta/sin\Theta) + (cos^2\Theta/\Theta)$, were also highly inaccurate. Consequently, the lattice parameters presented here are the values calculated from an appropriate low angle line position. For the hexagonal phases, the **c** and **a** values were calculated from the (00·2) and (20·0) lines, respectively. In the present study, it was the change in these values with temperature that was evaluated. Even though the accuracy of these values is only a couple of percent, their relative values are much more precise.

The temperature dependence of the lattice parameters was used to calculate the linear thermal expansion coefficients: (1/a)(da/dT) and (1/dc/dT). Because the linear expansion coefficients form a second rank

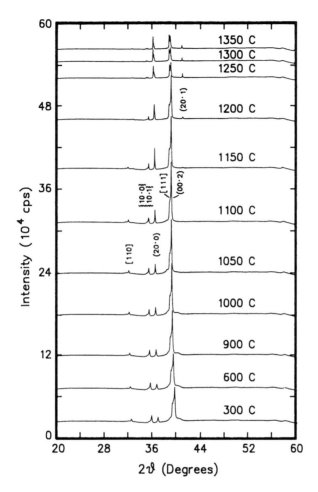

Fig. 5. X-ray intensity vs 2Θ data measured for $Ti_{55}Al_{45}$ (without mono-chromator) at the indicated temperatures during cooling (following measurement of the data in Fig. 4). Indices for the γ, $α_2$, and $Ti_2AlN$ lines are indicated by [ ], and ( ), and { } brackets respectively. The zero intensities for the spectra have been shifted; scales are the same.

tensor (15), the volume coefficient of thermal expansion could be calculated for a hexagonal structure as follows

$$(1/V) \bullet dV/dT = 2 \bullet (1/a) \bullet da/dT + (1/c) \bullet dc/dT \qquad (1)$$

where V is the volume.

For $Ti_{48}Al_{52}$ the intensities of the X-rays diffracted by the γ-TiAl phase were analyzed. The diffracted X-ray intensity, I, at a Bragg scattering position ($Θ_B$) is a function of the Lorentz Polarization factor, LP (Θ), and the scattering factor, F(H)

$$I = C \bullet m(θ) \bullet LP(θ) \bullet |F(H)|^2 \qquad (2)$$

where C is a constant of the experimental arrangement, m(Θ) is a multiplicity factor depending on the sample texture, and F(H) is expressed as

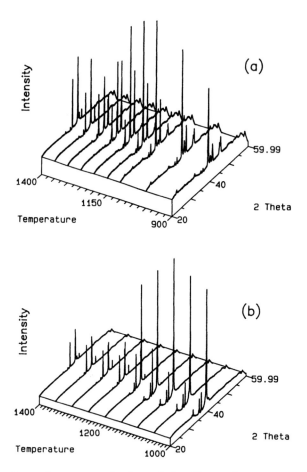

Fig. 6.   Three-dimensional plot of the X-ray data measured for $Ti_{55}Al_{45}$ during (a) heating and (b) subsequent cooling.

$$F(H)=\Sigma f_k\,T_k\exp[2\pi iH\cdot r(k)] \qquad (3)$$

In Eq. (3) the summation is over all atoms (k) in the unit cell, the bold letters designate vector quantities for the scattering vector (**H**) and the atom position vector (**r**(k)), $f_k$ is the atomic scattering factor, and $T_k$ is the temperature factor of scattering for atom k. For a monatomic structure, all the $T_k$ values are the same and F(**H**) becomes simply $T_k$ times the scattering factor without the temperature factor included: F′ (**H**) (i.e., the scattering factor usually considered in the evaluation of room temperature data). $T_k$, also called the Debye-Waller factor (16), is an exponential function of the mean squared displacement of the atom, $<u^2>$

$$T_k=\exp[-8\pi^2\sin^2\theta<u^2>/\lambda^2] \;=\; \exp[-B_k\sin^2\theta/\lambda^2] \qquad (4)$$

$$\text{with } B_k=8\pi^2<u^2>.$$

Conceptually, this effect may be understood as the effect of smearing out the electron charge distribution of an atom over a region in space rather than at a lattice point as the temperature is raised. The intensity of

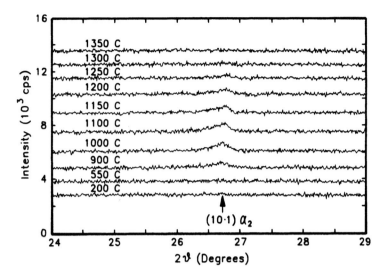

Fig. 7. Enlargement of the X-ray data shown in Fig. 4, showing the disappearance of the $(10 \cdot 1)$-$Ti_3Al$ superlattice line (near $2\Theta \simeq 26.7°$) on heating $Ti_{55}Al_{45}$. The zero intensities for the spectra have been shifted; scales are the same.

scattering is a function of the number of electrons around the lattice site (i.e., around the atom at the lattice point). Since the thermal motions of all the atoms are not coordinated, superposition of the diffracted X-rays will still be constructive only for those diffracted at the lattice sites (*16*). If the electron charge density at the lattice point is reduced, then the scattered intensity will also be decreased. Also, no broadening of the diffraction peak results from the temperature increase. Intensity measurements at elevated temperatures can, therefore, also be used to determine the mean squared displacements of the atoms in the crystals.

In the present study, it was assumed that the Ti and Al atoms have similar temperature factors (an assumption that, although not strictly correct, does not introduce large errors into the general temperature dependence of $<u^2>$). Consequently, $T_k$ could be taken outside of the summation in Eq. (3), simplifying the expression for the intensity

$$I_\theta = C \cdot m(\theta) \cdot LP(\theta) \cdot \exp[-2B_k \sin^2\theta/\lambda^2] \cdot \left| \Sigma f_k \exp[2\pi iH \cdot r(k)] \right|^2 \quad (5)$$

All of the angular dependent terms, except $\exp[-2B_k\sin^2\Theta/\lambda^2]$, in Eq. (5) may be eliminated by taking the ratio of intensities for the same diffraction line measured at two different temperatures, $T_1$ and $T_2$

$$I_\theta(T_1)/I_\theta(T_2) = \exp[-2B_k(T_1)\sin^2\theta/\lambda^2]/\exp[-2B_k(T_2)\sin^2\theta/\lambda^2] \quad (6)$$

$$= \exp\{-2[B_k(T_1) - B_k(T_2)] \cdot \sin^2\theta/\lambda^2\}$$

Note, this ratio also accounts for the sample texture factor, $m(\Theta)$. From the slope of $Ln[I_\theta(T_1)/I_\theta(T_2)]$ vs $\sin^2\Theta/\lambda^2$ graphs, values for

$[B_k(T_1) - B_k(T_2)]$ were determined. Only differences in mean squared displacements $<u^2(T_1)> - <u^2(T_2)>$, calculated from Eq. (5), are obtained by this technique. If the sample geometry allowed a good calculation of $m(\Theta)$ and $LP(\Theta)$, plots of the logarithm of $I/m(\Theta)LP(\Theta)$ vs $\sin^2\Theta/\lambda^2$ of the reference temperature data would provide a unique value for $<u^2(T_2)>$. However, if $T_2$ is kept constant at some reference temperature ($=T_R$) while $T_1$ is varied, then this technique provides a method for obtaining the temperature dependence of the mean squared displacements. In the analysis of the $Ti_{48}Al_{52}$ spectra herein, the reference X-ray spectrum was that measured at 850°C ($=T_R$).

Temperature increases are also accompanied by an increase in the background intensity of the scattered X-rays. This intensity, which occurs in all directions, is called thermal diffuse scattering (TDS). TDS essentially comes from those X-rays scattered by the smeared electron density located away from the mean lattice positions. This intensity will also be a function of the mean squared displacements, $<u^2>$, of the atoms and will increase with increasing values of the $\sin^2\Theta/\lambda^2$ (17). Consequently, in the Debye-Waller factor analysis of the diffraction peak intensities described above, the background contribution was first subtracted.

## RESULTS AND DISCUSSION

Figures 4 and 5 show a set of X-ray diffraction spectra sequentially measured (before installation of a monochromator) for $Ti_{55}Al_{45}$ as the sample was first heated from room temperature to 1400°C and then cooled back to room temperature. Intensity variations in the major peaks can be more readily discerned in a 3-dimensional composite of these spectra, as shown in Fig. 6. Initially, the sample was comprised of two phases: $Ti_3Al$ and TiAl (also denoted $\alpha_2$ and $\gamma$, respectively). However, note from the lowest temperature spectrum (200°C) in Fig. 4 that the (00·2) $Ti_3Al$ line is the most intense diffraction peak present for this phase. In the absence of sample texture, the (20 · 1) $Ti_3Al$ line would be the strongest (18). Even though this was a polycrystalline sample, the casting and homogenization processes have resulted in a preponderance of (00 · 2) $Ti_3Al$ planes parallel to the specimen surface. This is a normal occurrence, but does indicate that for these samples it is not correct to use intensity ratios based on the randomness assumption. A second feature of the room temperature spectrum is the broad linewidth. This is partially a consequence of cold working the specimen while cutting and polishing flat prior to the measurements. Strain relief annealing may be observed to occur during the first three spectra shown in Fig. 4—the lines sharpen until at 900°C the overlapping (002) and (200) $\gamma$-TiAl lines may be resolved near $2\Theta \simeq 45°$. The remaining large breadth after the strain relief at the base, especially of the high intensity lines, is a consequence of an incorrectly adjusted PSPC.

Between 900–1200°C there are unassigned diffraction lines at $2\Theta \simeq$ 35° and 40°. These are the (10·0) and (00·6) lines, respectively, for a ternary $Ti_2AlN$ compound (*19*), which, in a previous study of the Ti-Al system (*20*), was mistaken for a novel $Ti_2Al$ compound (*see* ref. *21* for a clarification of the situation). These low intensity lines (indicating the presence of only a very slight amount of atmospheric contamination) disappear on heating at temperatures near 1150°C. The disappearance of these lines after only a 250°C temperature increase suggests a relatively narrow temperature range of stability for $Ti_2AlN$ when the concentration of nitrogen is low. For all the spectra, the broad band of slightly increased intensity located between 48–52° is caused by scattering off the beryllium window of the furnace chamber.

Figure 4 shows that at low temperatures the diffraction peak for the (10 · 1) superlattice line of the ordered hexagonal $Ti_3Al$ phase was observed near $2\Theta \simeq 26.5°$. An enlargement of the X-ray spectra shown in Fig. 4 in this angular region are shown in Fig. 7. This superlattice peak (and, therefore, the $Ti_3Al$ order) disappeared when the temperature was raised in excess of 1200°C. All the fundamental lines, however, for the disordered hexagonal $(\alpha - Ti)$ structure in the 1–60° $2\Theta$ angular range [(20 · 0), (00 · 2), (20 · 1), (20 · 2), and (22 · 0) shown later] remained, proving the high temperature structure is hexagonal (for consistency, the line indices quoted in this work for the disordered $\alpha$ phase are with respect to the doubled **a**-parameter unit cell of $Ti_3Al$). Since no other superlattice lines were created simultaneous with the disappearance of the $Ti_3Al$-order (further substantiated by subsequent high temperature measurements down to $2\Theta \simeq 10°$), it is proven that the disordered hexagonal $\alpha$-Ti structure is the high temperature phase involved in the previously determined eutectoid reaction at $\simeq 1125°C$ near this alloy composition. This was originally suggested several years ago in the phase diagram of one of the authors (*6*), and consistent with the results of the earlier comprehensive studies of Lipsitt (*22*) and Blackburn (*23*). The fact that the (10 · 1) $\alpha_2$ superlattice line was observed to as high a temperature as 1200°C, however, indicates that the disordering reaction of the $\alpha_2$ structure is sluggish. This result is reasonable, since the reverse ordering reaction is extremely fast and short-range ordering in the $\alpha$ phase in advance of $\alpha_2$ phase boundaries has already been observed at lower aluminum compositions (*6,24*).

Figure 6a shows that on heating to 1100°C, the major diffraction effect is a growth in intensity of the $\alpha_2$ (20 · 0), (00 · 2), and (20 · 1) lines $(2\Theta \simeq 36.5, 39,$ and $41.5°$, respectively). Since in this temperature range the $\alpha_2 + \gamma/\gamma$ boundary is relatively vertical, whereas the $\alpha_2/\alpha_2 + \gamma$ boundary slopes toward higher Al contents (*see* Figs. 1 and 2), an increase in temperature of an alloy in the two phase $\alpha_2 + \gamma$ field would result in an increase in the $\alpha_2$ volume percent, consistent with the X-ray line intensity data. Figure 6b shows the reverse effect on cooling. Between 1100–1200°C, there is a gradual *decrease* in intensity of the $2\Theta \simeq 39°$ line on heating (Fig. 4), whereas both the (20 · 0) and (20 · 1) $\alpha_2$ lines continue to grow larger. This is a reflection of having crossed the $\alpha_2 + \gamma$ $\rightarrow \alpha$ eutectoid invariant at a hypereutectoid composition and subsequent motion on the phase diagram upward through the $\alpha + \gamma$ phase field. The diffraction line centered around $2\Theta \simeq 39°$ is a composite of the (111) line of $\gamma$ and the $(00 \cdot 2)$ line of $\alpha_2$ (or $\alpha$). Because temperature increases of an alloy located in the $\alpha + \gamma$ field result in large decreases in $\gamma$ volume percent as it transforms into $\alpha$, the (111)-$\gamma$ contribution to the 39° diffraction peak decreases quickly. During this reaction, only some of the lost intensity near $2\Theta \simeq 39°$ is replaced by that owing to the newly formed $\alpha$ phase possessing the $(00 \cdot 2)$ orientation. Since not all the newly formed $\alpha$ possesses the $(00 \cdot 2)$ orientation, the composite (111)-$\gamma$ and $(00 \cdot 2)$-$\alpha$ line decreases intensity, even though the volume percent of $\alpha$ in the alloy increases while raising the temperature between 1100–1200°C. This transformation is quite different from the lower temperature (T < 1100°C) readjustment of $\gamma$ volume percent for the alloy located in the $\alpha_2 + \gamma$ field where the disappearance of (111) $\gamma$ with increasing temperature was accompanied by a growth of the orientation related $\alpha_2$: $(00 \cdot 2)\alpha_2 //(111)\gamma$. On *cooling* the 45% Al alloy between 1200–1100°C, Fig. 6b shows a growth in the (111)-$\gamma$ line, again consistent with and opposite to the heating data results. The drastic *decrease* in intensity (shown in Fig. 6a) of the $2\Theta \simeq 39°$ diffraction line on heating from 1200–1250°C (and, conversely, the drastic increase in this line's intensity on cooling from 1250 to 1200°C, as shown in Fig. 6b) locates the $\alpha/\alpha + \gamma$ phase boundary. Above 1250°C, all $\gamma$ diffraction peaks had disappeared.

The present X-ray data on $Ti_{55}Al_{45}$, a sample located well within the composition limits of any $\beta + \gamma$ field, supports the conclusion of Mc-Cullough et al. *(10)* (from measurements on a 50% Al sample) that no $\beta + \gamma$ field exists in the Ti-Al system. In fact, as shown in Fig. 4, only hexagonal $\alpha$-Ti diffraction peaks (with the exception of the two very small additional lines due to $Ti_2AlN$ near $2\Theta \simeq 35.5°C$ and 39.9° in the highest temperature spectra) are observed for $Ti_{55}Al_{45}$ in the whole temperature range between 1250–1400°C. In this temperature range, the $\gamma$-TiAl phase is absent, as well as any BCC $\beta$-Ti phase. Since melting occurs at only slightly higher temperature, and since the $\beta$ phase was not observed at any lower temperature for this sample, it is concluded that the two phase $\beta + \gamma$ field, as drawn in most previous diagrams *(7,25–27)*, does not exist. Our required extension of the hexagonal $\alpha$-Ti phase to 1400°C for $Ti_5Al_{45}$ also supports the recent modified diagram of Mc-Cullough *(10)* shown in Fig. 2, but the present data would require a minor shift of the $\alpha + \gamma$ transus to slightly higher Al contents. Such changes are shown in our diagram for this system in Fig. 9.

Closer inspection of the high temperature diffraction spectra shown in Fig. 4 for $Ti_{55}Al_{45}$ during heating between 1250–1400°C, reveals a rather novel feature. In this temperature range, only hexagonal diffraction lines are observed. However, each of these peaks is not a single peak, but at least a doublet. As the sample was heated there was a gradual shift in the intensity of *all* three major $\alpha$-Ti hexagonal lines [$(20 \cdot 0)$, $(00 \cdot 2)$, and $(20 \cdot 1)$] from the right side of the doublet to the left side, as one would expect to occur if the sample was located in a two phase (both hexagonal) field. On cooling from 1400 to 1250°C, there was an intensity shift back to the right. In this sample, between 1250–1400°C, there are apparently two hexagonal phases, each characterized by a different lattice parameter ($a_1$ = 0.4937nm, $c_1$ = 0.4592nm and $a_2$ = 0.4951nm, $c_2$ = 0.4607nm at 1300°C). The approximate reversibility of the effect implies this is an equilibrium effect, and not some spurious result resulting from contamination or composition change. The low temperature diffraction spectra, Fig. 5, for this sample, after cooling to 300°C, also indicates the sample was unchanged (with the exception of possessing the different texture developed at high temperature). Subsequent high temperature measurements on additional samples down to $2\Theta = 10°$, incorporating the use of an incident beam monochromator (Fig. 8), also showed very clearly the 2 sets of hexagonal lines between 1250–1400°C, but failed to detect any low angle superlattice lines. The same results (not shown) were also obtained for a sample heated very quickly to 1400°C from room temperature and measured during cooling. Consequently, it is concluded that in the temperature range between 1250–1400°C, there exists a 2-phase field for Ti-45Al comprised of the hexagonal $\alpha$-Ti phase and a new second disordered hexagonal $\alpha'$ phase.

Figure 9 shows the proposed placement of the new $\alpha'$ phase by means of only slight modifications to the McCullough diagram. The existence of 2 very similar structures located close to each other in the phase diagram, differing only in lattice constants, is certainly unusual but not unique. An example is to be found in the In-Pb system (*28*), where two tetragonal phases of different lattice constants exist next to each other in the phase diagram at compositions near 86 at% In. In fact, these phases can coexist in a two phase field with a composition difference of only 2 at%. Both phases are even stable at room temperature. The 2 tetragonal phases also coexist all the way up to the melting point, similar to the present finding in the Ti-Al system.

An alternative interpretation of the hexagonal doublets detected at high temperature in $Ti_{55}Al_{45}$ might be inhomogeneity in composition at high temperature, owing to the nonequilibrium formation of the high temperature $\alpha$ at different times during heating from different reactions: (1) from the sluggish conversion of $\alpha_2$, and (2) from the final conversion of $\gamma$. If this were the explanation, diffusion of the species at high temperatures (where the diffusivities (*29*) are large) would act to equilibrate the material into a single composition. During the measurement of the data shown in Fig. 4 and 5, the sample spent over 3 h at temperatures above 1250°C, and no single peaks were observed.

A second possible interpretation is that the $\alpha'$ hexagonal phase is a ternary structure stabilized by the addition of interstitial elements (e.g., nitrogen and oxygen). This ternary structure would, then, either be a previously unknown ternary phase or an expanded $\alpha$ Ti caused by solid solution of the interstitials. The latter option, however, would not be expected to give rise to either distinct (e.g., as shown in Fig. 8), or compositionally broadened, diffraction lines. If distinct lines were ob-

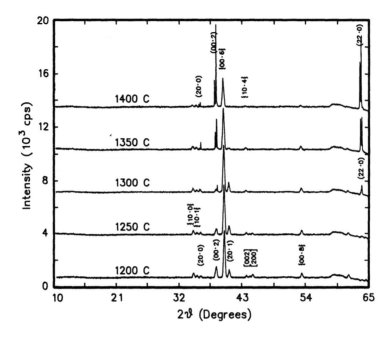

Fig. 8. X-ray intensity vs 2Θ data measured for $Ti_{55}Al_{45}$ (with mono-chromator) at the indicated temperatures during heating. Indices for the γ, $α_2$, and $Ti_2AlN$ lines are indicated by [ ], ( ), and { } brackets respectively. The zero intensities for the spectra have been shifted; scales are the same.

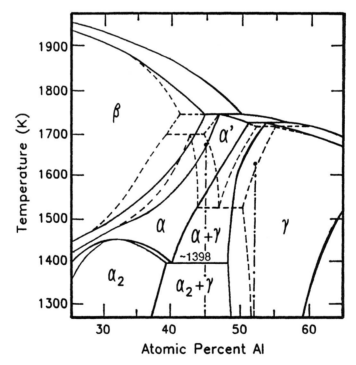

Fig. 9. Ti-Al phase diagram from ref. (*10*), modified to show the phase boundaries (dashed) suggested by the present data, the location of the new hexagonal phase α', and the temperature range and compositions (vertical dot-dashed lines) of the samples presently studied.

served as a result of a large amount of interstitials in solid solution, there would not be two sets of them. In the several high temperature experiments conducted in the present study on different $Ti_{55}Al_{45}$ samples with widely varying interstitial contents, the occurrence of the dual hexagonal structure did not appear to be connected with the qualitative amount of gaseous impurity present (30). For example, the hexagonal doublets were observed both in Fig. 4, for a sample where very small amounts of the ternary nitride were found, and also in Fig. 8, for a sample that contained a very large amount of $Ti_2AlN$. The most likely explanation for the additional set of hexagonal peaks, if resulting from the interstitial elements, is the presence of a new ternary compound. Such an occurrence would be expected to vary with the amount of interstitial elements available. Since this latter effect is not observed, even though the interstitial stabilization explanation has not been ruled out, it appears unlikely.

Further indication that the hexagonal phase experienced at least 2 transformations as it was heated from room temperature to 1400°C may be seen in the temperature dependence of its lattice parameters (Fig. 10). The data shown in Fig. 10 are those calculated from the line positions for the low temperature hexagonal phase only. These parameters determine the volume (V) of the material, and the volume is a function of the material's free energy (G): $V = (dG/dP)_T$, where $P$ is the pressure (31). Consequently, a phase transition of first order would be accompanied by a discontinuity in this parameter and its derivatives, e.g., the volume expansion coefficient. Note that both the **c** and **a** parameters follow the same trends and that there is a "contraction" of the lattice on heating between 1100–1300°C, followed by a large expansion above 1300°C. The volume coefficient of thermal expansion, calculated as described in the experimental procedure from the data in Fig. 10, is shown in Fig. 11, along with the linear coefficients of thermal expansion. There are two discontinuities, near 1100 and 1300°C, apparent in the thermal expansion coefficients for the $\alpha$ phase. These are exactly what would be expected to occur at the temperatures of the established $\alpha_2 + \gamma \rightarrow \alpha$ and proposed $\alpha + \gamma \rightarrow \alpha'$ transformations, respectively. If the $\alpha'$ phase was a non-equilibrium phase formed from $\gamma$ (i.e., unrelated to the prior-formed $\alpha$), it would be difficult to explain why the $\alpha'$ formation would be accompanied by a change in the thermal expansion coefficient of the previously formed $\alpha$. It is, consequently, concluded that the $\alpha'$ phase indeed forms as a result of a phase transition involving the low temperature hexagonal $\alpha$-Ti phase, as shown in Fig. 9.

The structural changes observed in a Ti-52 at% Al sample as it was heated to high temperatures were much different than those described above for $Ti_{55}Al_{45}$. Figure 12 shows the X-ray diffraction spectra measured for $Ti_{48}Al_{52}$ (without a monochromating crystal), as the sample was sequentially heated to 1350°C and then cooled back to room temperature. A 3-dimensional display of these data is shown in Fig. 13. At room temperature, this sample was primarily $\gamma$-TiAl, with a small admixture of $\alpha_2$, the presence of the latter being attested to by the appearance of the $(20 \cdot 0)$-$Ti_3Al$ diffraction peak at $2\Theta \simeq 36°$ in the 200°C spectra. Since this

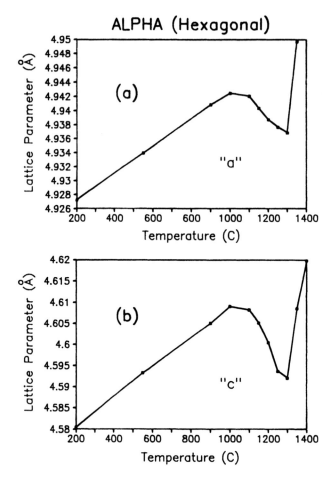

Fig. 10.   Lattice parameters measured for the hexagonal phase, from the Ti$_{55}$Al$_{45}$ data shown in Fig. 4. For consistency with the lower temperature data, above 1250°C, only the data from the higher angle lines of the hexagonal doublets are included.

sample had been well equilibrated (1000°C for ~11 d) to eliminate casting artifacts, our phase diagram of Fig. 9 reflects a location of the $\gamma/\alpha_2 + \gamma$ boundary at this higher Al content.

On heating Ti$_{48}$Al$_{52}$, Fig. 12 shows the $\gamma$ phase remained up to the highest temperatures, and no new additional phases formed in this alloy until above 1250°C. Between 1250–1350°C, the appearance of the hexagonal $(20 \cdot 1)$ and $(20 \cdot 2)$ lines indicates the alloy has crossed into a different equilibrium phase region. Since the $\gamma$ diffraction peaks are also observed in this high temperature range, the new phase field is identified as the $\alpha + \gamma$ region. The reverse effects (disappearance of $\alpha$) in the X-ray spectra are observed during cooling (Fig. 12). In this latter data, note from the $(20 \cdot 1)$-$\alpha$ diffraction peak (near $2\Theta \simeq 42°$) in the 300°C spectra that a small amount of the hexagonal phase exists in this sample after cooling to the lowest temperature, consistent with the initial sample condition.

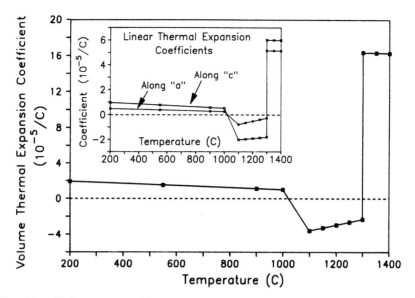

Fig. 11. Volumetric and linear (inset) coefficients of thermal expansion as a function of temperature for the hexagonal phase observed during heating $Ti_{55}Al_{45}$.

Figure 13 shows that on heating $Ti_{48}Al_{52}$ up to 1250°C, even though according to the phase diagrams of Figs. 1 and 2, there is no change in vol% of the $\gamma$ phase, the TiAl phase diffraction lines decrease in intensity. If it is assumed there is no simultaneous decrease in long range order of the $\gamma$ phase, this intensity decrease can be accounted for in terms of the thermal motions of the constituent atoms. It can also be used to verify that no phase transitions occur in this temperature region. After subtracting the background contributions, the intensities of the diffraction lines for the $\gamma$ phase measured at all temperatures were normalized to the data measured at 850°C ($= T_R$). The latter X-ray spectrum was chosen as a reference because at this temperature the sample was essentially single phase and strain relief effects had been completed. In accordance with the description in the experimental procedure, the difference in mean squared displacements, $<u^2 (T)> - <u^2 (T_R)>$, of the atoms in the $\gamma$ phase were determined for each temperature up to 1250°C. An example of this determination from the measured data at 1050°C is shown in the inset of Fig. 14. Within the accuracy of the data, Fig. 14 shows that the $<u^2 (T)> - <u^2 (850°C)>$ increased with temperature between 850–1250°C at a rate of approximately $2.3 \times 10^{-4}$ Å$^2$/C. Phase transitions in the alloy would be accompanied by discontinuities in this temperature dependence. In the present case, consistent with the disappearing phase method results described above, no phase transitions were detected, since $<u^2(T)>$ varied relatively smoothly in this temperature range. This is also consistent with the earlier phase diagram investigations of this system.

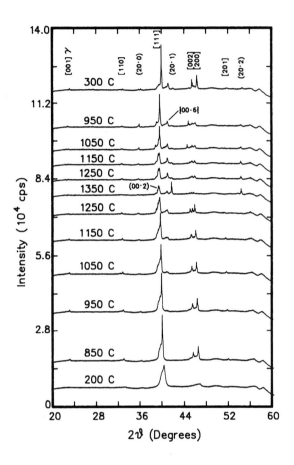

Fig. 12. X-ray intensity vs 2Θ data measured for $Ti_{48}Al_{52}$ (without mono-chromator) at the indicated temperatures in the order from bottom to top. Indices for the γ, $α_2$, and $Ti_2AlN$ lines are indicated by [ ], ( ), and { } brackets respectively. The zero intensities for the spectra have been shifted; scales are the same.

## CONCLUSION

In conclusion, it has been proven that the $α_2 + γ \rightarrow α$ eutectoid reaction near 1125°C and 45 at% Al, previously proposed by one of the authors several years ago, exists. The disordered hexagonal α-Ti phase indeed extends to high Al contents at elevated temperatures, and no β + γ phase field exists in $Ti_{55}Al_{45}$ up to 1400°C, consistent with other recent diffraction work on alloys of other compositions. The present diffraction data for $Ti_{55}Al_{45}$ suggest the presence in the Ti-Al system at high temperatures of an as yet unreported second disordered hexagonal structure (α'), similar in structure to α titanium, but with different lattice parameters. For $Ti_{48}Al_{52}$, the γ/α + γ phase boundary is located between 1250–1350°C, and no β-Ti phase was detected up to 1350°C.

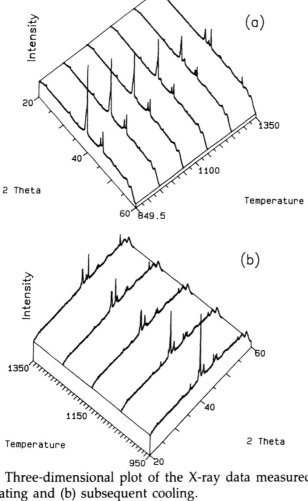

Fig. 13.   Three-dimensional plot of the X-ray data measured for $Ti_{48}Al_{52}$ during (a) heating and (b) subsequent cooling.

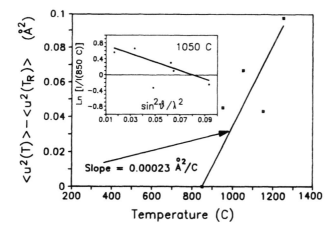

Fig. 14.   Temperature dependence of the difference in mean squared atomic displacements from that at 850°C for the γ-TiAl phase. Inset shows an example of the logarithm of $I(T)/I(850 C)$ vs $sin^2\Theta/\lambda^2$ data used to obtain $<u^2 (T)>$ - $<u^2 (T_R)>$ at 1050°C. Solid lines show least squares fits to the data.

## ACKNOWLEDGMENT

One of the authors (R. D. S.) would like to acknowledge the many helpful discussions on the Ti-Al system with M. J. Kaufman, J. C. Williams, H. A. Lipsitt, U. Kattner, J. L. Murray, J. H. Perepezko, W. J. Boettinger, D. Shectman, and H. L. Frazier. Special thanks are given to I. Ansara for bringing to the authors' attention the similarities to the In-Pb system and to R. M. Waterstrat and E. Prince for kindly providing the authors permission to quote some of their results prior to publication.

## REFERENCES

1. Vujic, D., Li, Z., and Whang, S. H., *Met. Trans.* **19A**, 2445 (1988).
2. Sircar, S., Narasimhan, K., and Mukherjee, K., *J. Mat. Sci.* **21**, 4143 (1986).
3. Mishurda, J. C., Lin, J. C., Chang, Y. A., and Perepezko, J. H., *High Temperature Ordered Intermetallic Alloys III*, Liu, C. T., Taub, A. I., Stoloff, N. S., and Koch, C. C., eds., Elsevier, MRS Symposium Proceedings, vol. 133 (1989).
4. Collings, E. W., *Met. Trans.* **10A**, 463 (1967).
5. Shull, R. D., Cuthill, J. R., Reno, R. C., Murray, J. L., and Mehrabian, R., *Proceedings of the Ti-6211 Basic Research Program*, MacDonald, B. A., Arora, O. P., and Rath B. B., eds., Office of Naval Research publishers, p. 1 (1982).
6. Shull, R. D., McAlister, A. J., and Reno, R. C., *Titanium Science and Technology*, Lutjering, G., Zwicker, J., and Bunk, W., eds., Deutsche Gesellschaft fur Metallkunde, Munich, p. 1459 (1985).
7. Murray, J. L., *Binary Alloy Phase Diagrams*, American Society for Metals, Metals Park, OH, p. 175 (1987).
8. Jones, S. A., Shull, R. D., McAlister, A. J., and Kaufman, M. J., *Scripta Metall.* **22**, 1235 (1988).
9. Waterstrat, R. M. and Prince, E., to be published.
10. McCullough, C., Valencia, J. J., Mateos, H., Levi, C. G., and Mehrabian, R., *Scripta Metall.* **22**, 1131 (1988).
11. Shull, R. D., Singhal, S. P., Mozer, B., and Maeland, A., *Rapidly Solidified Metastable Materials*, Kear, B. H. and Giessen, B. C., eds., MRS Conference Proceedings, vol. 28, Elsevier, New York, p. 279 (1984).
12. *1981 Annual Book of ASTM Standards*, Part 44, American Society for Testing and Materials Publ., Philadelphia, PA, p. 654 (1981).
13. Klug, H. P. and Alexander, L. E., *X-Ray Diffraction Procedures*, Wiley, New York, ch. 3 (1954).
14. Nelson, J. B., and Riley, D. P., *Proc. Phys. Soc.* (Lond.) **57**, 160 (1945).
15. Krishnan, R. S., Srinivasan, R., and Devanarayanan, S., *Thermal Expansion of Crystals*, Pergamon, New York, p. 48 (1979).
16. Willis, B. T. M. and Pryor, A. W., *Thermal Vibrations in Crystallography*, Cambridge University Press, Cambridge, UK, p. 81 (1975).
17. Ibid., p. 207.
18. Goldak, A. J. and Parr, J. G., *Trans. Met. Soc. AIME* **221**, 639 (1961).
19. Jeitschko, W., Nowotny, H., and Benesousky, F., *Monatsh. Chem.* **94**, 1198 (1963).
20. Loiseau, A. and Lasalmonie, A., *Acta Cryst.* **39**, 580 (1983); *Mat. Sci. Eng.* **67**, 163 (1984).
21. Kaufman, M. J., Konitzer, D. G., Shull, R. D., and Fraser, H. L., *Scripta Metall.* **20**, 103 (1986).

22. Martin, P. L., Lipsitt, H. A., Nuhfer, N. T., and Williams, J. C., *Titanium '80 Science and Technology*, Kimura, H. and Izumi, O., eds., TMS-AIME, New York, p. 1245 (1980).
23. Blackburn, M. J., *The Science, Technology, and Application of Titanium*, Jaffee, R. I. and Promisel, N. E., eds., Pergamon, New York, p. 633 (1970).
24. Namboodhiri, T. K. G., McMahon, C. J., and Herman, H., *Met. Trans.* **4,** 1323 (1973).
25. Farrar, P. A. and Margolin, H., *The Metals Handbook, 8th Edition*, Am. Soc. for Metals Publ., Metals Park, OH, vol. 8, p. 264 (1973).
26. Bumps, E. S., Kessler, H. D., and Hansen, M., *Met. Soc. AIME* **194,** 609 (1952).
27. Kornilov, I. I., Pylaeva, E. N., Volkova, M. A., Kripyakevich, P. I., and Markiv, V. Y., *Dokl. Akad. Nauk SSSR* **161,** 843 (1965).
28. Hansen, M., *Constitution of Binary Alloys, 2nd Edition*, McGraw-Hill, New York, p. 855 (1958).
29. Goold, D., *J. Inst. Met.* **88,** 444 (1959–60).
30. The relative amount of interstitial elements was determined qualitatively from the presence and intensity of X-ray diffraction peaks from known ternary oxygen- and nitrogen-containing compounds detected in the spectra for the samples and from the color and reflectivity of the sample surface following the high temperature measurements.
31. Denbigh, K., *The Principles of Chemical Equilibrium*, University Press, Cambridge, UK, p. 87 (1964).

# High Temperature Neutron Diffraction Studies of $YBa_{2-x}Sr_xCu_3O_{7-\delta}$

J. FABER, JR.* AND R. L. HITTERMAN

*Materials Science Division, Argonne National Laboratory, Argonne, IL 60439*

## ABSTRACT

*In situ*, high-temperature neutron diffraction experiments have been performed on the nonstoichiometric, Sr-doped, 123 superconducting material $YBa_{2-x}Sr_xCu_3O_{7-\delta}$ using a unique, restricted-angle, high temperature furnace constructed for use with time-of-flight scattering techniques. The furnace provides a completely isolated specimen chamber in the temperature range of $20 < T < 1400C$, and the oxygen partial pressure can be controlled in the range of $10^{-20} < P_{O_2} < 1$ atm ($10^5$ Pa). For $x = 0.33$ (in $YBa_{2-x}Sr_xCu_3O_{7-\delta}$) and $P_{O_2} = 1$ atm, an order-disorder transition is observed at $T = 650$ C. The crystal symmetry changes from orthorhombic (Pmmm) below the transition to tetragonal (P4/mmm) above. The mechanism of the phase transition involves the redistribution of anions on the O(1) and O(5) lattice sites; these sites become equivalent in the high temperature tetragonal phase. With $P_{O_2} = 1$ atm the nonstoichiometric state of the specimen changes with temperature, hence the experimental results must be mapped onto the $(T,\delta)$ plane. The transition temperature, $T_{ot}$, decreases with increasing Sr concentration. Under isothermal conditions with $T = 490C$, the transition occurs at $P_{O_2} = 3 \times 10^{-3}$ atm.

## INTRODUCTION

Mechanisms responsible for superconductivity in the high $T_c$ oxides are not well understood. However, it is known that the presence of oxygen vacancies strongly affect the transition temperature to the super-

*Author to whom all correspondence and reprint orders should be addressed.

High Temperature Science, Vol. 26 &copy; 1990 by the Humana Press Inc.

119

conducting state (1). The Y-123 compound ($YBa_2Cu_3O_{7-\delta}$, has been the subject of numerous experimental (2–7) and theoretical (4,7–11) investigations. The results of neutron diffraction studies (4,8,9,11,12) have substantially improved our understanding of this compound. The basic Y-123 structure is illustrated in Fig. 1. There seems to be general agreement that the two layers of Cu(2)-O(2)-O(3) form the superconducting planes that are separated by an yttrium ion (Fig. 1). The other prominent feature in Fig. 1 is the Cu(1)-O(1) chains that extend along the *b*-axis of the orthorhombic unit cell, and are considered to form the charge reservoir for the superconducting process and also participate in the defect behavior of the system. As we shall see, the O(5) lattice sites along the *a*-axis of the unit cell are essentially vacant at low temperature, and it is these sites that are the sources of oxygen vacancies in the non-stoichiometric state.

An important approach for characterizing oxygen stoichiometry and associated ordering of anion vacancies is to study the effects caused by controlled substitution of dopant ions on different cation sites. A number of studies have been carried out to test the response of the structure to dopant additions. One surprising result is that $T_c$ appears to be insensitive to rare-earth substitution on the yttrium sites in the lattice, even if the substitution is complete and the substituted ion is a rare-earth with a large magnetic moment. In contrast to Y ion site behavior, substitutions on the Ba sites (although not as extensively studied) have a much more pronounced effect on $T_c$. In particular, for Sr-substituted material, there is a depression of $T_c$ of approximately 0.45K/%Sr content (14).

An orthorhombic to tetragonal phase transition is observed in the Y-123 material, with $P_{O_2}$ = 1 atm at $T_{ot}$ = 700 C. In this paper, we will examine the structure of $YBa_{1.67}Sr_{0.33}Cu_3O_{7-\delta}$ using both isobaric and isothermal, *in situ* neutron diffraction techniques and compare these results with the behavior of Y-123. It is perhaps interesting to note that for undoped Y-123 materials, $T_c$ vanishes when the structure transforms to the tetragonal phase. Preliminary results from isobaric studies of this compound have been reported elsewhere (15). For our studies of the Sr-substituted material ($YBa_{2-x}Sr_xCu_3O_{7-\delta}$), we assume that $Sr^{2+}$ substitutes for $Ba^{2+}$, that the cations vibrate isotropically, and that the anion vacancies are localized, but randomly distributed among the O(1) and O(5) lattice sites.

## EXPERIMENT

The neutron scattering experiments were carried out on well-characterized samples of $YBa_{2-x}Sr_xCu_3O_{7-\delta}$ on the General Purpose Powder Diffractometer (GPPD) at Argonne's Intense Pulsed Neutron Source (IPNS) (16–18). Some detail concerning the method of sample preparation has been reported (14). To understand how Bragg scattering information is obtained with a pulsed neutron source, it is useful to note

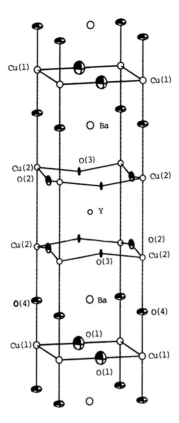

Fig. 1. Room temperature, $T = 20$ C, orthorhombic (Pmmm) crystal structure of $YBa_{1.67}Sr_{0.33}Cu_3O_{7-\delta}$ determined from neutron powder diffraction experiments. The thermal vibration ellipsoids define contours with 50% probability that the atoms will be found inside these surfaces. The cations are defined by the open ellipsoids and the anions by the segmented ellipsoids.

the relationship between the neutron wavelength, $\lambda$, and the Bragg angle, $\vartheta$.

$$\lambda = hmt/L = 2d\sin\vartheta \qquad (1)$$

In this expression, $h$ is Planck's constant, $m$ is the neutron mass, $t$ is the time-of-flight (TOF), $L$ is the total path length of the neutron from the source to the detector, and $d$ is the interplanar $d$-spacing of interest. The measurements yield Bragg integrated intensity information that is then used to recover the structural detail. For $2\vartheta = 90$ degrees, with $\lambda$ in the range $0.5 < \lambda < 5$Å, the accessible $d$-space range is $0.5 < d < 3.95$Å. The resolution, $\delta d/d$, determined at the full width at half maximum, is 0.004 at this scattering angle.

A unique, restricted-angle, high temperature furnace has been developed that provides a completely isolated specimen environment in the temperature range $20 < T < 1400°C$. A cross-sectional view of the furnace is schematically illustrated in Fig. 2. The furnace consists of three major component assemblies: a vacuum chamber with a heating assembly, an

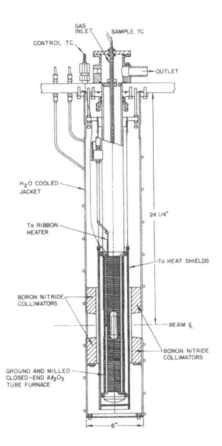

Fig. 2. Schematic illustration of the artificial environment, high-temperature furnace for *in situ* studies of materials in the temperature range 20 < $T$ < 1400 C. The heater is constructed from Ta ribbon and operates under high-vacuum conditions. The sample chamber is isolated from the vacuum system by a closed-end, high-purity alumina tube. The sample is inserted from the top of the furnace.

optical assembly with components that provide for neutron collimation, and the sample chamber. The restricted-angle feature is accomplished with the use of boron nitride collimator elements located around the furnace components. These collimator elements eliminate direct line-of-sight viewing of the detector banks at $2\vartheta = 90$ degrees. The detectors are positioned 1.5 m from the sample position and subtend a solid angle of 0.05 steradians. The cutouts in the boron nitride collimator provide an unimpeded path from sample to detector banks, while collimating out the interference Bragg scattering signals from the Ta heater winding, Ta radiation shields and $Al_2O_3$ closed-end sample chamber isolation tube. The power requirements for this furnace are quite modest, in part owing to the use of high-vacuum conditions and radiation shields. For example, to maintain a sample temperature of $T_s = 1000°C$ requires 700 watts, whereas at $T_s = 1400°C$, 2200 watts are required.

To provide for controlled sample environments, an open system design is employed in this furnace. The sample chamber configuration is

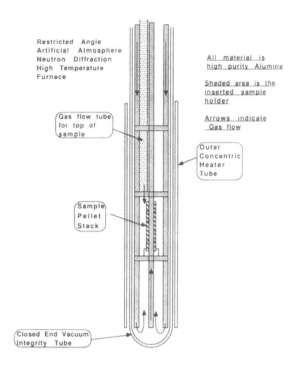

Fig. 3.   Sample chamber sample holder assembly. Two gas flow systems are employed. The first directs a stream of gas at the top of the sample. The second is accomplished with the use of hollow alumina support rods and circulates the gas from the bottom of the closed-end alumina tube that defines the sample chamber.

illustrated in Fig. 3. All of the sample support components are fabricated from high-purity polycrystalline $Al_2O_3$. Two systems are used to flow gases of appropriate composition over the sample. The main gas flow tube directs a stream of gas near the top of the specimen. A secondary gas flowstream is implemented with hollow alumina support rods. This system minimizes the effects of gas segregation in the specimen area. A typical sample is a stack of sintered cylindrically shaped pellets, 1 cm in diameter, stacked 5 cm high. Two thermocouples are used. The first is a sheathed couple located very near the Ta ribbon heater windings and is primarily used for temperature control (*see* Fig. 2). The second thermo-couple is located in the sample environment tube and is positioned in close proximity to the sample. The top and bottom of the specimen are held in place by ceramic supports. Typical temperature gradients at the sample position are <2.5°C for most ceramic materials.

## RESULTS AND DISCUSSION

The isobaric ($P_{O_2}$ = 1 atm) experiments were carried out in the temperature range $20 < T < 900$°C. The TOF diffraction data were analyzed using Rietveld profile refinement techniques. The results of a

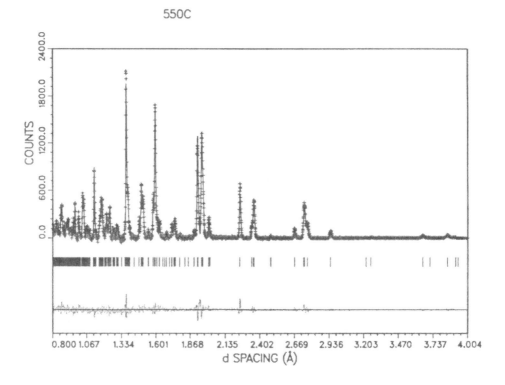

Fig. 4.    Results of Rietveld profile refinement on $YBa_{1.67}Sr_{0.33}Cu_3O_{7-\delta}$ at $T$ = 550 C and $P_{O_2}$ = 1 atm. The pluses are the experimental data points corrected for background, the solid line drawn through the points represents the least-squares best fit to the data, the tic marks define the positions of the Bragg peaks, and the lower portion of the figure illustrates the point-by-point differences between theory and experiment.

representative fit to the experimental data at $T = 550°C$ are illustrated in Fig. 4. There are 440 Bragg reflections in the $d$-space range illustrated in the figure, and 37 adjustable parameters were used to calculate the intensity at 2625 data points, yielding a weighted residual, $R_{wp}$ = 0.06 and $\chi^2$ = 1.9. The agreement between theory and experiment is very good (note the difference curve in Fig. 4). The unit cell thermal expansion behavior of $YBa_{1.67}Sr_{0.33}Cu_3O_{7-\delta}$ with $P_{O_2}$ = 1 atm is illustrated in Fig. 5. A phase transition is observed with $T_{ot}$ = 650°C. The transition is second-order. The structure in the tetragonal phase (P4/mmm) defined by $T_{ot} >$ 650°C is illustrated in Fig. 6. By comparison with Fig. 1, we see that the occupation of anions on the O(1) sites has become disordered, and O(1) and O(5) are equivalent lattice sites in the tetragonal phase. A quantitative description of O(1) and O(5) lattice site occupation are illustrated in Fig. 7. The vertical line marks $T_{ot}$ for the undoped, Y-123 material (12). Also, it is apparent from Fig. 7 that the average site occupation for O(1) and O(5) decrease with increasing temperature. This behavior is especially obvious for $T > T_{ot}$. Therefore, these results must be mapped onto

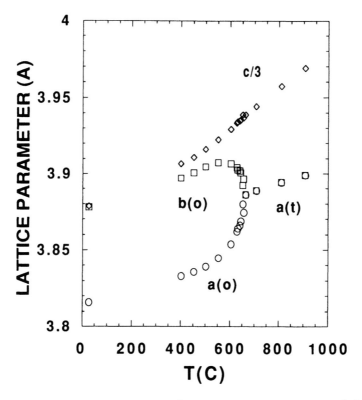

Fig. 5. Lattice parameter dependence upon temperature as defined from the positions of the Bragg peaks in $d$-space coordinates. The orthorhombic phase is denoted by the symbol (o) and the tetragonal phase by (t). For convenience in plotting, the $c$ lattice parameter has been divided by 3.

the $(T,\delta)$ plane. To examine the properties of the phase transition more clearly requires us to devise an experimental result that is a function of only one extensive parameter of the system.

To examine this behavior, we have carried out a series of isothermal experiments at $T = 490°C$ (*see* (12) for the corresponding behavior of Y-123), in which only the defect concentration is varied. The lattice response to changes in oxygen partial pressure is given in Fig. 8. As the $P_{O_2}$ decreases, the departure from stoichiometry increases ($\delta$ increases) and we are able to induce the orthohombic-to-tetragonal transition with $P_{O_2(ot)} = 3 \times 10^{-3}$ atm. The corresponding behavior of the anion occupation of O(1) and O(5) sites is illustrated in Fig. 9. A comparison of these results with those for the Y-123 compound (*12*) show that O(1) site occupation in YBa$_{1.67}$Sr$_{0.33}$Cu$_3$O$_{7-\delta}$ is lower than that for the Y-123 material under the same conditions. Moreover, O(5) site occupation in the former is higher than that for the Y-123. These results, and those illustrated in Fig. 7, lead us to conclude that the O(1)–O(5) anion repulsive interaction is smaller in the Sr-substituted compound than that for the Y-123.

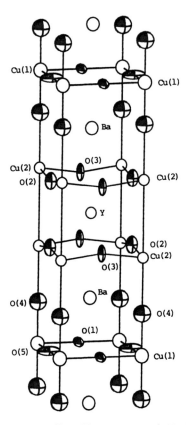

Fig. 6.  High temperature, $T > T_{ot}$ tetragonal (P4/mmm) crystal structure of $YBa_{1.67}Sr_{0.33}Cu_3O_{7-\delta}$ determined from neutron powder diffraction experiments. The thermal vibration ellipsoids are defined as in Fig. 1. The cations are defined by the open ellipsoids and the anions by the segmented ellipsoids. In the tetragonal phase, the O(1) and O(5) lattice sites are equivalent and the occupation numbers for these sites must be equal.

## CONCLUSIONS

High-temperature, *in situ* neutron scattering experiments have been carried out on $YBa_{1.67}Sr_{0.33}Cu_3O_{7-\delta}$ in the temperature range $20 < T < 900$ C. A second order, orthorhombic-to-tetragonal phase transition with $T_{ot} = 650°C$ was observed. The order–disorder transition involves the rearrangement of anion vacancies on the O(1) and O(5) lattice sites. In the high temperature tetragonal phase, O(1) and O(5) sites are equivalent and the site occupancies are equal. We have observed that the non-stoichiometric defect concentration increases with increasing temperature. Thus, the results of our isobaric experiments must be mapped onto the $(T,\delta)$ plane. To separate the effects of defect formation from those associated with thermal energy leads us to consider an isothermal investigation, and these are carried out at $T = 490°C$ by varying the oxygen partial pressure. In this case, $P_{O_2(ot)} = 3 \times 10^{-3}$ atm was found. A comparison of our results with those in the literature (*12*) shows that the O(1)–O(5) anion repulsive interaction is smaller in the Sr-substituted material than that found in the undoped Y-123 material.

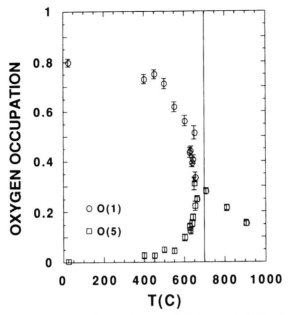

Fig. 7.   The temperature dependence of O(1) and O(5) site occupation. Note that δ (in Yba$_{1.67}$Sr$_{0.33}$Cu$_3$O$_{7-\delta}$) decreases with increasing temperature.

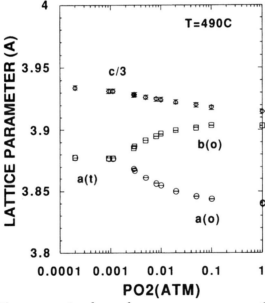

Fig. 8.   Lattice parameter dependence upon oxygen partial pressure under isothermal conditions with $T = 490$ C. The symbols are defined as in Fig. 4. Under these conditions, P$_{O_2(ot)}$ occurs at approximately $3 \times 10^{-3}$ atm.

To quantitatively describe the oxygen content and ordering in these materials at high temperature, a statistical thermodynamic model based on the quasichemical approximation has been developed (*19*). An analysis of the experimental results with this model is in progress. There are two important parameters in this theory. The first is the site binding energy for oxygen on the O(1) and O(5) lattice sites. The second is the

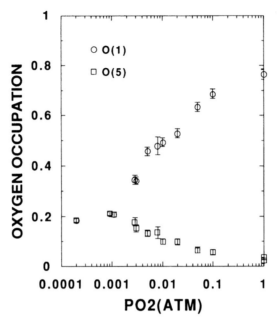

Fig. 9.   The oxygen partial pressure dependence of O(1) and O(5) site occupation. In this case, the defect concentration is being adjusted under conditions of constant thermal energy.

short-range order parameter defining the probability of near-neighbor O(1)–O(5) interactions. Further experiments on $YBa_{1.33}Sr_{0.67}Cu_3O_{7-\delta}$ are planned to test our understanding of the Sr dopant concentration dependence of these energy parameters. An understanding of these interaction energies may help us to better understand the suppression of $T_c$ that occurs at low temperature.

## ACKNOWLEDGMENTS

This work is supported by the US Department of Energy, BES-Materials Sciences under contract W-31-109-ENG-38. We thank B. W. Veal, A. P. Paulikas, L. J. Nowicki, and J. W. Downey who prepared and characterized the samples, and H. Shaked, J. D. Jorgensen, and B. W. Veal for many useful discussions.

## REFERENCES

1.  Jorgensen, J. D., Veal, B. W., Kwok, W. K., Crabtree, G. W., Umezawa, A., Nowicki, L. J., and Paulikas, A. P., *Phys. Rev.* **B36,** 5731 (1987).
2.  Jorgensen, J. D., Beno, M. A., Hinks, D. G., Soderholm, L., Volin, K. J., Hitterman, R. L., Grace, J. D., Schuller, I. K., Segre, C. U., Zhang, K., and Kleefish, M. S., *Phys. Rev.* **B36,** 3608 (1987).
3.  Fritas, P. P. and Plaskett, T, S., *Phys. Rev.* **B36,** 5723 (1987).
4.  Kubo, Y., Nakabayashi, Y., Tabuchi, J., Yoshitake, T., Ochi, A., Utsumi, K., Iagarashi, H., and Yonezawa, M., *Japn Jour. Appl. Phys.* **26,** L1888 (1987).

5. Kishio, K., Shimoyana, J., Hasegawa, T., Kitezawa, K., and Fueki, K., *Japn. Jour. Appl. Phys.* **26**, L1228 (1987).
6. Specht, E. D., Sparks, C. J., Dhere, A. G., Brynestad, J., Cavin, D. B., Kroeger, D. M., Oye, H. A., and Seiler, F. J., *Phys. Rev.* **B37**, 7426 (1988).
7. McKinnon, W. R., Post, M. L., Selwyn, L. S., Pleizier, G., Tarascon, J. M., Barboux, P., Greene, L. H., and Hull, W., *Phys. Rev. B* (submitted).
8. Bakker, H., Welch, D. D., and Lazareth, Jr., O. W., *Phys. Rev. B*, in press.
9. Salamons, E., Koeman, N., Brouwer, R., deGroot, D. G., and Greissen, R., *Solid State Commun.* **64**, 1141 (1987).
10. Willie, L. T., Berera, A., and deFontaine, D., *Phys. Rev. Lett.* **60**, 1065 (1988).
11. Nakanura, K. and Ogawa, K., *Japn. Jour. Appl. Phys.*, submitted.
12. Jorgensen, J. D., Shaked, H., Hinks, D. G., Dabrowski, B., Veal, B. W., Paulikas, A. P., Nowicki, L. J., Crabtree, G. W., Kwok, W K., and Nunez, L. H., *Physica* **C153–155**, 578 (1988).
13. Beno, M. A., Soderholm, L., Capone, D. W., Hinks, D. G., Jorgensen, J. D., Grace, J. D., Schuller, I. K., Segre, C. U., and Zhang, K., *Appl. Phys. Lett.* **51**, 57 (1987).
14. Veal, B. W., Kwok, W. K., Umezawa, A., Crabtree, G. W., Jorgensen, J. D., Downey, J. W., Nowicki, L. J., Mitchell, A. W., Paulikas, A. P., and Sowers, C. H., *Appl. Phys. Lett.* **51**, 279 (1987).
15. Faber, Jr. J., Shaked, H., Veal, B. W., Hitterman, R. L., Paulikas, A. P., Nowicki, L. J., and Downey, J. W., *Advanced Characterization Techniques for Ceramics*, McVay, G. L., Pike, G. E., and Young, W. S., eds., American Ceramics Society, Westerville, OH, 1989, in press.
16. Faber, Jr., J. and Hitterman, R. L., *Advances in X-Ray Analysis* **29**, Plenum, New York, 1985, pp. 119–130.
17. Jorgensen, J. D. and Faber, Jr., J., ICANS-VI Meeting, ANL, June 17–July 2, 1982, ANL Report ANL-82-80(1983), pp. 105–114.
18. MacEwen, S. R., Faber, Jr., J., and Turner, A. P. L., *Acta Met.* **31**, 657 (1983).
19. Shaked, H., Jorgensen, J. D., Faber, Jr., J., Hinks, D. G., and Dabrowski, B., *Phys. Rev. Rapid Commun.* in press.

# Atom Probe Field-Ion Microscopy Applications

Patrick P. Camus

*National Institute for Standards and Technology,*
*Surface Science and Metallurgy Divisions,*
*Gaithersburg, MD 20899*

## ABSTRACT

Field-ion microscopy is a real-space atomic resolution microscopy that has the unique ability of serial sectioning a specimen in a controlled manner. Thus, the true three-dimensional structure of defects and particles may be obtained. Atom probe analysis is a powerful chemical analysis technique that possesses both high spatial (<2 nm) and depth (0.2 nm) resolutions. Light elemental analyses are routine because there is no mass limitation for this mass spectrometric technique. This paper first presents a general description of the physics of these techniques followed by a selected review of applications.

**Index Entries:** Field-ion microscopy; field evaporation; field ionization; atom probe; FIM; APFIM; time-of-flight mass spectrometer.

## INTRODUCTION

Field-ion microscopy and atom probe chemical analysis are techniques that have been gaining in popularity in a variety of scientific fields owing to their powerful and sometimes unique capabilities. A general description of each technique will be presented to familiarize the reader with the most important details of their function. More detailed descriptions and quantitative analysis of the physics of the techniques may be found in the manuscripts by Müller and Tsong (1,2), Wagner (3), and Miller (4,5). The last portion of this manuscript will be devoted to description of studies in a variety of disciplines illustrating the capabilities of these powerful techniques.

High Temperature Science, Vol. 26    © 1990 by the Humana Press Inc.

# FIELD-ION MICROSCOPY

Field-ion microscopy produces atomic resolution images of the surface of a very sharply pointed specimen. A typical microscope consists of an ultra-high vacuum chamber, cryogenic specimen cooling stage which may be rotated approximately $\pm 45°$ about 2 axes, and a channel plate/phosphor screen imaging assembly. Images are recorded directly from the screen using either film or video tape. A high voltage, up to 25 kV, is applied to the specimen. Images are produced by the field ionization of an imaging gas introduced into the vacuum system. The interior of the specimen may be imaged by sequentially removing atomic layers via field evaporation.

## Field Ionization

Field ionization is a phenomenon where gas atoms in the vicinity of the specimen surface become ionized in a high electric field. To attain the necessary field (20– 45 V/nm) at moderate voltages, the preferred specimen geometry is a sharply pointed object with a radius of approximately 100nm. The imaging gas, usually a noble gas or hydrogen, is polarized in the field, attracted to the specimen, impacts and rebounds from the surface following ballistic trajectories, Fig. 1. If the specimen is cryogenically cooled, the kinetic energy of the gas atom is reduced upon each impact with the surface. When the atom travels through a region above the surface where the field is large enough to ionize the atom and there exists a high probability for electron tunnelling to the surface, ionization occurs. The distance from the surface at which the tunnelling probability is greatest is of the order of 0.2 nm. The tunnelling probability is the greatest in the highest field regions or regions of smallest radius of curvature. The positively charged gas ions radiate outward along the field lines where they form a bright spot on the imaging assembly.

Since the surface of the specimen is not smooth on the atomic scale but can be described as constructed of hard spheres, the surface actually consists of plateaus and steps forming the curvature, Fig. 1. Due to the atomic stacking within the crystal structure, each planar low field region of the hemispherical specimen corresponds to a given (hk1) plane. Because the gas atoms ionize predominantly in the high field regions above the atoms at the edges of the planar regions, only these atoms lead to spots in the image. Therefore, the field ion image consists of a series of concentric rings around each (hk1) plane and can be indexed as an approximately stereographic projection. A field-ion image, therefore, is a map of the ionization probability of the specimen surface. Any modification to any of the parameters affecting ionization will have a marked effect upon the image, Fig. 2. Some of these parameters are local work function, ionization potential of the gas, and local field.

Because of the radial trajectories of the gas ions and the absence of any lenses, the field-ion microscope is a point projection microscope. As

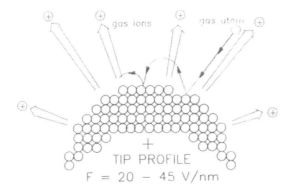

Fig. 1.   A schematic diagram of the field ionization process responsible for the formation of the field-ion image. Surface atoms which protrude slightly from the average surface have a greater probability of ionizing image gas atoms and generate bright spots in the image.

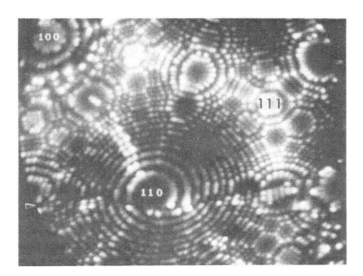

Fig. 2.   A field-ion micrograph of tungsten where each spot is the image of a single atom on the surface of the specimen. Note the characteristic ring pattern and the symmetry elements similar to a stereographic projection and the grain boundary running horizontally through the <110> pole. (Courtesy A. J. Melmed.)

such, the magnification, M, may be given as $M = R/\beta r$, where R is the specimen to screen distance, r is the average specimen radius, and $\beta$ is a geometric factor arising from the non-spherical specimen geometry. With typical values of $R = 10$ cm, $r = 50$ nm, and $\beta = 5$, magnifications of images are in the millions. Because the specimen radius and the geometric factor cannot be varied, the magnification may only be changed during the experiment by a motion of the imaging assembly, thus changing R.

### Field Evaporation

Field ionization is the process by which the surface structure is imaged in field-ion microscopy. If the third dimension is to be investigated, the surface atoms must be removed and a new internal surface exposed. A common technique to remove material from a specimen is to raise the temperature to thermally evaporate the atoms. However, evaporation may also occur by increasing the applied field and ionizing the surface atoms.

If the applied field is increased beyond that necessary for field ionization, the surface atoms in the highest field regions will become ionized. They will be removed from the surface and radiate along the field lines, exposing a new surface, Fig. 3. Since the atoms at the edge of the planes are in the highest field regions, they have the greatest likelihood of being removed. This has the effect of reducing the size of the planar ring until the atomic plane is completely removed. In this manner, surface atoms over the whole specimen may be removed in a controlled manner with a depth resolution of less than 0.2 nm.

This technique may be continued for as long as necessary to investigate internal features, especially grain boundary structure, true three-dimensional morphology, and average distances between phases. An additional benefit of the technique for surface studies is the preparation of a reproducibly clean surface. Any contamination layer present owing to specimen preparation or exposure to the atmosphere may be removed prior to investigation within the UHV chamber.

## ATOM PROBE CHEMICAL ANALYSIS

Instead of raising the standing field slowly, one may apply a pulsed field to the specimen, causing the surface ions to evaporate at a known instant. This pulse is used as the start signal of a time-of-flight mass spectrometer with charge pulses generated at a detector, owing to the arrival of individual surface ions, providing the stop signals. This technique is known as pulsed field evaporation or atom probe chemical analysis. Pulsed laser-assisted field evaporation may also be used to evaporate the atoms and commence the timing sequence. Because the initial potential energy is fully transferred into kinetic energy of the travelling ion, the mass-to-charge ratio of each ion may be determined by

$$neV_{tot} = \tfrac{1}{2}\, m(d/t)^2$$

$$m/n = 2e/d^2\, V_{tot}\, t^2$$

where $d$ is the flight distance of the ions, $V_{tot}$ is the total voltage applied to the specimen during the pulse, and $t$ is the measured time-of-flight. To determine the mass-to-charge ratio of each signal from the detector, only the time-of-flight must be measured, typically to a resolution better than 10 ns. Multiple signals at the detector may be processed

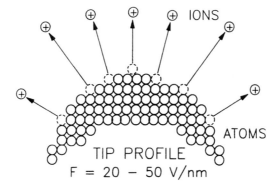

Fig. 3.  A schematic diagram of the field evaporation process responsible for the removal of surface atoms. This facilitates sectioning the material in the third dimension for imaging and chemical analysis.

by the timing system for each applied pulse for detection of simultaneous evaporation events.

Atom probe analysis is an absolute measure of the chemistry of a region of the specimen within the limits set by a detector efficiency less than unity. To determine the local composition of element A, the only requirement is to count the number of detected A ions and divide by the total number of detected ions. A microcomputer usually initiates the pulse, reads the time from the high speed electronics, calculates the mass of each signal, records the data, and displays some representation of the data on the screen.

Typically, an aperture is placed in the imaging assembly that defines the area to be analyzed, Fig. 4. As such, the diameter of the area can be changed by moving the aperture, i.e, the imaging screen. The diameter of the area of analysis may be varied from one atomic diameter to approximately 10 nm, although typical values range from 1–5 nm. If a feature of interest is not behind the aperture, a rotation of the specimen may be performed to move the feature under the aperture. As the specimen evaporates and ions are detected, material is being analyzed into the depth of the specimen. In this way, a composition profile through the feature of interest is obtained. As the field evaporation process occurs in a controlled manner, compositional analyses may be obtained with a depth resolution of one atomic plane, Fig. 5 (5).

Since the atom probe is a time-of-flight technique, there is no inherent mass limitation; light elemental species (Li, Be, B, C, O) are as readily detected as heavier species. *A priori* selection of individual species is not necessary during analysis as the detector is activated to identify the arrival of all possible species. Note that all isotopic species of the elements are obtained during the course of the experiment, which may lead to systematic errors in materials with isotopic overlap, Fig. 6.

There are a number of enhancements to and variations on the typical atom probe describe here (1– 4). Each instrument possesses its own

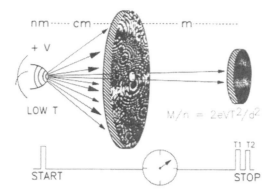

Fig. 4. A schematic diagram of field-ion microscopy and atom probe chemical analysis. All the surface atoms are imaged on the large screen, whereas only those surface atoms that travel through the small aperture and strike the small detector are analyzed by the mass spectrometer.

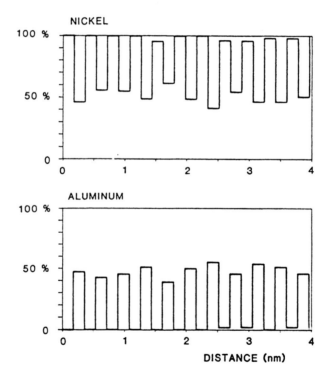

Fig. 5. Atom probe composition versus distance profiles taken perpendicular to the (100) planes of a $L1_2$ ordered $Ni_3Al$ alloy (5). The structure consists of alternating planes of pure nickel and planes of mixed nickel and aluminuim. The planar depth resolution is clearly visible. (Courtesy M. K. Miller.)

advantages which usually increase the area of analysis or increase the mass resolution. However, the fundamental ideas presented here are universal to all atom probes with further details presented in the references.

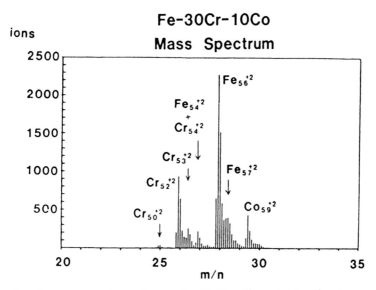

Fig. 6.   A mass spectrum for an Fe-Cr-Co alloy obtained using a conventional atom probe. Adequate resolution is obtained to resolve mass-to-charge ratios differing by 0.5 amu ($m/\Delta m > 100$). Extremely good mass resolution ($m/\Delta m > 1000$) is possible with additional instrumentation. Note also the difficulty that may arise with mass spectrometric techniques with the mass overlap at $m/n = 27$.

## INVESTIGATIONS

Field-ion microscopy provides atomic resolution images of the surface of specimens while the additional capability of serial sectioning is provided by field evaporation. Atom probe chemical analysis provides compositional determination of features on the nanometer scale. A few examples of investigations exploiting these techniques will be described.

### Field-Ion Microscopy

Individual atoms are visible in field-ion images, however, this area is not where the majority of the investigations concentrate. A field-ion image also consists of a series of concentric rings overlaid on a stereographic projection of the material. As such, the symmetry elements of the crystal structure may be determined. For example, images of the Al-Mn icosahedral material clearly show the 5-, 3-, and 2-fold rotation axes of the structure (6). Local atomic stacking sequences in ordered materials may also be observed. NiAl is a BCC-derivative B2 ordered material with alternating nickel and aluminum planes in the <100> directions. In the field-ion images, these planes appear as alternating brightly and dimly imaging rings (7).

Surface diffusion studies of adatoms on various crystal faces of metals have been performed (8). The observed atomic positions of diatomic or triatomic clusters are recorded after each heating of the specimen to a predetermined temperature. From this data, a plot of the energy potential and interaction energies may be determined in two dimensions. Surface reconstruction of metallic and semiconducting surfaces have been reported in atomic detail. Unreconstructed surfaces were produced at cryogenic temperatures by field evaporation, whereas reconstruction was observed after heating the specimens to higher temperatures (9,10).

Observations of segregants to defects in materials necessitates a large amount of contrast between the solute and the bulk material. This usually occurs when the solute atoms have either a much larger or a much smaller radius than the matrix. A few examples are Pd in steel (11), B in $Ni_3Al$ (12), and B in steel (13), Fig. 7. In each case, the solute atoms protrudes slightly from the surface and appear very brightly in the image. Note, however, that chemical identification from the images alone is only speculative, as the contrast may originate from other sources. In these examples, the chemical identity of the species was confirmed by atom probe analysis.

As in segregation studies, contrast must be present between phases for their identification in multiphase materials. In Cu-Ti materials, no contrast was observed in the image, even when transmission electron microscopy and atom probe analysis confirmed the presence of a two-phase mixture (14). Depending upon the physics of field evaporation and field ionization processes in the material, the minor phase may either be brightly imaging, as is the case of Guinier-Preston in Al-Cu (15), or darkly imaging, as is the case of Cr-enriched (16) and Cu-enriched (17) particles in iron-based alloys. The unique sectioning ability of field evaporation permits the true three-dimensional characterization of complex two-phase morphologies (18).

### Atom Probe Analysis

Atom probe analysis of catalytic products on the surfaces of refractory specimens has been investigated (19,20). These studies begin with a clean specimen in UHV conditions. The reactive gas of interest is introduced into the chamber continuously and a spectrum is obtained. Based on a variation of the specimen temperature, the pressure of the reactive gas, and the relative abundances of the detected species, a better understanding of catalytic reactions is obtained.

In many studies, the composition of phases present in the multiphase material is required on a scale less than 10 nm in diameter. In nickel-based superalloys, quantitative determination of all solute species in the multiple phases present would help with the understanding of the mechanical properties (21–23). Much work has been performed that showed that Mo, W, Cr, and Co are retained in the gamma matrix, whereas Al and Ti are enriched in the gamma-prime precipitate phase, Fig. 8. In Al-Cu alloys, a determination of the composition of GP-zones along with the microscopy observations is assisting in the determination of the complex precipitation sequence (24). In reactor pressure vessel steels, degradation of mechanical properties was attributed to the presence of very small (<5 nm) diameter precipitates of unknown composi-

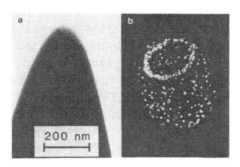
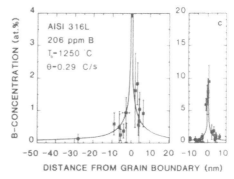

Fig. 7. A transmission electron micrograph (a), a field-ion micrograph (b), and corresponding composition profiles (c) of segregation to a grain boundary in stainless steel (13). The bright spots along the grain boundary were identified by atom probe analysis to arise from boron segregation. Note the narrow width of the boron distribution at the boundary. (Courtesy H. Nordén.)

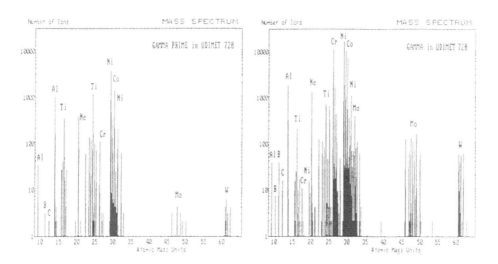

Fig. 8. Mass spectra obtained from a nickel-based Udimet 720 alloy in (a) gamma matrix and (b) gamma-prime precipitates (23). Note the excellent mass resolution, the analysis of low masses, and the wide range of masses that may be analyzed simultaneously. (Courtesy K. L. More.)

tion. Atom probe analyses showed that the particles were copper enriched, but also contained significant levels of other solute elements (17).

Two extremes in the modes of decomposition are spinodal decomposition (gradual composition change with ageing time) and nucleation and growth (discrete nuclei formation near equilibrium composition). To distinguish between these two modes of precipitation, the development of the second phase must be monitored form the earliest stages of particle formation. This usually occurs when the composition amplitude and the diameter of the particles are small. During spinodal decomposition, a gradual increase in the composition amplitude from the mean composition as a function of ageing time must be measured. For a nucleation and growth mechanism, the composition of the particle at the earliest ageing time must be far from the average composition or close to the equilibrium composition. The major change occurring during the process is an increase in the particle size. Extensive studies have been performed indicating the mode of decomposition to be spinodal decomposition in Fe-Cr-Co (16), Fe-Al-Ni-Co (25,26), Cu-Ti (13), and nucleation and growth in Fe-Cr-Co (27) and Ni-Al (28). Chemical analyses were usually performed at a depth resolution of 0.2 nm and a spatial resolution of 2 nm. Quantitative composition determinations were obtained, usually where the smallest particle sizes in each investigation was less than 5 nm.

In many practical alloy systems, segregation of elements to defects make a significant contribution to mechanical properties. However, quantifying the elemental content at the defects is usually difficult because of spatial resolution or light element constraints. Since atom probe analysis does not typically have these constraints, investigators have used this technique to examine many segregation systems. $Ni_3Al$ is typically an extremely brittle material, however the introduction of approximately 2000 wppm of boron was shown to increase ductility markedly while many other techniques failed to identify the position of the boron, atom probe analyses showed that the boron segregated to many types of planar defects in the material (12). The investigation of the segregation of boron to grain boundaries in stainless steels has been performed by many techniques. However, none of the composition versus distance results have been performed with sufficient spatial resolution to compare to theoretical calculations. Atom probe was then used to measure the boron distribution around grain boundaries (13). The distributions has sufficient spatial resolution (4 nm) that direct comparison to theory was possible, Fig. 7. A change in the aluminum concentration by only a few atomic percent in binary NiAl alloys is sufficient to change the mode of fracture from ductile to brittle. Grain boundary structure and/or chemistry was proposed as the controlling mechanism. Atom probe analyses of grain boundaries in a brittle alloy showed that a large portion of the boundaries are depleted of aluminum in a layer less than 0.5 nm in width (7), Fig. 9.

## CONCLUSIONS

Atom probe analysis and field-ion microscopy have been applied to a wide variety of problems including surface diffusion, gas-metal reactions, catalysis, precipitation, solute segregation in a wide variety of materials (metals and alloys, semiconductors, high temperature superconductors, and polymers). Field-ion microscopy is a real-space atomic resolution microscopy and, more importantly, has the unique ability for serial sectioning the specimen in a controlled manner. Thus, the true three-dimensional structure of defects and particles may be obtained.

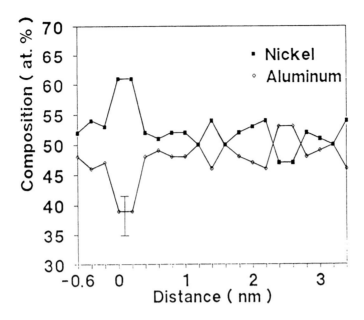

Fig. 9.   A composition profile across a grain boundary in binary NiAl (17). Note the narrow width (< 0.5 nm) of the aluminum depletion at the grain boundary.

Atom probe is a powerful chemical analysis technique that possesses both high spatial (<2 nm) and depth (0.2 nm) resolutions. There is no mass limitation for the technique, so light elemental analysis, which may be more difficult at high resolution for other techniques, present no problem for this mass spectrometric technique.

## REFERENCES

1.  Müller, E. W. and Tsong, T. T., *Field Ion Microscopy, Principles  and Applications*, Elsevier, Amsterdam, 1969.
2.  Müller, E. W., and Tsong, T. T., *Prog. Surf. Sci.* **4**, 1 (1973).
3.  Wagner, R., *Field Ion Microscopy in Materials Science*, Springer, Berlin, 1982.
4.  Miller, M. K., *Intl. Mater. Rev.* **32**, 1 (1987).
5.  Miller, M. K., and Smith, G. D. W., *Atom Probe Microanalysis: Principles and Applications to Materials Problems*, MRS, Boston, 1989.
6.  Melmed, A. J. and Klein, R., *Phys. Rev. Lett.* **56**, 1478 (1986).
7.  Camus, P. P., Baker, I., Horton, J. A., and Miller, M. K., *J. de Phys.* **49-C6**, 329 (1989).
8.  Watanabe, F., and Ehrlich, G., *J. de phys.* **49-C6**, 267 (1989).
9.  Tsong, T. T., Liu, H. M., Gao, Q. J., and Feng, D. L., *J. de Phys.* **48-C6**, 41 (1987).
10. Kellogg, G. L., *J. de Phys.* **48-C6**, 59 (1987).
11. Miller, M. K., Brenner, S. S., and Wilde, B. E., *Proc. 29th Intl. Field Emiss. Symp.*, H.-O. Andrén, H. Nordén, eds., Almqvist & Wiskell, Stockholm, 1982, 481.
12. Horton, J. A. and Miller, M. K., *Acta Met.* **35**, 133 (1987).

13. Karlsson, L. and Nordén, H., *Acta Met.* **36**, 13 (1988).
14. Biehl, K.-E. and Wagner, R., *Proc. Intl. Conf. Solid-Solid Phase Transform*, H. I. Aaronson et al., eds., Metall. Soc. AIME, Warrendale, PA, 1982, 185.
15. Mori, T., Wada, M., Kita, H., Uemori, R., Horie, S., Sato, A., and Nishikawa, O., *Jap. J. Appl. Phys.* **22**, 203 (1983).
16. Soffa, W. A., Brenner, S. S., Camus, P. P., and Miller, M. K., *Proc. 29th Intl. Field Emiss. Symp.*, H.-O. Andrén, H. Nordén, eds., Almqvist & Wiskell, Stockholm, 1982, 511.
17. Miller, M. K. and Burke, M. G., *J. de Phys.* **48-C6**, 429 (1987).
18. Camus, P. P., Soffa, W. A., Brenner, S. S., and Miller, M. K., *J. de Phys.* **45-C9**, 265 (1984).
19. Liu, W., Ren, D. M., Bao, C. L., and Tsong, T. T., *J. de Phys.* **48-C6**, 487 (1987).
20. Chuah, G.-K., Krose, N., Block, J. H., and Abend, G., *J. de phys.* **48-C6**, 493 (1987).
21. Delargy, K. M. and Smith, G. D. W., *Proc. Conf. on High Temp Alloys for Gas Turbines*, Liege, Belgium, R. Brunestand et al., eds., D. Reidel, Dordrecht, Holland, 705 (1983).
22. Blavette, D., Bostel, A., and Bouet, M., *J, de Phys.* **45-C9**, 379 (1984).
23. More, K. L. and Miller, M. K., *J. de phys.* **49-C9**, 391 (1989).
24. Hono, K., Sakurai, T., and Pickering, H. W., *J. de Phys.* **48-C6**, 349 (1987).
25. Cowley, S. A., Hetherington, M. G., Jakubovics, J. P., and Smith, G. D. W., *J. de Phys.* **47-C7**, 211 (1986).
26. Zhu, F., von Alvensleben, L., and Haasen, P., *Scr. Met.* **18**, 337 (1984).
27. Zhu, F., Wendt, H., and Haasen, P., *Decomposition of Alloys: the Early Stages*, P. Hassen et al., eds., Pergamon Press, Oxford, 1984, 139.
28. Wendt, H., *Proc. Intl. Conf. Solid-solid Phase Transform.*, H. I. Aaronson et al., eds., Metall. Soc. AIME, Warrendale, PA, 1982, 455.

# Emissivities and Optical Constants of Electromagnetically Levitated Liquid Metals as Functions of Temperature and Wavelength

S. KRISHNAN,*[1] G. P. HANSEN,[2] R. H. HAUGE,[3]
AND J. L. MARGRAVE[3]

[1]Intersonics, Inc., Northbrook, IL; [2]Texas Research Institute, Austin, TX;
and [3]Rice University and the Houston Area Research Center,
Houston, TX

## ABSTRACT

The development of a unique, noncontact, temperature measurement device utilizing rotating analyzer ellipsometry is described. The technique circumvents the necessity of spectral emissivity estimation by direct measurement concomitant with radiance brightness. Using this approach, the optical properties of electromagnetically levitated liquid metals Cu, Ag, Au, Ni, Pd, Pt, Zr were measured *in situ* at 4 wavelengths and up to 600 K superheat in the liquid. The data suggest an increase in the emissivity of the liquid compared with the incandescent solid. The data also show moderate temperature dependence of the spectral emissivity, particularly when the wavelength of light is close to an absorption edge of the material. The data for both solids and liquids show excellent agreement with available values in the literature for the spectral emissivities as well as the optical constants.

**Index Entries:** liquid metals; Cu, Ag, Au, Ni, Pd, Pt, Zr, emissivities; optical constants; non-contact pyrometry; ellipsometry.

## INTRODUCTION

The behavior of the spectral emissivities of liquid metals at elevated temperatures has not received much attention. However, it assumes

*Author to whom all correspondence and reprint orders should be addressed.

High Temperature Science, Vol. 26     © 1990 by the Humana Press Inc.

fundamental importance when it becomes necessary to carry out radiation thermometry with only an assumed emittance. The questions are addressed in this study are

1. How do the spectral emissivities depend on temperature above the melting point?
2. How do spectral emissivities vary as a function of wavelength? and
3. Is the temperature dependence different at different wavelengths?

The emissivities of liquid metals at elevated temperatures have been measured previously by several methods.

1. Comparison of the measured brightness temperature of the liquid at the accepted fusion temperature (1–3);
2. Comparison of the measured radiances by a two-color pyrometer with a single color pyrometer, and assuming the measurement of the first is independent of emissivity (4), and
3. Having the sample in thermal contact with a substrate of known temperature, measuring the radiance brightness temperature of the sample and comparing the 2 temperatures (5).

There are several difficulties associated with each of these techniques. For example, in the case of melting-point brightness measurements, the question arises as to whether the measured emissivity corresponds to the liquid, solid, or a mixture of both. If the measurement is made precisely at the melting point, equilibrium is established between the liquid and solid, so the measurement may correspond to a mixture. Even assuming one is making the measurement on the liquid, the errors resulting from brightness measurement can be substantial. Bonnell et al. (3) have calculated a possible error of 5–6% resulting from a 10 K uncertainty in brightness temperature, in estimating a 0.30 value of emissivity at 2000 K. If the temperature being measured is higher, where the brightness-temperature uncertainty is greater, the measured emissivity can be in error by more than 10%. Several workers have pointed out the difficulties associated with 2-color pyrometry as a technique for deriving thermodynamic temperatures (6–9). The third approach is limited by container contact interactions that can cause uncertainty in the spectral emissivity. This method is also unlikely to provide reliable emissivity estimates because of the chemical reactivity and solubility properties of liquid metals at high temperatures (>1800 K).

The use of ellipsometry for measurement of the optical properties of metals and alloys is not new (10–18). However, the use of this technique for emissivity measurements on very refractory liquid metals and alloys as a function of temperature above the melting point is certainly novel. The difficulty in handling very refractory liquid metals is that they tend to be highly reactive. However, this problem is completely eliminated in the current experiments with the use of the electromagnetic levitation system where the liquid drop is completely isolated from any containers.

The levitation technique also provides other attractive features, such as the ability to control and vary liquid temperatures in addition to being able to study solid emissivities by resolidifying the liquid (this generally tends to produce very smooth surfaces). With the use of this arrangement, it becomes possible to measure solid and liquid metal emissivities and optical constants in the temperature range 800–3000 K quite easily. The highest temperature is usually dictated by the vapor pressures of the material being studied (vapor obscuration) and the coil designs that are used.

In this paper, the results of optical property measurements, including spectral emissivity, as functions of temperature at wavelengths of 488, 514.5, 632.8, and 1064 nm for a number of electromagnetically levitated liquid metals including Cu, Ag, Au, Ni, Pd, Pt, and Zr are presented. The data include measurements on liquids, solids, and undercooled liquids. Perhaps, the most significant outcome of this work is that a technique has been developed to measure accurately the optical constants, and thus temperatures, of levitated liquid materials where sample contact is undesirable or impractical. This is of enormous importance for containerless processing technologies.

### Theoretical Background

According to Wien's approximation to Planck's law, the flux distribution from a black body per unit solid angle can be written as a function of wavelength, $\lambda$, and temperature, T

$$W\,(\lambda,T) = C_1 \pi^{-1} \lambda^{-5} \exp(-\frac{C_2}{\lambda T}) \tag{1a}$$

where $C_1$ and $C_2$ are the first and second Planck radiation constants, respectively. Equation (1a) has been shown to be accurate (6) to within 1% for $\lambda T < 2897.8$ μM K. A useful form of Wien's approximation is obtained by taking the ratio of the real-body radiation at wavelength, $\lambda$, to that of a blackbody at the same wavelength

$$\frac{1}{T_{TH}} - \frac{1}{T_B} = \frac{\lambda \ln (E_\lambda)}{C_2} \tag{1b}$$

where $T_B$ is the measured brightness temperature, $T_{TH}$ the thermodynamic (blackbody) temperature, and $E_\lambda$ is the normal spectral emissivity of the real body at the wavelength, $\lambda$.

Assuming the liquid metal surface is specular, Kirchoff's law can be stated as

$$E_\lambda + R_\lambda = 1 \tag{2}$$

where $R_\lambda$ is the reflectivity and $E_\lambda$ is the spectral emissivity. The subscript denotes the wavelength dependence of these quantities. Equation (2) holds for any given or specified angle of incidence or emission.

$\varepsilon_1$ and $\varepsilon_2$, the real and imaginary parts of the dielectric constant, are related to the refractive index, $n$, and extinction coefficient, k, through the relations:

$$\varepsilon_1 = n^2 - k^2 \quad \text{and} \tag{3a}$$

$$\varepsilon_2 = 2nk \tag{3b}$$

If $\varepsilon_1$ and $\varepsilon_2$ are experimentally determined, then the remaining optical constants of the material in question can be obtained for the wavelength, $\lambda$.

The normal incidence reflectivity is calculated from the Beer equation:

$$R_\lambda = \frac{(n-n_o)^2 + k^2}{(n+n_o)^2 + k^2} \tag{4}$$

where $n_o$ is the refractive index of the ambient, transparent medium. Using Eq. (2), the spectral emissivity is determined. Simultaneous measurement of radiance brightness at normal incidence provides the thermodynamic temperature from Eqs. (1b, 2,4).

A schematic of the ellipsometric measurement is illustrated in Fig. 1. In this figure, it can be seen that the light reflected by the sample is passed through a rotating analyzer (polarizer) which discriminates the orthogonal components of the beam. In Fig. 2, we have illustrated the change in amplitude and phase suffered by an incident, linearly polarized beam that becomes elliptically polarized on reflection.

When collimated monochromatic light, plane polarized with an initial azimuth $\Psi_o$ to the plane of incidence is reflected from a metal surface, the P and the S components of the incident electric vector experience different phase changes and different reflectivities, producing a new azimuth $\Psi$. If $\Psi_o = \pi/4$, then the complex reflectance ratio, $\rho$, is:

$$\rho = \frac{r_p}{r_s} = \tan(\Psi)\, e^{i\Delta} \tag{5}$$

where, $r_p$ and $r_s$ are the amplitude reflection coefficients of the P and S components, respectively, $\Psi$ is the restored azimuth and $\Delta$ the relative phase difference between the two components of the reflected electric vector ($\Delta = \delta_{rp} - \delta_{rs}$). If the reflected intensity is measured at four azimuths (obtained at two positions of an analyzer) - $I_1$ (90°), $I_2$ (0°), $I_3$ (45°, $I_4$ (135°), then the fundamental ellipsometric parameters $\Psi$ and $\Delta$ are given by (14)

$$\tan(\Psi) = \left(\frac{I_2}{I_1}\right)^{1/2} \quad \text{and} \tag{6}$$

$$\cos(\Delta) = \tfrac{1}{2}\left[\left(\frac{I_2}{I_1}\right)^{1/2}t + \left(\frac{I_1}{I_2}\right)^{1/2}\right] \times \left[\frac{(1 - I_4/I_3)}{(1 + I_4/I_3)}\right] \tag{7}$$

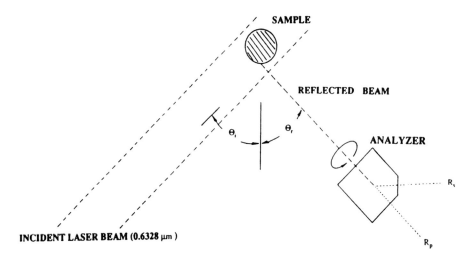

Fig. 1. Schematic illustration of ellipsometric measurement.

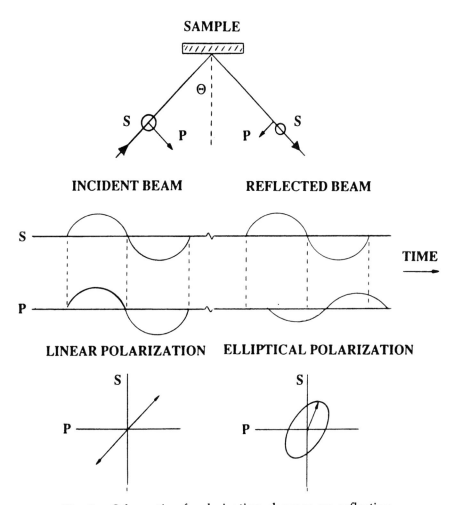

Fig. 2. Schematic of polarization changes on reflection.

The real and imaginary parts of the dielectric constant are then calculated from $\Psi$, $\Delta$, and the angle of incidence, $\Theta$ using (31)

$$\varepsilon_1 = \sin^2\theta \, \tan^2\theta \frac{\cos^2 2\psi - \sin^2 2\psi \, \sin^2\Delta}{(1 + \cos\Delta \, \sin 2\psi)^2} + \sin^2\theta \qquad (8a)$$

and

$$\varepsilon_2 = \frac{2 \sin 2\psi \, \cos 2\psi \, \sin\Delta}{(1 + \cos\Delta \, \sin 2\psi)^2} \sin^2\theta \, \tan^2\theta \qquad (8b)$$

Once $\varepsilon_1$ and $\varepsilon_2$ are obtained, one calculates n and k. Then, the normal incidence reflectivity is obtained from Eq. (4). It is important to note that the entire measurement essentially consists of two ratio measurements ($I_2/I_1$ and $I_4/I_3$). Thus, systematic errors tend to cancel.

## Optical Constants

In addition to spectral emissivities, the current technique provides also provides values of related optical constants. The optical properties of liquid metals that melt at accessible temperatures have been extensively investigated in the wavelength range 3 μm to about 0.3 μm by Arakawa (19), Comins (20), Miller (21), Schulz (22), Hodgson (23–27), and Smith (28). The results for these metals (Cu, Ag, Au, Bi, Sn, Sb, Ga, Al) are in reasonable agreement with the values calculated from the free electron theory (29) and have usually been expressed in terms of the real and imaginary parts of a complex dielectric constant $\varepsilon = \varepsilon_1 - i\varepsilon_2$. The product $\varepsilon_0\varepsilon_2\omega$ is usually denoted $\sigma$ ($\omega$) and is called the optical conductivity; $\varepsilon_0$ is the permittivity of vacuum. These results have fitted quite well with the Drude expressions (29)

$$\varepsilon_1(\omega) - 1 = \frac{n^* e^2 \tau}{m\varepsilon_0 (1 + \omega^2\tau^2)} \qquad (9a)$$

and

$$\varepsilon_2(\omega) = \frac{n^* e^2 \tau}{\omega m\varepsilon_0 (1 + \omega^2\tau^2)} \qquad (9b)$$

after some adjustment of the effective carrier density $n^*$ and the relaxation time $\tau$.

The dc conductivity is given by

$$\sigma_0 = \frac{n^* e^2 \tau}{m} \qquad (10)$$

The closeness of the fits achieved by previous work (19–28), and the absence of any sort of structure in their curves for $\varepsilon_1$ and $\varepsilon_2$ that is associated with interband absorption in solid metals, have provided much support for the nearly-free-electron electron model of liquid metals, upon which Ziman's (29) theory of their electrical properties is based.

In our study, we compared our results with those predicted from the free electron model. Such an exercise was carried out only for the noble metals. In order to perform such a fit, it was necessary to adjust the relaxation time so as to obtain the correct dc conductivity $\sigma_0$. The value of $n^*$ was obtained from previous studies on noble metals by some of the workers listed.

## EXPERIMENTAL

Figure 3 shows the experimental arrangement used to make emissivity measurements by Rotating Analyzer Ellipsometry (RAE) on electromagnetically levitated materials. A chopped (2000 Hz), focused, plane polarized laser beam impinged on the sample at 67.5° angle of incidence. The laser used at the 488 and 514 nm wavelengths was a Lexel (Model 95) 2 watt argon-ion laser that was lasing multiline. The measurements at 633 nm were carried out using the Aerotech 5 mW He-Ne laser, whereas the measurements at 1064 nm were carried out using an ALC 1064-50P (Amoco Laser Corporation) diode-pumped YAG laser with an output of 50 mW polarized light. The experiments that were performed at the YAG and He-Ne wavelength required the laser to be focused, whereas no focusing was needed for measurements with the argon-ion laser. The plane of polarization of the beam was rotated 45° (clockwise for a viewer looking at the sample from the laser) with respect to the plane of incidence (plane of the figure). Also shown in the figure is a laser shutter positioned between the chopper and the sample chamber. Closure of this shutter allowed measurements of radiance brightness of the sample without error owing to the laser. A Leeds & Northrup, Model 8641, Automatic Optical Pyrometer was used to monitor the brightness temperature of the sample via another port on the chamber situated at 67.5° from the incident laser port. Prior to the experiments, this pyrometer was calibrated using a NIST traceable strip lamp obtained from the Pyrometer Instrument Company, Northvale, NJ. The thermodynamic temperatures of the liquid were calculated using Wien's approximation to Planck's law and using values of spectral emissivity measured at $\lambda = 633$ nm. Thermodynamic temperatures were also determined using the radiance detector situated on the ellipsometer based on measured brightness and the optical constants measured by the ellipsometer. To calculate the temperatures using the radiance detector, it was necessary to recompute an angular (67.5°) spectral emissivity from the measured dielectric constants and then using Wien's approximation to Planck's law. The two temperatures usually agreed to within ± 5 K. The second measurement was not feasible in the blue and green regions of the spectrum since samples usually emitted very little light.

Light from the sample (radiated and reflected) was imaged (focused) on an aperture by an objective lens. A second lens and aperture generated parallel light from this image. The f number of the first lens and

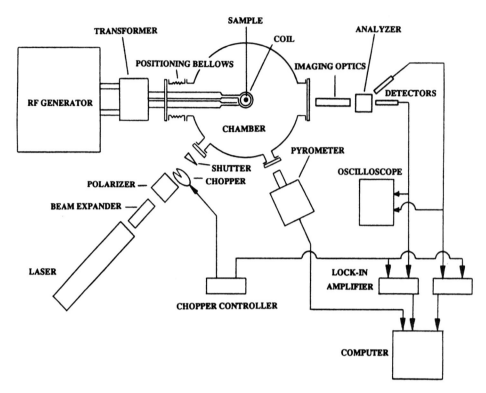

Fig. 3. Schematic of the levitation system, ellipsometric arrangement, data acquisition, and temperature measurement.

aperture fixed the maximum solid angle from which light was collected to about 1°. Since the study involved several wavelengths, spectral discrimination was achieved using laser line interference filters obtained from Corion and Ealing. Light that passed through the hole was then analyzed by a beam-splitting Thompson calcite prism that was rotatable around the beam axis, thus facilitating RAE.

Shown in Fig. 2 is the flow diagram for the data acquisition system. The outputs of the polarization detectors were monitored individually with separate lock-in amplifiers which were phase-locked with the reference signal from the chopper. The lock-in amplifiers were obtained from EG&G PARC and had 20 KHz bandwidths. The outputs of the lock-in amplifiers were monitored continuously by the computer, as were the outputs of the radiance detector and the Leeds & Northrup pyrometer.

In a previous paper (30), the dynamics of electromagnetically levitated liquid metals and alloys at elevated temperatures were described. These results showed that the levitated liquid metal droplets underwent rotations, shape oscillations, and translations in the coil. This behavior causes oscillations in the intensity of the emitted as well as reflected light from the sample. The intensity variations were roughly 10–20% in magnitude. The frequency was typically 20–40 Hz.

It was necessary to minimize the effect of these surface perturbations by carrying out signal averaging. Since the perturbation frequencies from previous studies were established to be on the order of 50–200 Hz, the output time constant on the lock-in amplifiers was set at 0.1 s. The outputs of the lock-in amplifiers were monitored for 10–20 s, until a reliable mean value could be extracted. The analyzer was then rotated for the second set of measurements and the signal averaging repeated. Finally, the signal intensities were compared for the two analyzer positions and the criterion of validity elucidated by Beattie (15), was verified. Any data not meeting this criterion were rejected. A detailed description of the calibration procedure for the optics and electronics has been presented elsewhere (32) and will not be presented here.

All samples used had purities in the range 99.9–99.999%. The samples were cut from rod or bar stock and had masses in the range 0.5–1.5 g. After melting in the levitation coil, residual oxide surface contaminants were allowed to vaporize and be removed from the chamber before the measurements were started.

The experiments were performed at pressures on the order of 1 atmosphere and gas flowrates of 10–20 cc/min. The gases used were argon and helium with 3% hydrogen added to help inhibit surface oxide formation. Temperature control was achieved to a limited extent by regulating the gas mixture and flowrate. Helium, with its higher thermal conductivity, was used to obtain lower temperatures, including the undercooled state, whereas argon was used to obtain the higher temperatures. The upper temperature limit for each sample was set by the vaporization rate of the metal. At higher temperatures, smoke near the metal surface scattered the reflected laser light, which resulted in much larger measurement errors.

## RESULTS

Tables 1 to 4 give a summary of some of the results of the experiments reported here. Values are given for the dielectric constants and spectral emissivities of Cu, Ag, Au, Ni, Pd, Pt, and Zr over wide ranges of temperature above the nominal freezing point of the liquid at the 4 wavelengths, respectively. Since much of the data reported here showed moderate to strong temperature dependence, an interpolated value based on a linear fit to the data is provided. A value of the confidence interval is listed in the parenthesis below each of the tabulated values. These intervals correspond to a confidence level of 95% for the data listed in the Table. For purposes of clarity, the symbol ± has been omitted. Also listed parenthetically below the temperature range is the number of determinations of the optical properties over the range. The temperatures for interpolation (in some cases, extrapolation) were chosen on the basis of the temperature range over which most of the data were acquired. The data for liquid zirconium represent an average of the optical properties, as is the case with the data for solid palladium. Correspondingly, the error values represent a standard deviation and not standard errors of estimate. In a few cases, extrapolation of the data to the chosen temperature was necessary, and, consequently, the error bars were slightly higher.

Table 1
Optical Constants of Liquid Metals at 488 nm[a]

| Metal | Temp. range (K)[b] | $\varepsilon_1$ | $\varepsilon_2$ | $E_\lambda$ |
|---|---|---|---|---|
| Copper | 1499–1877 | −4.92 | 7.63 | 0.45 |
|  | (13) | (0.13) | (0.12) | (0.01) |
| Silver | 1398–1552 | −9.28 | 1.78 | 0.11 |
|  | (17) | (0.38) | (0.31) | (0.01) |
| Gold | 1420–1907 | −2.77 | 7.57 | 0.522 |
|  | (16) | (0.28) | (0.62) | (0.018) |
| Nickel | 1789–2109 | −2.66 | 13.17 | 0.486 |
|  | (24) | (0.35) | (0.30) | (0.007) |
| Palladium (s) | 1424–1699 | −5.66 | 12.23 | 0.43 |
|  | (11) | (0.24) | (0.31) | (0.01) |
| Palladium (l) | 1911–2120 | −3.47 | 13.51 | 0.468 |
|  | (27) | (0.39) | (0.65) | (0.004) |
| Platinum | 2266–2646 | −1.86 | 15.72 | 0.48 |
|  | (33) | (0.66) | (0.90) | (0.01) |

[a]The confidence interval for 95% limits of the data is listed parenthetically below each value. The number of determinations is listed parenthetically below the temperature range.
[b]Temperatures chosen for interpolation of optical constant values are: Cu (1400 K), Ag (1300 K), Au (1400 K), Ni (1800 K), Pd (1925 K), and Pt (2250 K).

Table 2
Optical Constants of Liquid Metals at 514.5 nm[a]

| Metal | Temp. range (K)[b] | $\varepsilon_1$ | $\varepsilon_2$ | $E_\lambda$ |
|---|---|---|---|---|
| Copper | 1361–1903 | −8.81 | 7.11 | 0.30 |
|  | (19) | (0.16) | (0.42) | (0.01) |
| Silver | 1452–1541 | −10.86 | 0.84 | 0.05 |
|  | (17) | (0.52) | (0.14) | (0.01) |
| Gold | 1449–1903 | −6.34 | 8.07 | 0.47 |
|  | (20) | (1.08) | (0.47) | (0.02) |
| Nickel | 1776–2063 | −6.35 | 15.42 | 0.412 |
|  | (31) | (0.39) | (1.20) | (0.008) |
| Palladium (s) | 1486–1810 | −11.63 | 13.68 | 0.32 |
|  | (11) | (0.37) | (0.26) | (0.005) |
| Palladium (l) | 1869–2134 | −7.83 | 16.95 | 0.381 |
|  | (25) | (0.42) | (0.83) | (0.006) |
| Platinum | 2307–2649 | −2.46 | 25.44 | 0.45 |
|  | (19) | (0.37) | (0.60) | (0.01) |

[a]The confidence interval for 95% limits of the data is listed parenthetically below each value. The number of determinations is listed parenthetically below the temperature range.
[b]Temperatures chosen for interpolation of optical constant values are: Cu (1400 K), Ag (1300 K), Au (1400 K), Ni (1800 K), Pd (1925 K), and Pt (2250 K).

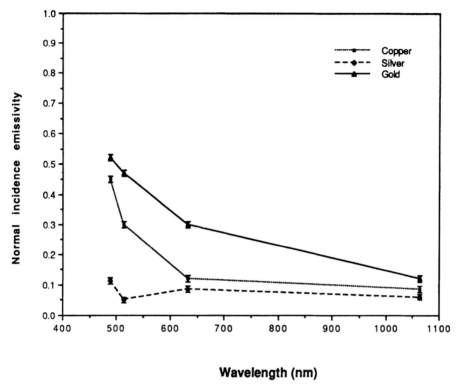

Fig. 4. Normal incidence emissivity of liquid metals in Cu group as a function of wavelength. Error bars represent 95% precision limits.

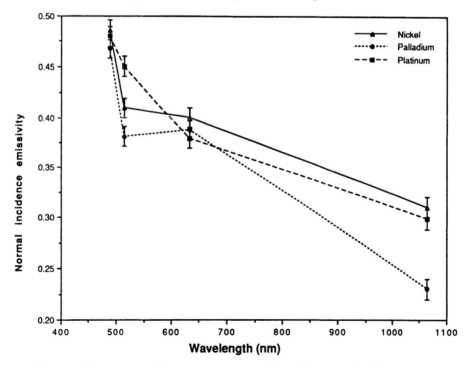

Fig. 5. Normal incidence emissivity of liquid metals in Ni group as a function of wavelength. Error bars represent 95% precision limits.

Table 3
Optical Constants of Liquid Metals at 633 nm[a]

| Metal | Temp. range (K)[b] | $\varepsilon_1$ | $\varepsilon_2$ | $E_\lambda$ |
|---|---|---|---|---|
| Copper | 1315–1727 | −14.11 | 3.83 | 0.13 |
| | (27) | (0.45) | (0.39) | (0.01) |
| Silver | 1291–1481 | −16.31 | 3.34 | 0.086 |
| | (41) | (0.36) | (0.44) | (0.01) |
| Gold | 1270–1932 | − 9.58 | 8.20 | 0.30 |
| | (37) | (0.79) | (1.36) | (0.01) |
| Nickel | 1822–2143 | − 6.60 | 18.55 | 0.40 |
| | (21) | (0.90) | (1.24) | (0.01) |
| Palladium (s) | 1297–1577 | −12.12 | 19.35 | 0.32 |
| | (11) | (0.43) | (0.39) | (0.005) |
| Palladium (l) | 1807–2126 | − 7.32 | 18.93 | 0.38 |
| | (23) | (0.72) | (0.73) | (0.01) |
| Platinum | 20062475 | − 6.04 | 25.20 | 0.38 |
| | (33) | (1.38) | (1.63) | (0.01) |
| Zirconium | 2175–2402 | − 4.81 | 21.7 | 0.41 |
| | (3) | (0.87) | (2.46) | (0.005) |

[a]The confidence interval for 95% limits of the data is listed parenthetically below each value. The number of determinations is listed parenthetically below the temperature range.
[b]Temperatures chosen for interpolation of optical constant values are: Cu (1400 K), Ag (1300 K), Au (1400 K), Ni (1800 K), Pd (1925 K), and Pt (2250 K).

Table 4
Optical Constants of Liquid Metals at 488 nm[a]

| Metal | Temp. range (K)[b] | $\varepsilon_1$ | $\varepsilon_2$ | $E_\lambda$ |
|---|---|---|---|---|
| Copper | 1434–1729 | −33.19 | 11.18 | 0.085 |
| | (11) | (1.25) | (1.02) | (0.008) |
| Silver | 1236–1489 | −38.13 | 7.68 | 0.06 |
| | (12) | (1.60) | (0.44) | (0.005) |
| Gold | 1417–1976 | −34.82 | 14.25 | 0.12 |
| | (10) | (1.88) | (1.02) | (0.01) |
| Nickel | 1888–2031 | −12.71 | 15.31 | 0.31 |
| | (9) | (1.33) | (0.92) | (0.016) |
| Palladium | 1947–2073 | −15.21 | 7.91 | 0.23 |
| | (5) | (2.05) | (1.08) | (0.01) |
| Platinum | 2106–2386 | −12.05 | 13.44 | 0.298 |
| | (10) | (1.62) | (2.80) | (0.006) |

[a]The confidence interval for 95% limits of the data is listed parenthetically below each value. The number of determinations is listed parenthetically below the temperature range.
[b]Temperatures chosen for interpolation of optical constant values are: Cu (1400 K), Ag (1300 K), Au (1400 K), Ni (1800 K), Pd (1925 K), and Pt (2250 K).

Since the chosen temperatures were at least 100° or more above the melting point, some variation from available literature data is expected. Each of these four tables corresponds to measurements at a single wavelength. The 3 noble metals Cu, Ag, and Au, and the three transition metals Ni, Pd, and Pt were studied at all 4 wavelengths, whereas Zr was studied only at the He-Ne wavelength. A salient feature of the data listed in Tables 1–4 is that almost all spectral emissivities were measured with an error of ± 0.01 corresponding to 95% precision limits of the data at the chosen temperatures.

Table 5 is a compilation of emissivity data found in the literature, together with the data obtained in this study, to illustrate the changes in emissivity as a function of temperature. In this table we list emissivity data for these metals at room temperature (RT), incandescent temperatures (IT) and in the liquid regime. The spectral emissivities calculated from the data reported by Miller (21) and our study have been obtained at a photon energy of 1.97 eV ($\lambda$ = 0.63 $\mu M$) whereas the other emissivity data pertaining to the elevated temperatures were measured at $\lambda$ = 0.65 − 0.665 $\mu M$ (1.91–1.87 eV). The room temperature emissivity data were obtained at 1.97 eV. It can be seen that the agreement of the present results with data found in the literature is excellent in all cases, except, perhaps, in the case of liquid zirconium, where there appears to be a large discrepancy in the literature data. The lack of spectral emissivity data in the literature at the other wavelengths made a similar compilation at other wavelengths impossible.

To illustrate the overall wavelength dependence of the spectral emissivity, the spectral emissivities of the copper and nickel groups are plotted in Figs. 4 and 5, respectively. The spectral emissivity data used for these plots are listed in Tables 1–4. The error bars in Figs. 4 and 5 correspond to a 95% confidence interval. The spectral emissivities correspond to the previously listed interpolation temperatures. In these two plots. it is seen quite clearly that the spectral emissivity increases gradually as a function of wavelength, until the blue region where a sharp rise is seen. The behavior of Cu and Au are similar, although silver does not show such a dramatic rise. The corresponding plot in Fig. 5 for the nickel group has two points of interest. First, all three metals behave identically in the visible region of the spectrum. Second, they seem to show very little dependence between the red and green regions of the spectrum and then show a rise in the blue. They also converge to almost the same value of spectral emissivity in the blue.

The optical constants of liquid copper were measured in the temperature range 1315–1903 K(0.97–1.40 $T_m$), where $T_m$ is the melting point ($T_m$ = 1357 K). Measurements of spectral emissivity and optical properties at the 4 wavelengths were conducted in this temperature range. This included a measurement on the undercooled liquid and 1 measurement after the sample resolidified at the He-Ne wavelength.

Figure 6 is a plot of the spectral emissivity of liquid copper in the above temperature range at the 4 wavelengths. The solid lines in the

Table 5
Comparison of Normal Incidence Spectral Emissivities
with Selected Literature Values for $\lambda = 650$ nM

| Metal | RT[a] | Ref. | IT[a] | Ref. | Liquid | Ref. |
|-------|-------|------|-------|------|--------|------|
| Copper | 0.08 | 32 | 0.10 | *This work* | 0.13 | *This work* |
|  | 0.12 | 31 | 0.10 | 5 | 0.13 | 21 |
|  |  |  |  |  | 0.14 | 3 |
| Silver | 0.03 | 32 | 0.05 | 5 | 0.086 | *This work* |
|  |  |  |  |  | 0.08 | 3 |
|  |  |  |  |  | 0.09 | 21,23 |
| Gold | 0.17 | 31 | 0.15 | 5 | 0.30 | *This work* |
|  |  |  | 0.22 | 33 | 0.31 | 21 |
|  |  |  |  |  | 0.22 | 5 |
| Nickel | 0.35 | 32 | 0.36 | 5 | 0.40 | *This work* |
|  |  |  | 0.38 | 33 | 0.35 | 3 |
|  |  |  |  |  | 0.37 | 5 |
| Palladium | 0.23 | *This work* | 0.32 | *This work* | 0.38 | *This work* |
|  |  |  | 0.32 | 5 | 0.37 | 5 |
| Platinum | 0.32 | 32 | 0.30 | 31 | 0.38 | *This work* |
|  | 0.27 | 31 | 0.30 | 34 | 0.39 | 1,4 |
|  |  |  |  |  | 0.38 | 5 |
| Zirconium |  |  | 0.32 | 5 | 0.41 | *This work* |
|  |  |  |  |  | 0.30 | 5 |
|  |  |  |  |  | 0.47 | 4 |
|  |  |  |  |  | 0.32 | 3 |

[a]RT represents room temperature data; IT represents incandescent temperatures.

figure represent the results of a least squares fit to the data at each of the wavelengths. The results of the least squares fit, $E_\lambda = a + bT$, are also shown in the figure.

It can be seen from Fig. 6 that there is moderate temperature dependence in the blue and green regions of the spectrum where the spectral emissivity decreases with temperature above the melting point. However, there appears to be an increase in spectral emissivity as a function of temperature in the red (633 nm), whereas in the infrared (1064 nm), the spectral emissivity is relatively constant.

The value of spectral emissivity measured in the infrared for liquid copper was 0.085, which compares favorably with that computed from the data of Miller (21). A comparison of our liquid copper spectral emissivity (1400 K) in the red with those obtained using the melting point method shows our value to be slightly lower (Table 5). On the other hand, our solid state data agree well with the data reported in the literature for copper, near the melting point. In addition, a value of spectral emissivity computed from the optical constants obtained by Miller (21) agrees well with our result. The comparison of spectral emissivity data in the green and blue with those of Miller (21) and Otter (35) suggest that our values are slightly higher.

The optical constants and spectral emissivity of liquid gold were measured in the temperature range 1270–1980 K (0.95–1.48 $T_m$, $T_m$ = 1338 K).

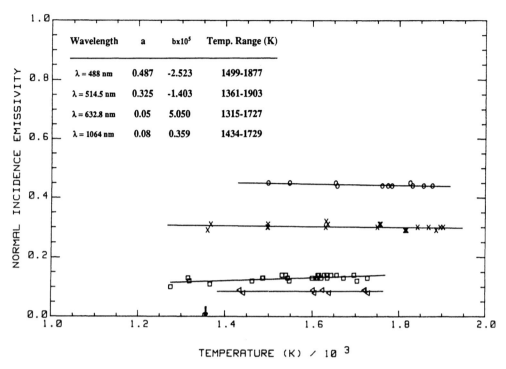

Fig. 6. Normal incidence spectral emissivity of liquid copper as a function of temperature at 1064 (△), 632.8 (□), 514.5 (x), and 488 nm (o). Solid line represents the least squares fit $E_\lambda = a + bT$ to the data. Melting point indicated by arrow. The results of the fit are indicated on the plot.

These data include 3 measurements on the undercooled liquid (He-Ne wavelength).

A plot of the spectral emissivity of liquid gold as a function of temperature is shown in Fig. 7, including the results of the least squares fit to the data. A small increase in spectral emissivity is seen in the blue, whereas there is a very large decrease in the green. The spectral emissivity in the green close to the melting point is 0.47, whereas at 1900 K it decreases to about 0.40. This corresponds to about 17.5% decrease in spectral emissivity. The spectral emissivity is constant in the red and has a small positive slope in the infrared.

The spectral emissivity of gold (at 1400 K) in the blue is 0.52 and compares well with that calculated from the data of Miller (21) (0.52). The spectral emissivity is 0.40 in the green and is a little lower than that of Miller. The emissivity of liquid gold in this range for the He-Ne wavelength was determined to be 0.30. This value is in agreement with that computed from the data of Miller (21), but is higher than that reported by Burgess and Waltenburg (49). There appear to be no data available in the infrared for liquid gold, and the value reported here is 0.12.

The optical properties and spectral emissivities of liquid platinum were measured in the temperature range 2000–2650 K (0.98–1.30 $T_m$, $T_m$ = 2045 K). The data obtained in this range included 6 measurements on the undercooled liquid (for the He-Ne wavelength).

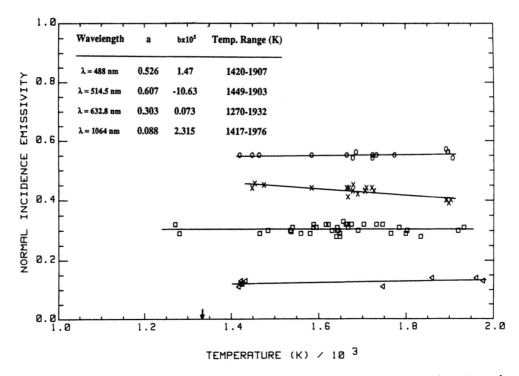

| Wavelength | a | $b \times 10^5$ | Temp. Range (K) |
|---|---|---|---|
| $\lambda = 488$ nm | 0.526 | 1.47 | 1420-1907 |
| $\lambda = 514.5$ nm | 0.607 | -10.63 | 1449-1903 |
| $\lambda = 632.8$ nm | 0.303 | 0.073 | 1270-1932 |
| $\lambda = 1064$ nm | 0.088 | 2.315 | 1417-1976 |

Fig. 7. Normal incidence spectral emissivity of liquid gold as a function of temperature at 1064 ($\triangle$), 632.8 ($\square$), 514.5 (x), and 488 nm (o). Solid line represents the least squares fit $E_\lambda = a + bT$ to the data. Melting point indicated by arrow. The results of the fit are indicated on the plot.

Figure 8 is a plot of the spectral emissivity of liquid platinum as a function of temperature at the 4 wavelengths together with the results of a least squares fit to the data at each of the wavelengths. It is seen quite clearly that the spectral emissivity decreases with increasing temperature in the blue, green, and in the infrared, whereas it appears to increase in the red. It must be pointed out that the data in the green would probably fit better with a curve rather than a straight line, although no such fit was attempted. An unusual feature of the data in the green was the dramatic change in the values of the dielectric constants in the neighborhood of 2550 K. It is apparent that some major change took place on the sample surface at around 2550 K. When the dielectric constants were plotted against temperature, a sudden drop in the value of $\varepsilon_1$ and an almost unaffected value of $\varepsilon_2$ was observed above 2550 K. Ten measurements were performed above this temperature, and 9 were below this temperature. The data for $\varepsilon_1$ above this temperature are all bunched up, whereas below this temperature, there is some scatter in the values of $\varepsilon_1$. Since platinum is unreactive with likely contaminant gases such as oxygen, water, and nitrogen, the observed change is not likely to be a result of chemical reactivity. Furthermore, these data were collected after several hours of levitation, which eliminates the possibility of surface active impurities appearing just after melting. Additionally, the flow gases consisted essentially of argon + hydrogen mixtures, and no impurity

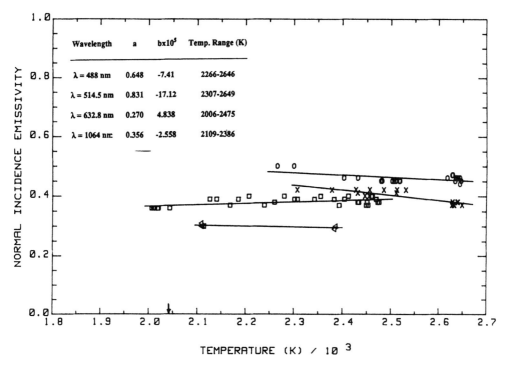

Fig. 8. Normal incidence spectral emissivity of liquid platinum as a function of temperature at 1064 (△), 632.8 (□), 514.5 (x), and 488 nm (o). Solid line represents the least squares fit $E_\lambda = a + bT$ to the data. Melting point indicated by arrow. The results of the fit are indicated on the plot.

introduction from this is likely. Its is speculated that such a change may be owing to some real effects in the electronic properties of liquid platinum at these elevated temperatures. Since the same feature did not appear in the blue, the behavior in the green is very puzzling.

Since most of the data for liquid platinum were measured at temperatures substantially higher than the melting point, extrapolation of the data to temperatures close to the melting point would result in larger errors. Therefore, the temperature of interpolation was chosen as 2250 K, which is about 200 K above the melting point. In carrying out the extrapolation, the data for the undercooled liquid was ignored. The spectral emissivity in the red was determined as 0.38 and compares very well with the data in the literature (*see* Table 5). The spectral emissivity of undercooled platinum in the red was measured as 0.36. The mean values in the blue, green and infrared were determined as 0.48, 0.45, and 0.30, respectively. It was not possible to assess the agreement with literature data since such no data are available at these wavelengths.

In order to illustrate the wavelength dependence of the optical constants, the polarization ($\varepsilon_1 - 1$) and the optical conductivity $\delta(\omega)$ for liquid copper are plotted as a function of photon energy in Figs. 9 and 10. The symbols *x* represent the data of Miller (*21*) and Otter (*35*), whereas the dashed line represents the data for the room temperature solid obtained from Palik (*31*). The dashed line represents the prediction from

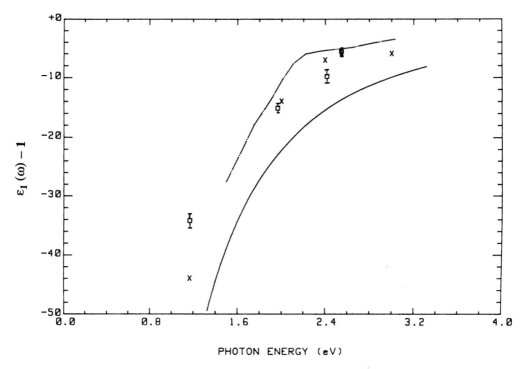

Fig. 9.    Plot of $\epsilon$, $-1$ vs hv for liquid copper ($\square$) at 1400 K. The data from Miller (21) and Otter (35) denoted by (x). The dashed line is the data for the room temperature solid from ref. (31). The solid line represents the Drude prediction with n*/n = 0.80.

the Drude theory using an effective carrier density, n*/n = 0.80, and a relaxation time of h/2$\pi\tau$ = 0.24 eV. For a detailed discussion on these parameters, see the papers by Miller (21) and Comins (20).

The reason the Drude curve does not overlap the data in Fig. 9 is because no account was taken in the fitting procedure for the contribution from the *d*-electrons. In Fig. 10, the data clearly show the strong absorption feature associated with the 2 eV excitation present in liquid copper. This suggests that the *d*-electron excitation to states in the conduction band above the Fermi level is scarcely affected on melting of the solid. The presence of *d*-electron excitation was also observed in the case of liquid gold at 2 eV. Further, in similar plots for liquid nickel, palladium, and platinum, the excitations that set in at very low energies (0.4–0.8 eV) for the room temperature solid, are seen to broaden out in the case of the liquid.

## DISCUSSION

One of the goals of this work was the development of a pyrometric technique by which one could measure both spectral emissivities and true temperatures of liquid metals *in situ*, since an error is introduced in temperature measurement of incandescent solids or liquids when spec-

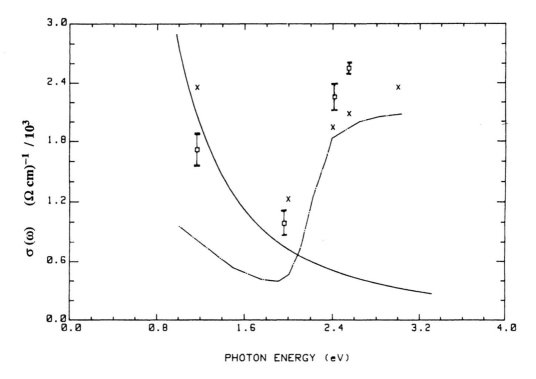

Fig. 10. Plot of σ (ω) vs hμ for liquid copper (□) at 1400 K. The data from Miller (*21*) and Otter (*35*) are denoted by (x). The dashed line is the data for the room temperature solid from ref. (*31*). The solid line represents the Drude prediction with $n^*/n = 0.80$.

tral emissivities are only estimated. Also, the paucity of emissivity data for liquid metals at temperatures above the melting point has made reliable temperature measurement in that regime difficult. As a result, the accuracy of thermophysical property data such as specific heats, expansion coefficients, and so forth, have always been limited by this uncertainty. In order to measure the thermodynamic temperature of these high temperature liquids accurately, it is necessary to know the surface spectral emissivity as a function of temperature and independent of radiance brightness measurements.

Three important results reported here are

1. The observations of the spectral emissivities at the 4 wavelengths of the metals in the Ni and Cu groups show a moderate temperature dependence in the liquid over the temperature ranges studied;

2. The spectral emissivities of the liquids reported here are *higher* than those of the corresponding solids; and

3. In those liquid metals where moderate supercooling was observed, the measured spectral emissivities of the supercooled liquid were essentially the same as the those of the liquid above the melting point.

Corrections may need to be made to thermophysical data obtained previously that have made the assumption of constant emissivity. For instance, Bonnell (36) observed an anomaly in measurements of the specific heat, $C_p$ (l), of liquid zirconium at elevated temperatures. He attributed the behavior to either a real change in $C_p$ (l) or a monotonic decrease in emissivity. The latter would compensate for the nonlinear rise in total enthalpy, H(T)–H(298), making $C_p$(l) constant. The few measurements on liquid zirconium suggested a constant spectral emissivity at $\lambda$ = 633 nm. At the present time, it appears that the data of Bonnell (36) may not need to be corrected, therefore suggesting that $C_p$ (l) does increase, but only further work can confirm this observation.

The possibility that spectral emissivities change appreciably with temperature has been suggested by Nordine (37) and Margrave (38), but have never really been established. However, indirect support for such an observation can be found in the literature. For instance, the contributions to $C_p$ (l) in the range 3000–5000 K, owing to expansion coefficients, compressibility factors, and electronic parameters, have been studied using exploding wires by Cezariliyan (39) and by Gathers et al. (40), and major deviations from the classical 3R value have been observed. Similar variations in spectral emissivities can be expected, especially at temperatures of 3000 K and higher.

The second important result concerns the increase in the spectral emissivity on melting. In these experiments, measurements of the spectral emissivities of solid copper and palladium were performed just below their melting point by allowing them to freeze within their levitation coil. The emissivity measured for solid copper was 0.10, whereas for palladium the value was 0.32 (for the He-Ne wavelength). These are 15–20% lower than the emissivity of the corresponding liquids. A similar increase of about 13% was observed in the green on melting for palladium. This increase in emissivity on melting was observed previously for these metals by others (3,5), but no explanations have been forthcoming. Intuitively, one expects the smoother, liquid surface to have a lower emissivity than the solid. This would be the case if factors affecting emissivity were the macroscopic physical and chemical structure of the surface; however, the optical properties are also governed by the electronic properties of the material. Factors that affect the electronic properties will play a role in determining the spectral emissivity of the material. For example, significant destruction of the long-range order of the lattice occurs on melting. This will modify the band structure and may provide at least part of the basis for the changes observed in the optical properties upon melting of these metals.

Measurements of the optical properties of undercooled liquids carried out in this study (palladium, platinum, gold) reveal that the spectral emissivities of the undercooled liquids are virtually the same as that of the liquid above the melting point, although they are just slightly lower. Since large undercoolings were not attained, the assumption of this behavior to large undercoolings must still be made with caution.

## CONCLUSIONS

Rotating analyzer ellipsometry has been used successfully to measure the optical constants of Cu, Ag, Au, Ni, Pd, Pt, and Zr and obtain normal incidence spectral emissivities at 488, 514.5, 632.8, and 1064 nm over a wide range of temperatures in the liquid state. The technique was developed as a noncontact diagnostic method for our electronmagnetic levitation system. It was possible to measure the dielectric functions and complex indices of refraction data for liquid metals at elevated temperatures.

The spectral emissivities at 633 nm of the liquid metals obtained in this study agreed well with most of the data available in the literature. In the cases of palladium and copper, the emissivities were also measured in the solid state, with the sample levitated. The data for all metals studied showed that the spectral emissivities of the liquids were higher than those of the solids. Additionally, the spectral emissivities of all the metals studied here showed moderate to strong temperature dependence in the liquid over large temperature ranges. This fact, therefore, suggests that corrections may need to be made to thermophysical property measurements that have been based on the assumption of constant emissivity.

Comparisons of the dielectric functions and complex index of refraction data with those available in the literature showed excellent agreement. Since only data for the low melting materials were available, it was not possible to compare the results for liquid Pd, Pt, and Zr.

## ACKNOWLEDGMENTS

This work performed for the Jet Propulsion Laboratory, California Institute of Technology, sponsored by the National Aeronautics and Space Administration.

## REFERENCES

1. Jones, F. O., Knapton, A. G., and Savill, J., *J. Less Comm. Met.* **1**, 80 (1959).
2. Treverton, J. A. and Margrave, J. L. *Proc. 5th Symp. Thermophysical Properties* ASME, Boston, MA, 1970, pp. 489–494.
3. Bonnell, D. W., Treverton, J. A., Valerga, A. J., and Margrave, J. L., *Presented at 5th Symp. on Temperature*, Washington, DC, 1971, pp. 483–487.
4. Koch, R. R., Hoffman, J. L., and Beall, R. A., *Rep. Inv. # 7743*, US Dept. of Interior, Bureau of Mines, Washington, DC, 1973.
5. Burgess, G. K., and Waltenburg, R. G., Bureau of Standards Scientific Paper # 242, 1914, pp. 591–605.
6. McElroy, D. L. and Fulkerson, W. *Temperature Measurement and Control*, J. Wiley, New York, 1968, R. F. Bunshah, ed., p. 489.
7. Coates, P. B., *Metrologia* **17**, 103 (1981).
8. Nordine, P. C., *High Temp. Sci.* **21**, 97 (1986).

9. Nutter, G. D., *Applications of Radiation Thermometry*, ASTM STP 895, Richmond, J. C. and DeWitt, D. P., eds., pp. 3–24.
10. Hunderi, O. and Ryberg, R., *Surf. Sci.* **56**, 182 (1976).
11. Liljenvall, H. G., Mathewson, A. G., and Myers, H. P., *Phil. Mag.* **22**, 243 (1970).
12. Smith, N. V., *Phys. Rev.* **183**, 634 (1969).
13. Budde, W., *Appl. Optics.* **1**, 201 (1962).
14. Givins, M. P., *Solid State Physics*, Seitz, F. and Turnbull, D., eds., vol. 6, pp. 313–353.
15. Beattie, J. R., *Phil Mag.* **46**, 235 (1955).
16. Hauge, P. S. and Dill, F. H. *IBM J. Res. Dev.* **14**, 472 (1973).
17. Hauge, P. S. *Surf. Sci.* **56**, 148 (1976).
18. Azzam, R. M. A. and Bashara, N. M., *Ellipsometry and Polarized Light*, North Holland, Amsterdam, 1987, pp. 1–267.
19. Arakawa, E. T., Ingaki, T., and Williams, M. W. *Surf. Sci.* **96**, 248 (1980).
20. Comins, N. R., *Phil. Mag.* **25**, 817 (1972).
21. Miller J. C., *Phil. Mag.* **20**, 1115 (1969).
22. Schulz, L. G., *ADV Phys.* **6**, 102 (1957).
23. Hodgson, J. N., *Phil. Mag.* **5**, 272 (1960).
24. Hodgson, J. N., *Phil. Mag.* **6**, 509 (1961).
25. Hodgson, J. N., *Phil. Mag.* **4**, 183 (1959).
26. Hodgson, J. N., *Phil. Mag.* **7**, 229 (1962).
27. Hodgson, J. N., *Proc. Phys. Soc. B* **68**, 593 (1955).
28. Smith, N. V., *Adv. Phys.* **16**, 629 (1967).
29. Ziman, J. M., *Phil. Mag.* **6**, 1013 (1961).
30. Hansen, G. P., Krishnan, S., Hauge, R. H., and Margrave, J. L., *Met Trans.* **19A**, 1939 (1988).
31. Handbook of Optical Constants, Palik, E. D., ed., Academic, 1985, pp. 275–368.
32. Hansen, G. P. Krishnan, S., Hauge, R. H., and Margrave, J. L., *Met Trans.* **19A**, 1889 (1988).
33. Worthing A. G., *Phys. Rev.* **28**, 174 (1926).
34. Barnes, B. T., *J. Opt. Soc. Am.* **56**, 1546 (1966).
35. Otter, M., *Z. Phys.* **161**, 539 (1960).
36. Bonnell, D. W., Ph.D. Thesis, Rice University, 1972.
37. Nordine, P. C. personal communication.
38. Margrave, J. L.,personal communication.
39. Cezariliyan, A., *High Temp. High Press.* **11**, 9 (1979).
40. Gaithers, G. R., Shaner, J. W., and Hodgson, W. M., *High Temp. High Press.* **11**, 529 (1979).

# Ion Equilibria

## A New Technique for Measurement of Low $O_2$ and Alkali Partial Pressures

E. B. RUDNYI, M. V. KOROBOV, O. M. VOVK,
E. A. KAIBICHEVA, AND L. N. SIDOROV*

Chemistry Department, Moscow State University,
Moscow, 119899, USSR

## ABSTRACT

Ion-molecule equilibria with participation of oxygen-containing negative ions were used to measure oxygen, sodium, and potassium partial pressures in high temperature systems. The pressure range from $10^{-10}$ to 1 Pa (where $1.013 \times 10^5$ Pa = 1 atm) and the temperature range 1200–1500 K were covered.

Application of the method is described for various high temperature systems:

Incongruent vaporization of potassium chromate;
Vaporization of sodium metaphosphate from cells made of different metals;
Nickel oxide dissociation pressure;
Sodium oxide activity in $Na_2O$-$SiO_2$ glasses and melts;
Alkali oxide activity in $Na_2O$-$K_2O$-$SiO_2$ melts.

**Index Entries:** High temperature mass spectrometry; ion-molecule equilibria; partial oxygen pressure; alkali oxide activity; oxygen-containing negative ions.

## INTRODUCTION

High-temperature studies of negative ions in the vapors of oxygen-containing salts of alkali metals have been carried out at our laboratory since 1981, and the results are given in Table 1.

*Author to whom all correspondence and reprint orders should be addressed.

Table 1
Enthalpy of Negative Ion Formation and Electron Affinity

| ion | $\Delta_f H_0^\circ$ | EA[a] | ion | $\Delta_f H_0^\circ$ | EA[a] |
|---|---|---|---|---|---|
| | kJ·mol$^{-1}$ | | | kJ·mol$^{-1}$ | |
| $PO_2^-$ | $-645 \pm 18$[b] | $367 \pm 21$ | $KCrO_4^-$ | $-995 \pm 16$ | |
| $PO_3^-$ | $-943 \pm 16$[b] | $433 \pm 51$ | $MoO_3^-$ | $-637 \pm 11$ | $277 \pm 19$ |
| $SO_3^-$ | $-602 \pm 6$ | $212 \pm 7$ | $MoO_4^-$ | $-800 \pm 11$ | |
| $SO_4^-$ | $-743 \pm 9$ | | $NaMoO_4^-$ | $-1037 \pm 17$ | |
| $LiSO_4^-$ | $-1041 \pm 12$ | | $KMoO_4^-$ | $-1065 \pm 15$ | |
| $NaSO_4^-$ | $-976 \pm 13$ | | $CsMoO_4^-$ | $-1060 \pm 17$ | |
| $KSO_4^-$ | $-992 \pm 12$ | | $WO_3^-$ | $-695 \pm 11$ | $380 \pm 19$ |
| $RbSO_4^-$ | $-982 \pm 12$ | | $WO_4^-$ | $-881 \pm 11$ | |
| $CsSO_4^-$ | $-998 \pm 13$ | | $NaWO_4^-$ | $-1139 \pm 15$ | |
| $CrO_3^-$ | $-669 \pm 9$ | $351 \pm 17$ | $KWO_4^-$ | $-1170 \pm 18$ | |
| $CrO_4^-$ | $-781 \pm 11$ | | $CsWO_4^-$ | $-1181 \pm 15$ | |
| $NaCrO_4^-$ | $-972 \pm 14$ | | | | |

[a]Relates to a corresponding neutral molecule.
[b]Data from ref. *(1)*, other data from ref. *(2)*.

The presence in salt vapors of ion pairs similar to $PO_3^-$ and $PO_2^-$, and $KCrO_4^-$ and $CrO_4^-$ has made it possible to devise a new method for determination of partial pressures of oxygen and of alkali metals in high-temperature systems. Since the ions are in equilibrium with neutral vapor species, the pressures of oxygen and of alkali metal can be calculated from the negative ion pressure ratios

$$p(O_2) = \{p(PO_3^-)/p(PO_2^-) \cdot 1/K_1\}^2 \qquad p(K) = p(KCrO_4^-)/p(CrO_4^-) \cdot 1/K_2$$

where $K_1$ and $K_2$ are equilibrium constants of reactions

$$PO_2^- + \tfrac{1}{2} O_2 = PO_3^- \tag{1}$$

$$CrO_4^- + K = KCrO_4^- \tag{2}$$

The K's are known from previously established negative ion thermodynamic functions [e.g., *see (5)*], together with the standard functions for $O_2$ and K. The ion pressure ratio is obtained from that of the ion currents [*see (3)*].

$$p(PO_3^-)/p(PO_2^-) = I_e(PO_3^-)/I_e(PO_2^-) \cdot \{M(PO_3^-)/M(PO_2^-)\}^{1/2}$$
$$= I_m(PO_3^-)/I_m(PO_2^-) \cdot M(PO_3^-)/M(PO_2^-)$$

where p is partial pressure; $I_e$ is the ion current measured by an electrometer, $I_m$ is the ion current measured by an electron multiplier, and M is the molecular mass. Numerical values of equilibrium constants of reactions similar to *(1)* and *(2)* are calculated from thermodynamic properties of ions.

The basic source of error that occurs when pressures of oxygen and of alkali metal are determined is introduced by the error in the equilibrium constants of the (1) and (2) type reactions. It may be noted that knowing the errors for formation enthalpies (*see* Table 1) and for the thermodynamic functions of negative ions, calculated according to the error propagation law, leads to a considerable overestimation of errors for equilibrium constants caused by a strong correlation of errors for the thermodynamic properties.

A correct estimation of the error in the reaction equilibrium constant can be obtained by applying the error propagation law directly to the product/ratio of the equilibrium constants. In this case the (1) and (2) type equilibrium constant error, within 1000–1500 K, will not exceed coefficients equal to two.

In this paper, application of the method for various high-temperature systems is described.

# EXPERIMENTAL

The experiment was carried out on a MX-1303 type mass spectrometer (200 mm, 60°) adapted for studying ion-molecular equilibrium (4). Temperature was measured by a Pt/Pt-Rh (10%) thermocouple with an accuracy of 3 K. The ion currents were measured by the channel electron multiplier in all cases unless otherwise specified.

## Vaporization of Potassium Chromate

In 1985, we published an article on the vaporization of potassium chromate using both Knudsen cell mass spectrometry (KCMS) with ionization by electron impact and the method of ion-molecular equilibria (5). It has been found that potassium chromate evaporates basically in the form $K_2CrO_4$ and partially is subject to decomposition. Negative ions, such as $CrO_3^-$, $CrO_4^-$, $KCrO_4^-$, $Cr_2O_6^-$, $KCr_2O_7^-$, and $K_3Cr_2O_8^-$ were found in the potassium chromate vapors. The ratio of ion currents for $CrO_4^-$ and $CrO_3^-$ and the equilibrium constants for the reaction

$$CrO_3^- + \tfrac{1}{2} O_2 = CrO_4^- \tag{3}$$

($\Delta H_0^\circ = -112$ kJ·mol$^{-1}$, $\Delta\phi_{1000}^\circ = -86.5$ J·mol$^{-1}$·K$^{-1}$, $\Delta\phi_{1200}^\circ = -86.9$ J·mol$^{-1}$·K$^{-1}$ $\Delta\phi_{1400}^\circ = -87.0$ J·mol$^{-1}$·K$^{-1}$) have made it possible to determine the oxygen partial pressure. The ratio of ion currents of $KCrO_4^-$ and $CrO_4^-$ and the equilibrium constant of reaction (2) ($\Delta H_0^\circ = -303.9$ kJ·mol$^{-1}$, $\Delta\phi_{1000}^\circ = -105.2$ J·mol$^{-1}$·K$^{-1}$, $\Delta\phi_{1200}^\circ = -104.6$ J·mol$^{-1}$·K$^{-1}$, $\Delta\phi_{1400}^\circ = 103.9$ J·mol$^{-1}$·K$^{-1}$) have allowed the determination of the potassium partial pressure in the vapor of chromate.

In 1987 Brittain et al. (6) studied the vaporization of potassium chromate by torsion effusion and the KCMS method with ionization by electron impact. It has been demonstrated that potassium chromate evaporates mainly in the form of $K_2CrO_4$ but is subject to decomposition partly into chromium oxide according to the reaction,

$$K_2CrO_4(s) = 2K + \tfrac{3}{4} O_2 + \tfrac{1}{2} Cr_2O_3(s) \tag{4}$$

The calculated and measured values of partial pressures of oxygen and potassium for reaction (4) are given in Table 2. The essential thermodynamic values were taken from handbook (7) and papers (8,5). It is assumed that $p(K) = \frac{5}{5}p(O_2)$.

Runs III and IV show a fair agreement between experimental and calculated pressures. Potassium pressure in the run I is four times higher, which may be owing to the presence of potassium carbonate impurities [for carbonate effect on vaporization of potassium chromate see (6)]. The results of Table 2 generally testify to the fact that potassium chromate dissociates partly with formation of chromium oxide.

## Crucible Effects on the Vaporization of Sodium Metaphosphate

Vapor composition over sodium phosphate was studied using both the KCMS with ionization by electron impact (9–11) and infrared spectroscopy with matrix isolation (12,13). It has been found that the vapor composition is affected by the crucible material. In platinum, the vapor is primarily $NaPO_3$, in molybdenum a mixture of $NaPO_3$ and $NaPO_2$, whereas after addition of tantalum powder the vapor contains only $NaPO_2$.

The ion-molecular equilibrium method was used for studying sodium metaphosphate vaporization in platinum, nickel, or molybdenum crucibles (1). The following mass spectra of negative ions (in relative units) at 1164 K have been recorded:

| crucible | $PO_2^-$ | $PO_3^-$ | $NaP_2O_4^-$ | $NaP_2O_5^-$ | $NaP_2O_6^-$ |
|---|---|---|---|---|---|
| platinum | 0.01 | 100 | — | — | 0.19 |
| nickel | 5.2 | 100 | — | 0.01 | 0.04 |
| molybdenum | 13 | 100 | 0.01 | 0.05 | 0.07 |

The ratios of ion currents of $PO_3^-$ and $PO_2^-$ and equilibrium constants of reaction (1) ($\Delta H_0^\circ = -198$ kJ·mol$^{-1}$, $\Delta\phi_{1000}^\circ = -90.3$ J·mol$^{-1}$·K$^{-1}$, $\Delta\phi_{1200}^\circ = - -90.4$ J·mol$^{-1}$·K$^{-1}$, $\Delta\phi_{1400}^\circ = -90.4$ J·mol$^{-1}$·K$^{-1}$) were used to determine the oxygen pressure in the vapor of sodium metaphosphate using different crucibles. The ratio of ion currents of $NaP_2O_6^-$ and $PO_3^-$ were used in the calculation of the sodium metaphosphate activity. The results are given in Table 3.

The data obtained explain the crucible effects of vaporization of sodium metaphosphate. Reaction with the nickel and molybdenum crucibles reduce the sodium metaphosphate activity by half. But the main effect of nickel and molybdenum is that they reduce the oxygen pressure almost a million times. It, in turn, makes measurable the concentration of $NaPO_2$ according to the equilibrium constant of the reaction

$$NaPO_3 = NaPO_2 + \frac{1}{2} O_2 \qquad (5)$$

Table 2
Pressure of Oxygen and Potassium in the Vapor of Potassium Chromate

| $N^a$ run | T(K) | $\ln \dfrac{I(CrO_4^-)}{I(CrO_3^-)}$ | $n_i^b$ | $p(O_2)^c$ $10^2Pa$ | $p(O_2)^d$ | $\ln \dfrac{I(KCrO_4^-)}{I(CrO_4^-)}$ | $n_i^b$ | $p(K)^c$ $10^2Pa$ | $p(K)^d$ |
|---|---|---|---|---|---|---|---|---|---|
| I | 1210 | −7.43 | 4 | 1.2 | 0.58 | 2.69 | 4 | 4.7 | 1.0 |
| III$^c$ | 1210 | −7.34 | 15 | 1.3 | 0.58 | 1.57 | 15 | 1.3 | 1.0 |
| IV$^c$ | 1219 | −7.71 | 3 | 0.72 | 0.73 | 1.82 | 3 | 2.1 | 1.3 |

[a]Experimental condition are given in (5).
[b]Number of measurements of ion currents ratio.
[c]Experimental pressure values which differ by 30% from those listed in (5) owing to corrections in ion formation enthalpies.
[d]Calculated pressures according to reaction (4).
[e]Ion currents are measured by an electrometer.

Table 3
Oxygen Pressure Over Sodium Metaphosphate

| Crucible | T(K) | $\ln \dfrac{I(PO_3^-)}{I(PO_2^-)}$ | $n_i^a$ | $p(O_2)$ Pa | $\ln \dfrac{I(NaP_2O_6^-)}{I(PO_3^-)}$ | $n_i^a$ | Activity, $(NaPO_3)$ |
|---|---|---|---|---|---|---|---|
| platinum | 1070 | 9.88 | 1 | $1.1 \cdot 10^{-6}$ | −7.53 | 3 | 1 |
|  | 1118 | 9.78 | 4 | $1.6 \cdot 10^{-5}$ | −6.93 | 4 | 1 |
|  | 1164 | 9.18 | 3 | $6.1 \cdot 10^{-5}$ | −6.28 | 4 | 1 |
| nickel | 1023 | 3.25 | 2 | $8.6 \cdot 10^{-14}$ |  |  |  |
|  | 1070 | 3.27 | 3 | $2.0 \cdot 10^{-12}$ |  |  |  |
|  | 1118 | 3.11 | 3 | $2.6 \cdot 10^{-11}$ | −8.07 | 3 | 0.31 |
|  | 1164 | 2.96 | 3 | $2.4 \cdot 10^{-10}$ | −7.73 | 3 | 0.24 |
| molybdenum | 1023 | 2.56 | 1 | $2.2 \cdot 10^{-14}$ |  |  |  |
|  | 1070 | 2.26 | 3 | $2.6 \cdot 10^{-13}$ | −8.14 | 3 | 0.56 |
|  | 1118 | 2.30 | 4 | $5.1 \cdot 10^{-12}$ | −7.70 | 4 | 0.44 |
|  | 1164 | 2.05 | 3 | $3.9 \cdot 10^{-11}$ | −7.21 | 3 | 0.40 |
|  | 1210 | 2.04 | 3 | $4.0 \cdot 10^{-10}$ | −6.88 | 3 | 0.32 |

[a]Number of measurements of ion current ratio.

## Nickel Oxide Dissociation Pressure

In contrast to the alkali metal salts, vapors of the Ni-NiO system have no measurable concentrations of negative ions. To make usable the ion-molecular equilibrium method we have proposed to generate suitable negative ions by introducing small amounts of easily ionized additives (14).

Three mg of $NaPO_3$ were added to 40 mg of Ni and 180 mg of NiO and the mixture was placed in the nickel effusion chamber. Only $PO_2^-$ and $PO_3^-$ were detected in vapors at 1300–1473 K. The experiment was run for 5 h. During this period of time the intensity of ion currents dropped continuously, but their ratios within the limits of error remained constant. This indicates that the influence of sodium metaphosphate on the oxygen pressure is insignificant. A more detailed description of the experiments is given in (14). Oxygen pressure given in Table 4 is calcu-

Table 4
Nickel Oxide Dissociation Pressure

| T(K) | $\ln \dfrac{I(PO_3^-)}{I(PO_2^-)}$ | $n_i^a$ | $p(O_2)^b$ | $p(O_2)^c$ |
|------|------|------|------|------|
| | | | $10^4 Pa$ | |
| 1300 | 4.72 | 9 | 0.06 | 0.11 |
| 1344 | 4.56 | 6 | 0.28 | 0.46 |
| 1387 | 4.19 | 5 | 0.68 | 1.7 |
| 1430 | 4.22 | 9 | 3.4 | 5.8 |
| 1473 | 3.98 | 9 | 9.1 | 18 |

[a]Number of measurements of ion current ratio.
[b]Experimental pressure.
[c]Calculated according to data bank IVTAN-TERMO.

lated from the ratio of $PO_3^-$ and $PO_2^-$ and the equilibrium constant of reaction (1). Pressure of oxygen in the Ni-NiO system calculated according to the data bank IVTAN-TERMO (*15*) (1986 version) is also listed there. There is fair agreement between experimental and calculated data.

## Sodium Oxide Activity in Na₂O-SiO₂ Glasses and Melts

The sodium oxide activity in the $Na_2O-SiO_2$ system was determined in a number of papers by the emf method, the KCMS method with electron impact ionization, and by the transpiration method. Comparison of the results reported by different authors was carried out by Sanders and Haller (*16*) and by Hastie and Bonnell (*17*). The results of different experiments on sodium oxide activity in the $Na_2O-SiO_2$ system differ by about a factor of four, which is common for measurements of this kind.

As in the case with the Ni-NiO system, vapors of $Na_2O-SiO_2$ system exhibit no measurable concentration of negative ions. In this case when ion-equilibria were employed a small amount of potassium chromate was added to the $Na_2O-SiO_2$ glasses (*18*). The mixture was thoroughly ground in an agate mortar and placed in the platinum effusion chamber. Ions such as $CrO_3^-$, $CrO_4^-$, and $NaCrO_4^-$ were found in the mass spectrum at 1255–1515 K. It has to be noted that all efforts to employ potassium sulphate additives were unsuccessful: No negative ions were found in the vapor.

The $Na_2O-SiO_2$ phase composition changes continuously in the course of the experiment owing to evaporation of sodium oxide. Therefore, the duration of the experiment was specified by the time during which the composition of the system did not change by more than 0.3 mol%. To examine the influence of an additive on activity, the experiments differing in the amount of the additive (e.g., 0.3 and 2 mol%) were carried out on the $Na_2O-SiO_2$ system of every composition. A more detailed description of the experiment is given in (*18*).

Oxygen pressure was determined from the ratio of ion currents $CrO_4^-$ and $CrO_3^-$ and the equilibrium constant of reaction (3). Sodium pressure was determined from the ion currents of $NaCrO_4^-$ and $CrO_4^-$ and the equilibrium constant of reaction

$$CrO_4^- + Na = NaCrO_4^- \tag{6}$$

($\Delta H_0^\circ = -298.8$ kJ·mol$^{-1}$, $\Delta\phi_{1000}^\circ = -108.4$ J·mol$^{-1}$·K$^{-1}$, $\Delta\phi_{1200}^\circ = -108.0$ J·mol$^{-1}$·K$^{-1}$, $\Delta\phi_{1400}^\circ = -107.4$ J·mol$^{-1}$·K$^{-1}$). Then the activity of sodium oxide was calculated using the equilibrium constant (19) of the reaction

$$Na_2O(l) = 2Na + \tfrac{1}{2} O_2 \tag{7}$$

Corresponding data are listed in Table 5, where, for sake of comparison, the activity values for sodium oxide determined by the transpiration method (16) and calculated according to Hastie and Bonnell's model of ideal mixing of complex components (17) are incorporated; see also the article by Bonnell and Hastie elsewhere in this volume.

The results demonstrate that addition of potassium chromate has a substantial influence on the pressure of sodium and oxygen but no effect on sodium oxide activity within the limits of the experimental error. That is $p(Na)^2 \cdot p(O_2)^{0.5}$ is constant.

This can be explained as follows. The amount of potassium chromate (several mol%) is too small to markedly affect the activity of sodium oxide, which is the main component in the $Na_2O$-$SiO_2$ system. However, the additive creates an oxygen pressure of its own in the system vapor. In this case the sodium pressure is adjusted in order to keep the equilibrium constant of reaction (7) unchanged.

Thus the method can be used to determine the activity of sodium oxide. This is confirmed by the fair agreement of sodium oxide activity data with those reported elsewhere in other communications.

## Alkali Metal Oxide Activity in $Na_2O$-$K_2O$-$SiO_2$ Melts

At the present time we have investigated the $Na_2O$-$K_2O$-$SiO_2$ system using the ion-molecular equilibrium method. This section reports preliminary data on the $0.116Na_2O$ - $0.223K_2O$ - $0.661SiO_2$(melt 1) and $0.172Na_2O$ - $0.161K_2O$ - $0.667SiO_2$(melt 2).

In this case, the approach is absolutely similar to that applied to the $Na_2O$-$SiO_2$ system. The melts were supplied with additives generating $CrO_3^-$, $CrO_4^-$, $NaCrO_4^-$, and $KCrO_4^-$ and pressures of sodium, potassium, and oxygen were determined from ion current ratios. Sodium oxide activity was calculated from the pressure of sodium and oxygen and the equilibrium constant of reaction (7). Potassium oxide activity was calculated from the pressure values of potassium and oxygen and the equilibrium constant (17) of reaction

$$K_2O(l) = 2K + \tfrac{1}{2} O_2 \tag{8}$$

To examine the influence of an additive on activity its composition was varied. Thus, chromium oxide, and lithium and cesium chromates were used as additives (*see* Table 6).

Table 5

$Na_2O$ Activity[a] in $Na_2O$-$SiO_2$ System at 1430 K

| $Na_2O$ mol% | $N$[a] run | $\ln \dfrac{I(CrO_4^-)}{I(CrO_3^-)}$ | $\ln \dfrac{I(NaCrO_4^-)}{I(CrO_4^-)}$ | $n_i$[b] | $p(O_2)$ | $p(Na)$ | | $\ln a(Na_2O)$ | |
| | | | | | Pa | | [c] | [d] | [c] |
|---|---|---|---|---|---|---|---|---|---|
| 10.3[f] | I | −8.15 | −3.22 | 19 | 0.091 | 0.024 | −21.8 | | −23.4 |
| | II | −8.38 | −3.04 | 5 | 0.058 | 0.028 | −21.8 | | |
| 25.4 | III | −7.30 | −3.34 | 7 | 0.50 | 0.021 | −21.3 | −19.3 | −21.1 |
| 30.3 | V | −7.10 | −2.29 | 5 | 0.75 | 0.060 | −19.0 | −18.1 | −20.0 |
| | VI | −8.10 | −1.85 | 4 | 0.10 | 0.094 | −19.1 | | |
| 40.2 | VII | −7.21 | −0.94 | 2 | 0.60 | 0.23 | −16.4 | −15.7 | −17.2 |
| | VIII | −7.87 | −0.44 | 2 | 0.16 | 0.38 | −16.0 | | |
| 50 | IX | −7.01 | 1.03 | 2 | 0.89 | 1.7 | −12.2 | −13.3 | |
| | X | −8.09 | 1.74 | 2 | 0.10 | 3.4 | −11.9 | | |

[a] The experiment is described in (18). At the same composition the runs differ only in the amount of potassium chromate.
[b] Number of measurements of ion currents ratio.
[c] Reference to this paper. The activity values differ from those obtained in (18) by a factor of two. This difference is due to the fact that in (18) the equilibrium constant of reaction (7) was calculated on the basis of handbook (7), whereas in this paper it was based on data given in (19). Ion current ratios were the same.
[d] Data from Sanders and Haller (16).
[e] Calculated according to Hastie and Bonnell's model (17).
[f] Glass, and other melts.

Table 6
Mass Spectrum of Negative Ions over $Na_2O$-$K_2O$-$SiO_2$ Melts
(relative units, 1387 K)

| run | composition | additive (mol%) | $CrO_3^-$ | $CrO_4^-$ | $NaCrO_4^-$ | $KCrO_4^-$ |
|---|---|---|---|---|---|---|
| I | melt 1 | $Cr_2O_3$ (3.2) | 100000 | 4.5 | 2.0 | 8.4 |
| II | melt 1 | $Li_2CrO_4$ (4.9) | 100000 | 18.8 | 4.6 | 20.8 |
| III | melt 2 | $Cr_2O_3$ (2.0) | 100000 | 31.9 | 9.6 | 22.0 |
| IV | melt 2 | $Li_2CrO_4$ (2.0) | 100000 | 52.1 | 11.6 | 22.9 |
| V | melt 2 | $Cs_2CrO_4$ (3.2) | 100000 | 14.8 | 7.1 | 16.5 |

Table 7
Alkali Oxide Activities in $Na_2O$-$K_2O$-$SiO_2$ Melts

| T(K) | N run | $p(O_2)$ Pa | $p(Na)$ Pa | $p(K)$ Pa | ln a ($Na_2O$) _a_ | ln a ($Na_2O$) _b_ | ln a ($K_2O$) _a_ | ln a ($K_2O$) _b_ |
|---|---|---|---|---|---|---|---|---|
| | | | Melt 1 | | | | | |
| 1344 | I | 0.00014 | 0.090 | 0.25 | −19.5 | −20.5 | −22.5 | −22.9 |
| | II | 0.013 | 0.028 | 0.073 | −19.6 | | −22.6 | |
| 1387 | I | 0.0013 | 0.12 | 0.26 | −19.3 | −19.9 | −22.5 | −22.2 |
| | II | 0.022 | 0.068 | 0.16 | −19.1 | | −22.2 | |
| 1430 | I | 0.016 | 0.17 | 0.35 | −18.9 | −19.4 | −21.9 | −21.6 |
| | II | 0.046 | 0.14 | 0.30 | −18.8 | | −21.7 | |
| | | | Melt 2 | | | | | |
| 1255 | V | 0.0011 | 0.011 | 0.015 | −19.2 | −21.7 | −24.1 | −25.3 |
| | III | 0.0093 | 0.019 | 0.023 | −18.9 | | −23.8 | |
| 1300 | IV | 0.026 | 0.013 | 0.015 | −19.2 | −21.0 | −24.1 | −24.5 |
| | V | 0.0021 | 0.026 | 0.037 | −19.1 | | −23.6 | |
| | III | 0.021 | 0.041 | 0.047 | −18.6 | | −23.3 | |
| 1344 | IV | 0.069 | 0.030 | 0.033 | −18.7 | −20.5 | −23.4 | −23.7 |
| | V | 0.0060 | 0.059 | 0.079 | −18.5 | | −22.9 | |
| | III | 0.062 | 0.083 | 0.097 | −18.1 | | −22.6 | |
| 1387 | IV | 0.17 | 0.062 | 0.062 | −18.2 | −19.8 | −22.9 | −23.0 |
| | V | 0.013 | 0.13 | 0.16 | −18.0 | | −22.4 | |
| 1430 | V | 0.42 | 0.14 | 0.13 | −17.5 | −19.3 | −22.2 | −22.3 |

_a_Determined by ion-molecular equilibrium method.
_b_Calculated according to Hastie and Bonnell's model.

In contrast to alkali chromates, chromium oxide is not an easily ionized substance. Yet, used as an additive, it produced the same negative ions in measurable concentration. It is owing to the fact that in the $Na_2O$-$K_2O$-$SiO_2$ vapors there are large concentrations of atoms of sodium and potassium and correspondingly a relatively large concentration of electrons.

Table 7 lists pressure values for sodium, potassium, and oxygen and activity values for the oxides of sodium and potassium. For the sake of comparison there are also activity values for the oxides of sodium and potassium calculated according to Hastie and Bonnell's model of ideal mixing of complex components (17). The comparison is very good for the

$K_2O$ activities and satisfactory for the $Na_2O$ activities. As in the case with $Na_2O$-$SiO_2$ systems, the additives exert a substantial influence on pressures of sodium, potassium, and oxygen, but do not influence the activity values of sodium and potassium oxides.

## REFERENCES

1. Rudnyi, E. B., Vovk, O. M., Sidorov, L. N., Sorokin, I. D., and Alikhanyan, A. S., *Teflofiz, Vys. Temp.* **24**, 62 (1986); *High Temp. USSR* (Engl. transl.) **24**, 56 (1986).
2. Rudnyi, E. B., Vovk, O. M., Kaibicheva, E. A., and Sidorov, L. N., *J. Chem. Thermodyn.* **21**, 247 (1989).
3. Sidorov, L. N., Zhuravleva, L. V., and Sorokin, I. D., *Mass Spectr. Rev.* **5**, 73 (1986).
4. Rudnyi, E. B., Sidorov, L. N., and Vovk, O. M., Teplofiz. Vys. Temp. 23, 291 (1985), *High Temp. USSR* (Engl. transl.) **23**, 238 (1985).
5. Rudnyi, E. B., Sidorov, L. N., Kuligina, L. A., and Semenov, G. A., *Int. J. Mass Spectr. Ion Proc.* **64**, 95 (1985).
6. Brittain, R. D., Lau, K. H., and Hildenbrand, D. L., *J. Electrochem. Soc.* **134**, 2900 (1987).
7. Termodinamicheskie svoistva individual'nykh veshestv. (*Handbook on thermodynamic properties of individual substances*), Glushko, V. P., ed., Moscow, Nauka, 1978–1982, in 4 vol.
8. O'Hare, P. A. G. and Boerio, J., *J. Chem. Thermodyn.* **7**, 1195 (1975).
9. Alikhanyan, A. S., Steblevski, A. V., Gorgoraki, V. I., and Sokolova, I.D. Dokl. *AN SSSR* **222**, 629 (1975).
10. Rat'kovskyi, I. A., Ashuiko, V. A., Urich, V. A., and Kris'ko, L. Ya., Izv. Vyssh, Uchebn. Zaved., *Khim. Khim. Tekhnol.* **19**, 675 (1976).
11. Gingerich, K. A. and Miller, F., *J. Chem. Phys.* **63**, 1211 (1975).
12. Ogden, J. S. and Williams, S. J., *J. Chem. Soc.*, Dalt. Trans. **N4**, 825 (1982).
13. Bencivenni, L. and Gingerich, K. A., *J. Mol. Struct.* **98**, 195 (1983).
14. Rudnyi, E. B., Vovk, O. M., Kappe, E. L. R., Kaibicheva, E. A., and Sidorov, L. N., *Izv. AN SSSR*, Metally (1989).
15. Gurvich, L. V., *Vestn. AN SSSR* **N3**, 54 (1983).
16. Sanders, D. M. and Haller, W. K., NBS Spec. Publ. 561, 1979, p. 111.
17. Hastie, J. W. and Bonnell, D. W., *High Temp. Sci.* **19**, 175 (1985).
18. Rudnyi, E. B., Vovk, O. M., Sidorov, L. N., Stolyarova, V. L., Shakhmatkin, B. A., and Rakhimov, V. I., *Fiz. Khim. Stekla.* **14**, 218 (1988).
19. JANAF Thermochemical Tables, 2nd ed., NSRDS-NBS 37, Washington DC, 1971.

# Thermal Desorption Kinetics of Water from Glass Powders Studied by Knudsen Effusion Mass Spectroscopy

MARGARET A. FRISCH* AND EDWARD A. GIESS

*IBM T. J. Watson Research Laboratory, Yorktown Heights, NY 10598*

## ABSTRACT

Water desorbing from cordierite-type glass powders, with a narrow particle size distribution, was studied by Knudsen effusion mass spectrometry. The thermal desorption spectra show both low temperature (300–800 K) and high temperature (800–1500 K) peaks that represent surface and bulk effects, respectively. Carbon dioxide desorbs in two low temperature peaks below 1000 K; the lower peak is attributed to the decomposition of a bicarbonate. However, above the glass transition temperature ($\approx$1050 K), water diffusing from the bulk desorbs in a first-order process with an activation energy of 260 kJ/mol. During the onset of bulk water desorption, the rate is dominated by bulk diffusion of water. All these processes are related to the glass particle size. Smaller particles store and, consequently, desorb more "surface" water. As the particle size gets larger, the temperature increases, at which the maximum in the bulk water desorption rate occurs. This change corresponds to an increase in the frequency factor, which correlates inversely to the mean particle size diameter.

**Index Entries:** Mass spectrometry; Knudsen effusion; thermal desorption; kinetics, water; glass particles.

## INTRODUCTION

The importance of dissolved water and hydroxyl ions in breaking the silicon–oxygen network bonds, and thereby increasing the reactivity and decreasing viscosity of silicate glasses, is well known (1). Most glasses do contain significant amounts of water, primarily in the form of hydroxyl ($-SiOH$) when the concentration is less than 2 at.%, viz., the following equilibrium reaction (2,3).

*Author to whom all correspondence and reprint requests should be addressed.

High Temperature Science, Vol. 26      © 1990 by the Humana Press Inc.

$$- \text{Si--O--Si} - + H_2O \leftarrow - \text{Si--OH} + - \text{SiOH} \qquad (1)$$

McMillan and Chlebik (4) showed that a soda lime–silicate glass contained 50–60 ppm hydroxyl ion after dry nitrogen was bubbled through the melt and 750–780 ppm after steam was used. In order to characterize any glass then, it is necessary to determine the effects of water (and other volatiles) dissolved in the glass during its formation.

In these studies, the thermal desorption of water has been measured by Knudsen effusion mass spectrometry for powders of a cordierite-type (magnesium alumino-silicate) glass containing the equivalent of about 400 ppm dissolved water. In this technique, the vapor species evolved upon heating may exist in other forms within the bulk and on the surface of powder particles.

## EXPERIMENTAL METHODS

A cordierite-type glass with excess magnesia and silica (plus less than 5% boric and phosphoric oxides) in the cordierite primary phase field was chosen for study. A series of particle-sized crushed glass powders was made where each sample of powder had passed the coarser, but was retained on the finer wire mesh screen of a pair of screens with only a small difference in the size of their openings. The samples are designated by the average diameter (μm) and are listed in Table 1.

The samples, contained in a Knudsen cell, are heated at a linear rate in a UHV vacuum furnace, whereas the modulated molecular beam is measured with an automated quadrupole mass spectrometer. This system has been described elsewhere (5), but the important details are given here. The following operating parameters of the system were under computer control: mass position, ionizing energy, integration time (modulation cycles), scaler readout, temperature setpoint, and temperature readout. The furnace was resistively heated with a tungsten mesh element that was surrounded by 13 concentric tantalum radiation shields, providing an excellent approximation to a blackbody cavity. The mass spectrometer and furnace were in separately pumped UHV chambers that were isolated by a valve and a tube 5 mm in diameter by 30 mm long. During heating, the pressure of the mass spectrometer chamber was $<2 \times 10^{-10}$ Torr $(= 2.7 \times 10^{-8}$ Pa), whereas the furnace chamber was $<2 \times 10^{-8}$ Torr $(= 2.7 \times 10^{-6}$ Pa). These very high vacuums were necessary in order to make the mass spectrometric measurements on the permanent gases, e.g., $H_2O$, CO, and $CO_2$, with high sensitivity and wide dynamic range. The Knudsen cell, with an effusion orifice of 1 mm in diameter by 5 mm in length, was fabricated from molybdenum and lined with platinum. The sample cup was also made from platinum. In early experiments, a quartz liner was used which absorbed a large percentage of the water desorbing from the samples and had to be abandoned. Temperatures were measured with a reference grade thermocouple (100% Pt vs 90% Pt–10% Rh), secured in the cell wall. This same thermocouple was also used for temperature control, the precision of

Table 1
Particle Size Ranges of Glass Powders Studied

| Particle Size, μm | Average, μm |
|:---:|:---:|
| 45–38 | 42 |
| 53–45 | 49 |
| 62–53 | 58 |
| 88–74 | 81 |
| 49–125 | 137 |
| 297–250 | 274 |
| 1190–1000 | 1100 |

which was better than 0.2 degrees. The molecular beam from the Knudsen cell was modulated at 20 Hz, giving continuous dynamic background correction, which as especially important for the permanent gases. An ionizing electron energy of 30 eV was used for obtaining all ion signals. The ion detector was operated in the pulse-counting mode.

For each sample, 100 mg was weighed into the platinum cup and then loaded into the furnace. This chamber was pumped overnight to reduce the residual pressures of the permanent gases. This procedure was especially important for $H_2O$ to increase the signal to background for the microgram quantities of water desorbing from the samples. The sample was then heated from room temperature up to 1500 K at 4 K/min. The temperature was held constant at this temperature for 100 min and then cooled. The ion intensities, $m/z = 18$, 23, and 44 ($H_2O$, Na, and $CO_2$, respectively), were counted using two separate scalers for the open and closed shutter positions, for 10 s integration time, each in sequence, over the entire temperature program.

## RESULTS AND DISCUSSION

For $m/z = 18$, water, we observe in the thermal spectrum (Fig. 1), a feature in the low temperature region ($\approx$300–800 K), consisting of at least two overlapping peaks (*L1 and L2*). In the high temperature region, beginning about 1000 K, a prominent peak ((*H*)$\approx$1110 K) dominates the spectrum. There is also a shoulder (*S*) that joins these low and high temperature peaks. Although the low temperature peaks appear to be associated with surface-adsorbed water and $CO_2$, the high temperature peak is probably from lattice (bulk)-derived water, as is the shoulder. The 660 K peaks (*L2*, dashed vertical line in Fig. 1) for $H_2O$ and $CO_2$ spectra are obviously correlated and best described by a bicarbonate decomposition reaction. Sodium effusion, on the other hand, occurs at higher temperatures, beginning about 1300 K, and at the isothermal temperature of 1500 K, the rate drops gradually, showing that the reaction is not in equilibrium, but most probably is controlled by bulk diffusion.

The amount of water in these glasses was measured by integrating the ion signal for water over two regions in the thermal spectra: 300–800 K for the surface and 800–1500 K for the bulk. These integrated ion values

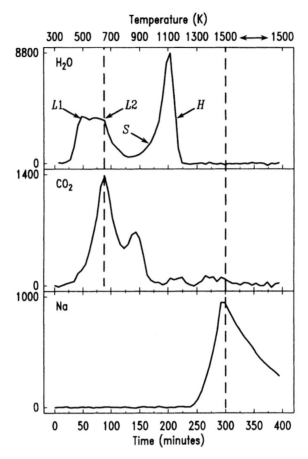

Fig. 1. The relative desorption rates of the principal gases from MgO–
Al$_2$O$_3$–SiO$_2$ glass powder (42 μm) when heated at 4K/min.

were calibrated by heating in the same Knudsen cell 10 mg of calcium
oxalate (CaC$_2$O$_4$·H$_2$O), which releases water stoichiometrically along
with an equal molar amount of carbon dioxide. The water concentrations
for surface and bulk water, measured by mass spectrometry, are given in
Table 2. The precision of these measurements is better than ±5%. One
notes that as the particle size gets larger, the surface water decreases as
expected. The amount, relative to the other particle sizes, can be affected
by the extent of surface area in cracks generated by the grinding process.
The amount of water released from the bulk was only partially complete
for the larger particle sizes, under the thermal treatment used in these
experiments, where the temperature was ramped into the crystallization
region before the water was fully desorbed from the glass. Slower heating ramp or isothermal anneal below 1150 K would allow a more complete release of the bulk water.

In Fig. 2, the high temperature desorption peak (H) shifts to higher
temperatures with increased particle size. However, the position of the
low temperature peak is invariant over this extremely wide range in
particle sizes. The assignment of these desorption processes to surface
(low T) and bulk (high T) is supported by the experimental evidence

Table 2
Water Desorbed in Surface and Bulk Regimes

| d (av.), μm | Surface, ppm | Bulk, ppm |
|---|---|---|
| 42 | 354 | 519 |
| 49 | 351 | 446 |
| 58 | 112 | 385 |
| 81 | 87 | 314 |
| 137 | 67 | 255 |
| 274 | 53 | 233 |

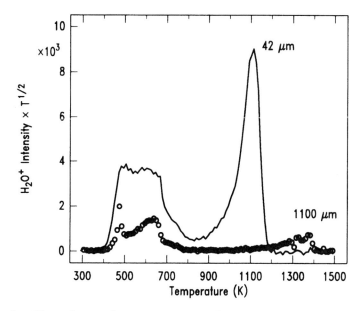

Fig. 2. The relative desorption rates of $H_2O$ from glass powders for two particle sizes, 42 and 1100 μm.

presented in this figure. Furthermore, the quantity of water released from the bulk for the 1100 μm particles is considerably less than the 42 μm ones even though the sample weights were identical. That portion of the bulk water released during crystallization must still reside in the particles, perhaps in voids or bubbles, escaping at a very slow rate to the surface. There is also visible in the high temperature peak some 'pulses' that probably correspond to bursting of these bubbles, suddenly releasing the water vapor trapped inside. The low temperature peak is smaller in intensity for the 1100 μm vs the 42 μm particles, because the larger ones have less surface area and a corresponding lesser amount of adsorbed atmospheric water than the smaller particles.

In the high temperature region (800–1300 K), there are three distinct features discernible in the data. The thermal desorption spectra, measured for particle sizes 42 through 274 μm, were analyzed for the best fit to several possible rate-controlling mechanisms: equilibrium reaction, first-and second-order kinetics, and diffusion. Because the peak temperature increased with increasing particle size, the possibility of this reaction being in equilibrium was immediately ruled out. In deciding which of the possible processes was operative, we used the analytical techniques described in a previous publication that modeled the effusion rates, as defined by the orifice of a Knudsen cell, both in respect to reaction kinetics and diffusion-controlled reactions (6).

Assuming diffusion in spherical or 'euhedral' type particles, a desorption curve, which satisfactorily fit the data, could not be generated. We then applied reaction kinetics to the analysis of the curve, according to the following equation

$$-dN/dt = N^n A e^{-E_a/RT} \tag{2}$$

where $N$ is the number of molecules unreacted, $n$ is the reaction order, $A$ is the frequency factor, and $E_a$ is the activation energy. In matching the calculated curve to the data, the overall shape determines the value for the activation energy, whereas the frequency factor determines the position of the peak temperature. We start with the initial quantity of water as unity and the mass spectrometric signal, which is proportional to $-dN/dt$, is normalized to the calculated curve at the maximum rate.

Using first-order, 260 kJ/mol for the activation energy and $2.90 \times 10^9$/s for the frequency factor, we were able to generate a theoretical curve that overlaid the experimental data for the 42 μm sample, with excellent precision spanning three orders of magnitude, as shown in Figs. 3a and b. We also show a theoretical curve for a second-order reaction to demonstrate the goodness of the first-order fit. The spectrum obtained for the 42 μm sample (the smallest particles) was used to calculate the best value for the activation energy of the reaction because, in this case, the desorption process was complete before the effect of the glass crystallizing became significant during heating. The precision with which we could fit the experimental data is ±10 kJ/mol for the activation energy and ±3% for the frequency factor, the latter corresponding to an uncertainty of about one degree.

This description of water desorbing from the surface of glass particles by a first-order process requires that the rate of desorption is proportional to the number of surface sites remaining occupied by water or its precursor. The surface activity of water must also be very close to the bulk because the reaction follows the first-order curve with such precision, showing no indication of diffusion limiting the final stages of desorption. This first-order reaction also infers that the surface water is 'molecular' prior to desorption even though the transport of bulk water to the surface is most likely by diffusion of ionic species (viz. $OH^-$ and $H^+$). As stated previously, at these very low water concentrations (400 ppm), water primarily exists in the form of hydroxyl ions ($-Si-OH$) (1–3). Lattice diffusion of these ions to the particle surface would also have to be fast enough so that desorption, not lattice diffusion, is the rate-limiting step in the reaction sequence.

The high temperature peak (1110–1163 K, depending on particle size) for this glass is about 50 degrees above the glass transition temperature, $T_g$, where, in general, any glass on heating begins to exhibit viscous

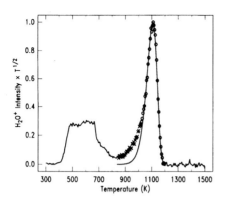
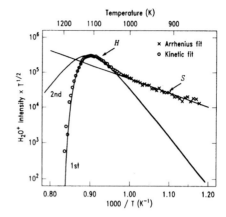

Fig. 3a.   Desorption rate of $H_2O$ from glass powder of particle size 42 μm. The high temperature points (○) were fitted with a first-order desorption curve, using $E_a$ = 260 kJ/mol and A = 2.90 × $10^9$/s (1110 K), whereas the low temperature points (x) were fitted with the Arrhenius equation, as shown in Fig. 3b.

Fig. 3b.   Desorption rate of $H_2O$ from glass powder of particle size 42 μm. The low temperature points (x) were fitted by linear regression that gave $E_a$ = 77 kJ/mol. The high temperature points (○) were fitted with a first-order desorption curve, using $E_a$ = 260 kJ/mol and A = 2.90 × $10^9$/s (1110 K).

behavior, as well as the elastic only behavior that exists below $T_g$. This high temperature peak, following the first set of peaks where all surface adsorbed water is believed to evolve, is probably associated with bulk (lattice)-derived water.

The bulk-derived water, first detected, is believed to be the beginning shoulder of the high temperature peak, clearly seen as points labeled "X" in Fig. 3a. Beginning at about 800 K to somewhat above 1000 K, the data were analytically best described by a straight line as a function of 1/T in the form of an Arrhenius equation, yielding a slope of 77 kJ/mol, as shown in Fig. 3b. We interpret this water desorption as being supplied by the diffusion of the molecular species through the bulk. This diffusion reaction dominates the desorption rate until the first-order surface reaction rate has accelerated owing to its much higher activation energy and becomes the major contributor to the water release (about 1050 K). For the different sample sizes, this activation energy of diffusion ranged from 70 to 80 kJ/mol. For other silicate glasses, Tomozawa (7) estimated this activation energy, $E_a$, to be in the range 80–100 kJ/mol. Moulson and Roberts (8) found an $E_a$ of 72 kJ/mol for the removal of hydroxyl ions by diffusion from pure silica glass slabs. In addition to silicon, this glass contains magnesium and aluminum, which would probably increase $E_a$ in the manner observed.

For particle sizes 48 through 274 μm, we used the same activation energy, 260 kJ/mol, to define the peak shape. However, as the particle size increased, the peak temperature also increased, which required a change in the frequency factor to match the calculated curve to the experimental data. The measured reaction rate of the highest temperatures in the thermal scan (above 1150 K) shows a departure from the theoretical curve, which is discernible even for the 48 μm sample. For the 137 μm sample, this experimental deviation from the theoretical curve actually results in a "double" peak. The experimental data and theoretical curve are shown in Fig. 4. This high temperature shoulder is attributed to the onset of glass crystallization, the rate of which becomes significant above 1150 K. In Fig. 5, the relative reaction rates for the 42 and 137 μm samples are compared on the log plot, which clearly shows the excellent agreement of the 42 μm sample to the simple first-order kinetic model and the breakdown of the model in the region of the crystallization for the 137 μm sample.

In Table 3, the experimentally-measured peak temperatures ($T_p$) and frequency factors (A) are presented for the six samples having the smallest particle diameters (d). One notices that a simple quantitative expression correlates the particle diameter and the frequency factor, i.e., that the latter are inversely proportional to the average diameter of the particles. Thus, in the last column of Table 3, we have shown this correlation by the quantity k × d, which is a constant for the five smallest particle sizes measured, averaging $117 ± 7 × 10^9$ μm/s. This correlation of particle size with desorption rate also follows the general relationship between the surface area and volume of particles

$$N_v/N_s = d/6 \tag{3}$$

Therefore, the number of surface sites is a major factor in the control of the desorption rate for bulk water. Although water diffuses at the same rate in large and small particles, it is the smaller surface area, not the longer diffusion path in the larger particles, that is the principal reason for the upward shift of the peak temperature.

If we assume a value for the spacing of water sites in the bulk, we can then compute an upper limit for the frequency factor. If we choose 0.4 nm as a reasonable value for the average cation–anion polyhedra bulk (or surface) unit spacing, then the upper limit for the frequency factor is $≈10^{14}$/s. For particle sizes 274 μm and larger, the onset of crystallization overlaps the early phase of the water desorption process and alters the peak temperature that is used to calculate the corresponding frequency factor from the thermal spectrum. Note the deviation of k × d for the particle size 274 μm sample, whose peak temperature is above the crystallization temperature.

This picture of the desorption process requires that the diffusion of water through the bulk be fast enough to supply the first-order desorption process, which becomes the dominant reaction (Fig. 3b), as the temperature is increased. Several phenomena could account for the two very different regimes. For most glasses, the diffusivity of hydroxyl ions does not change markedly upon going through $T_g$; however, our glass was quenched and is highly strained. Diffusivity has been found to increase below $T_g$ in strained glasses by Drury and Roberts (9) and Haider and Roberts (10). However, annealing the crushed glass to re-

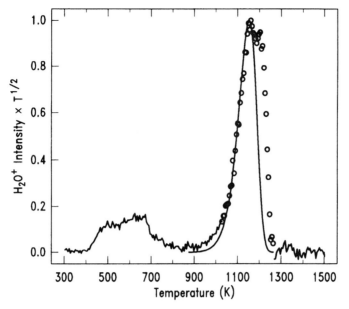

Fig. 4. Desorption rate of $H_2O$ from glass powder for particle size 137 μm. The high temperature points (○) were fitted with a first-order desorption curve, using $E_a$ = 260 kJ/mol and A = 0.90 × $10^9$/s (1154 K). Note the high temperature shoulder in the data, which deviates from the theoretical curve and coincides with the onset of crystallization.

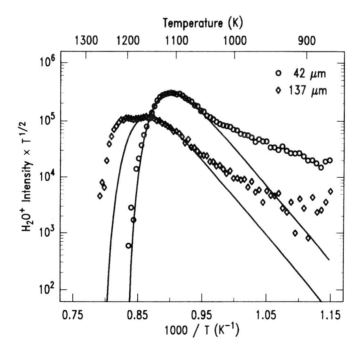

Fig. 5. The desorption rates of $H_2O$ from glass powders are compared for two particle sizes. The experimental data were fitted with first-order kinetics using $E_a$ = 260 kJ/mol and A = 2.90 × $10^9$/s (1110 K) for 42 μm and A = 0.90 × $10^9$/s (1154 K) for 137 μm.

Table 3
Rate Constants for Various Particle Sizes

| d (av.), μm | $T_p$, K | k, $10^9$/s | $k \times d$, $10^9$ μm/s |
|---|---|---|---|
| 42 | 1110 | 2.90 | 122 |
| 49 | 1117 | 2.35 | 115 |
| 58 | 1128 | 1.80 | 104 |
| 81 | 1136 | 1.50 | 122 |
| 137 | 1154 | 0.90 | 123 |
| 274 | 1163 | 0.73 | 200 |

move strain did not change the high temperature desorption peak, which is above $T_g$ where no strain can exist.

The activated process responsible for the shoulder at the beginning of the high temperature peak (*see* Fig. 3b) is clearly not controlled by the first-order reaction that dominates the water desorption rate above 1050 K, coinciding with $T_g$. Therefore, another desorption mechanism must exist for the shoulder. The desorption rate is faster below $T_g$ which may be the result of more reaction sites. The glass particles have corroded and microcracked surfaces which could heal or sinter smooth above $T_g$. However, significant surface healing and an attendant reduction in the desorption rate probably does not occur because the product of the frequency factor with particle size ($k \times d$) is invariant at temperatures above $T_g$. The higher rate at 800–1050 K is best explained by the fact that the activation energy for the diffusion of molecular water is rather small ($\approx 80$ kJ/mol) which, for this temperature range, yields a larger value for the exponential term. Thus, the diffusion of molecular water is more mobile than hydroxyl. Furthermore, water already in the molecular form is not impeded from direct desorption at the surface. However, the activity of molecular water in the bulk is very small at the 400 ppm concentration, and consequently, this diffusion process fails to deplete the bulk water before the first-order surface reaction takes over above 1050 K. The amount of water desorbed in this "shoulder" regime is only about 10% of the total released above 800 K.

The high temperature peak most likely has a source of water other than molecular water feeding the surface desorption process. The surface sites can be populated from the bulk, both by molecular water and hydroxyl ($-Si–OH$) diffusion (7). Also, molecular water and hydroxyl ions could have a higher diffusivity in the glass-transition regime than at lower temperatures because of the increased free volume that opens up the glass structure above $T_g$ (11). For example, Johnson et al. (12) found that with heating, the diffusivity of sodium ions in glass discontinuously increases at $T_g$ and continues increasing, but to a lesser degree, with continued heating above $T_g$. Such a phenomenon associated with water or hydroxyl ions could explain the present results. Alternatively, another source of water, such as might be bonded specifically with magnesium or aluminum ions, possibly becomes available above $T_g$. However, we believe this additional source is the hydroxyl ions.

## ACKNOWLEDGMENTS

For helpful discussions during the course of these studies, we are indebted to J. Acocella, G. H. Frischat, R. Ghez, S. H. Knickerbocker, R. A. Lussow, P. W. McMillan, A. Nowick, W. Reuter, M. W. Shafer, G. B. Stephenson, E. M. Stolper, M. Tomozawa, and D. Turnbull. C. F. Guerci assisted with sample preparation, and G. Pigey with Knudsen effusion-mass spectrometric measurements.

## REFERENCES

1. Scholze, H., *Glass Ind.* **47**, 546–554, 622–628, 670–675 (1966).
2. Stolper, E., *Contrib. Mineral Petrol.* **81**, 1–7 (1982).
3. Bartholomew, R. F., *Glass III, Treatise on Materials Science and Technology*, **22**, Tomozawa, M. and Doremus, R. H., eds., Academic, NY, (1982) pp. 75–127.
4. McMillan, P. W. and Chlebik, A., *J. Non-Cryst. Solids* **38/39**, 509 (1980).
5. Frisch, M. A. and Reuter, W. J., *Vac. Sci. Technol.* **16**, 1020 (1979).
6. Frisch, M. A., *Advances in Mass Spectrometry*, Quayle, A., **8A**, ed., Heyden, London, (1980) pp. 391–401.
7. Tomozawa, M., *J. Am. Ceram. Soc.* **68**, C251, C252 (1985).
8. Moulson, A. J. and Roberts, J. P., *Trans. Farad. Soc.* **57**, 1208–1216 (1961).
9. Drury, T. and Roberts, J. P., *Phys. Chem. Glasses* **4**, 79–90 (1963).
10. Haider, Z. and Roberts, G. J., *Glass Technol.* **11**, 153–157 (1970).
11. Kingery, W. D., Bowen, H. K., and Uhlmann, D. R., *Introduction to Ceramics*, 2nd edition, Wiley, New York, (1976) pp. 91, 758ff.
12. Johnson, J. R., Bristow, R. H., and Blau, H. H., *J. Am. Ceram. Soc.* **34**, 165–172 (1951).

# Twin-Chamber Knudsen Effusion Cell Mass Spectrometer for Alloy Thermodynamic Studies: II

M. J. STICKNEY, M. S. CHANDRASEKHARAIAH,
AND K. A. GINGERICH*

Chemistry Department, Texas A&M University,
College Station, TX 77843

## ABSTRACT

Additional modifications to the twin-chamber Knudsen effusion cell source mass spectrometer with a precision positioning device reported earlier are described. With this apparatus, the activities of palladium in a few palladium-niobium alloys are determined. The advantages of this design over the usual Knudsen effusion cell mass spectrometry for alloy thermodynamic studies are discussed.

## INTRODUCTION

The thermodynamic properties of alloys are generally deduced from the experimentally determined activity data of its components. The activity of a component of the alloy as a function of the composition and the temperature is sufficient in general to deduce other thermodynamic properties of the system. If one of the components is relatively more volatile than the rest, the measurements of the equilibrium partial pressures of that component as a function of the alloy composition becomes the method of choice (1). However, the accuracy of the activity data for solid alloys obtained from the usual methods of vapor pressure measurements, (viz., Knudsen effusion, Langmuir evaporation, or transpiration) suffers from the surface depletion problem (2). The high sensitivity and the wide dynamic range of the Knudsen cell mass spectrometry alleviate this problem of surface depletion and is an alternate method for the activity measurements. The use of a single chamber Knudsen cell source

*Author to whom all correspondence and reprint orders should be addressed.

High Temperature Science, Vol. 26        © 1990 by the Humana Press Inc.

in the mass spectrometer for the determination of the high temperature thermodynamic properties of solids is well documented (3,4) and does not require further elaboration here.

With the common single sample chamber Knudsen cell mass spectrometer, it is generally necessary to carry out at least two separate runs, one with the alloy and the other with the pure component, to obtain the activity value of the component. In terms of the conventional notations, the activity of the component, $a_i$, is given by the relation

$$a_i = [I^+_iT]_A \ S_R[I_e\eta \ \sigma_i(E)\gamma_i]R/[I^+_iT]_R \ S_A[I_e\eta \ \sigma_i(E)\gamma_i]_A$$

$$A = \text{alloy; } R = \text{pure component} \tag{1}$$

$S_A$ and $S_R$ are the respective geometrical factors such as the orifice size and the shape, the distance of the orifice from the entrance slit of the ion source, alignment of the orifice with respect to the entrance slit, and so on, of the run with the alloy and that with the pure component. All other terms in Eq. (1) have the usual meaning (4). The accuracy of the activity datum, $a_i$, derived from the measured ion currents, $(I^+_i)_A$ and $(I^+_i)_R$ depends not only on the accuracy with which these ion currents are measured but also on the ability to maintain the values of all the other factors in Eq. (1) the same in the two successive runs. It is difficult to maintain all the parameters of the instrument identical in two successive runs. Thus, the advantages of the better sensitivity of the mass spectrometric method may not result in better data for the activities unless some modification is introduced into the usual single chamber Knudsen cell mass spectrometer.

Among several innovations of the Knudsen cell mass spectrometry reported in the literature for the activity determination, the use of the multichamber cell as vapor source offers the greatest advantages (5,6). For example, it is relatively easy to maintain several factors indicated in Eq. (1) constant during the short time required to make the two successive measurements of ion currents, $(I^+_i)_A$ and $(I^+_i)_R$. This eliminates one source of errors. The errors attributed to the uncertainties of temperature values also are minimized. As it takes only about a minute or two to measure the two ion currents in succession in a twin-chamber cell mass spectrometer, it is easy to maintain the temperatures of both the chambers the same during the two measurements. If the temperature gradients within the Knudsen cell body is eliminated (a requirement demanded in both the single chamber and in the multichamber Knudsen cell sources) through proper design of the furnace, then Eq. (1) becomes

$$a_i = [I^+_i]_A/[I^+_i]_R \ S_R/S_A \tag{2}$$

Thus, the accuracy of the activity data does not depend on the uncertainties in the measurements of absolute values of the temperature.

There are a few reports in the literature describing the use of multiple Knudsen cells for the activity determination. Buchler and Stauffer (7) have described a twin crucible source in their vaporization study of the LiF-BeF$_2$ system. In their design, the twin cell source with the heater assembly was moved through an angle to bring one or the other orifice into focus. Jones et al. (8) have used a graphite Knudsen cell with two separate compartments for their study of the Pt-Au binary system. The molecular beam from either one or the other orifice was selectively admitted into the ion source by means of a movable beam-defining shutter. De Maria and Piacente (9) reported the use of a multiple cell for activity measurements. Their design consisted of a cell with five sample chambers drilled into a cylindrical block and the whole block was supported on a tungsten rod away from the axis of symmetry of the ion source. Each of the five effusion orifices, in turn, was brought into line with the entrance to the ion source by the rotation of the whole block. Chatillon et al. (5,6) have discussed at length the multiple cell technique for the activity measurements and pointed out the importance of temperature uniformity and the precise alignment of orifice positioning. They described an automatic positioning device for a four chamber Knudsen cell (5) in which the different orifices were brought into focus by means of two micrometer jacks. This positioning device was based on the polar coordinate system. It could control the polar and the azimuthal angles precisely but it had a fixed radial vector. Hence the movements of the positioning jacks resulted in a sweeping of a spherical area of about 40 mm square area by the orifices.

The temperature uniformity problem mentioned above is not unique to the twin-chamber or multiple chamber Knudsen cell techniques but is common to any high temperature thermodynamic study of materials. The use of a heat pipe device (10) for this purpose has certain advantages but is not necessary. Knudsen cells (~19 mm dia. and 25 mm height) have been maintained without any significant temperature gradients up to 2700 K with properly designed radiation shields (11). Though ascertaining the uniformity of the temperature within the whole Knudsen cell is essential for the determination of accurate activity data.

In the present design, a micrometer actuated X-Y alignment positioning device of a simple design enabled us to position the two orifices accurately under the entrance slit of the ionization chamber *in situ*. The cross-over problem was eliminated with the symmetrical positioning of two tungsten or alumina tubes (18). Further, the original ion source of Nuclide mass spectrometer Model HT 12-90 was modified to obtain more accurate appearance potentials and ionization efficiency curves. A microprocessor interfacing of the P.C. improved the data acquisition and data analysis. With these modifications, the accuracy of the activity data is improved. In this communication, some results obtained for the activities of a few Pd-Nb alloys are presented.

## EXPERIMENTAL

A single focusing mass spectrometer (Nuclide Model 12-90 HT) was used in this study, the details of which have been published previously (12). The twin-chamber Knudsen cell was mounted on the X-Y precision positioning device. The cell was radiatively heated with a thoriated tungsten rod coil heater. The Knudsen cells of about 25 mm total height were fabricated from a 19 mm diameter hot swaged molybdenum rod. Two cavities, each of 6 mm dia. and 15 mm depth, were drilled side by side, which served as the alloy and the reference chambers. The two effusion orifices were two cylindrical holes (0.70 mm dia. and 0.60 mm depth) drilled at the centres of these two cavities. Two one end closed, recrystallized alumina crucibles, one containing the alloy and the other the reference material, were introduced into the two chambers of the cell body through the bottom. Then the bottom of the cavities were sealed tightly into position with tight fitting molybdenum plugs. This sealed off the two chambers completely.

Any possible vapor phase crossover between the vapors effusing from the alloy and the reference orifices was eliminated with the symmetrical positioning of two tungsten or alumina tubes (4.0 mm I.D. and 30 mm long) above the two effusion orifices. The effectiveness of this arrangement was confirmed with a separate experiment (*18*).

The original ionization source was replaced by a new ionization source of modified design. A view of the fully assembled unit is shown in Fig. 1. The details of the assembly diagrams are shown in Figs. 2 & 3. The entire ionization source with the exception of the filament and its supports was fabricated using the hot swaged molybdenum plates (1/8th and 1/4th inch thick). A thoriated tungsten strip (0.030 inch wide and 0.001 inch thick with about 2% thoria) held tautly between the slits in the two tungsten rods, heated resistively, formed the source of electrons. The applied potential difference between the centre of the filament and the shield accelerated these electrons to the desired energies which enter the ionization chamber through a slot cut in the shield in front of the filament. The permanent magnet of the original source was removed.

The collimator arrangement and the shutter system was also redesigned. A view of this unit is shown in Fig. 4. It is possible with this new design to bring the orifice of the Knudsen cell, the collimating orifice and the entrance slit to the ionization chamber into a verticle axis precisely. The shutter could be fully opened or closed by means of a pneumatic device operated through the vacuum feedthrough shown in Fig. 4.

The data acquisition has been simplified by direct computer interfacing with the detector. The electron multiplier transmits the signal to a Keithley Model 617 electrometer which is equipped with a IEEE-488 interface port. A Zenith Model 159-12 computer receives the data through a customized program at the rate of about 3 transmissions per second. Firstly, 50 data points are collected with the shutter open fully to the ion source and then 100 data points are taken with the shutter closed, followed by collection of another 50 data points with the shutter open. These data points are averaged and stored on floppy disk. From the collected data points, both the average ion current and the standard deviation of it are calculated for each of the measurement. All the necessary programs are written in the basic.

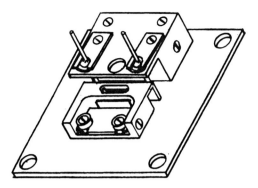

Fig. 1.  Fully Assembled View of the New Ion Source.

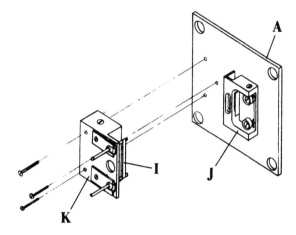

Fig. 2.  Partially Assembled Ion Source showing Sub-assemblies A-Mounting Plate, I-Filament, J-Shield Assembly, K-Filament Assembly

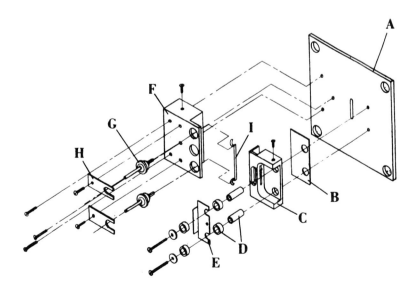

Fig. 3.  Assembly Diagram A-Mounting Plate, B-Glass Insulating Plate, C-Shield, D-Ceramic Insulators, E-Trap, F-Filament Block, G-Filament Support, H-Filament Support Retainer, I-Filament.

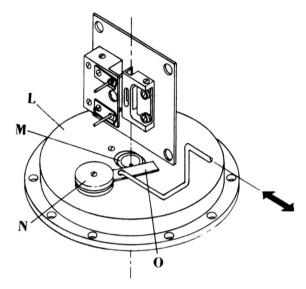

Fig. 4.  Source Entrance and Shutter Assembly Shown in Location with respect to the Ion Source L-Copper Plate, M-Movable Source Entrance, N-Shutter Assembly Bearing, O-Shutter Assembly.

Temperatures are measured with a calibrated Leeds and Northrup visual optical pyrometer, Model 8634-C, sighting at the blackbody cavity at the bottom of the cell. The common practice of drilling a tapped hole of about 10 mm depth and 1–1.5 mm dia. at the bottom of the cell as the black body cavity is not always satisfactory (*11*). Therefore, the blackbody design of Das and Chandrasekharaiah (*11*) is adopted in the present work. With a carefully designed thermal radiation shield pack of tantalum sheet surrounding the Knudsen cell assembly, temperatures up to 2000 K are obtained with insignificant temperature gradients in the cell body. The temperature readings obtained by sighting the blackbody orifice at the bottom did not differ from the temperature readings observed by sighting the effusion orifice.

The activities of palladium in two Pd-Nb alloys (76 and 97 at % Pd) were determined. Experiments for the measurement of the palladium activities in additional Pd-Nb alloys are in progress. All alloy samples were prepared by melting the mixtures of appropriate amounts of pure metals on a water cooled copper hearth arc furnace. The melted buttons of the alloys in general were remelted several times for homogenization of the composition.

## RESULTS AND DISCUSSION

The importance of minimizing the crossover of the vapor species between the two vapor fluxes effusing from the two orifices on the accuracy of the activity data obtained from the twin-chamber Knudsen cell source needs no further elaboration. The location of two tungsten

Table 1
Activities of Palladium in Pd-Nb Alloys

| Temperature T/K | $-\ln(I^+_{Pd}T)_{alloy}$ | $-\ln(I^+_{pd}T)_{Pd}$ | Activity of palladium |
|---|---|---|---|
| Run Dec. 88; 76Pd24Nb | | | |
| 1698 | 15.88 | 14.19 | $0.16_5$ |
| 1645 | 16.69 | 15.00 | $0.16_2$ |
| 1684 | 16.19 | 14.47 | $0.15_7$ |
| 1630 | 17.01 | 15.31 | $0.16_1$ |
| 1575 | 18.07 | 16.32 | $0.15_3$ |
| 1627 | 17.08 | 15.38 | $0.16_1$ |
| 1548 | 18.45 | 16.71 | $0.15_3$ |
| 1602 | 17.47 | 15.72 | $0.15_2$ |
| 1561 | 18.22 | 16.48 | $0.15_4$ |
| 1491 | 19.55 | 17.8 | 0.15 |
| Run Jan. 89; 97Pd3Nb | | | |
| 1696 | 14.01 | 13.94 | 0.94 |
| 1631 | 15.17 | 15.07 | 0.91 |
| 1674 | 14.51 | 14.51 | $0.91_3$ |
| 1605 | 15.72 | 15.61 | 0.90 |
| 1651 | 14.91 | 14.79 | 0.89 |
| 1576 | 16.23 | 16.13 | 0.90 |
| 1634 | 15.22 | 15.13 | 0.91 |
| 1561 | 16.48 | 16.39 | 0.91 |
| 1587 | 16.07 | 16.01 | 0.89 |
| 1615 | 15.55 | 15.46 | 0.91 |

tubes symmetrically above the two orifices eliminated this interference almost completely. The assumption that this arrangement of tubes affects the two effusion streams to the same extent was shown to be true from the results of the two experiments described previously (*18*).

The new ionization source gave consistently reproducible ionization efficiency curves as well as more accurate values for the appearance potentials. For example, the appearance potential (A.P.) values measured for the gold, palladium, silver and gallium are respectively 9.8 ± 0.3 eV, 8.3 ± 0.3 eV, 7.5 ± 0.4 eV and 6.14 ± 0.2 eV. The agreement with the corresponding values reported from the spectroscopic data 9.225 eV, 8.34 eV, 7.576 eV. and 5.999 eV. is good.

The determination of the $a_{Pd}$ in Pd-Nb alloys with the twin-chamber Knudsen cell source is in progress. The preliminary results for the two Pd-Nb alloys (76Pd 27Nb and 97Pd 3Nb) are presented in Table 1. The plots of $\ln(I_{Pd}T)$ vs 1/T for both the runs show good precision of measurements. The values $\Delta H^0_v$ (Pd, 298 K) = 91.2 ± 2.1 and 90.5 ± 1.1 kcal/mol, respectively, are calculated from the experimentally determined slopes and the thermal functions given by Hultgren et al. (*13*). The agreement with the corresponding value of 90.3 ± 0.2 kcal/mol reported recently by Chandrasekharaiah et al. (*14*) and with the value of 90.0 ± 0.5 kcal/mol recommended by Hultgren et al. (*13*) is good. The palladium activities of

$a_{Pd} = 0.14 \pm 0.005$ for 76Pd 24Nb and $a_{Pd} = 0.92 \pm 0.005$ for 97Pd3Nb at 1273 K are calculated assuming the corresponding partial molar enthalpies and entropies are temperature independent. Cima (17) has reported a value of $a_{Pd} = 0.14$ for the 75Pd25Nb alloy at 1273 K from his measured niobium activities using the solid oxide electrolyte galvanic cells. Both the 75Pd25Nb and the 76Pd24Nb alloys belong to the same homogeneous $Pd_3Nb$ phase. The agreement is good considering the difference in the method and the small compositional differences.

The twin-chamber or multichambers Knudsen cell mass spectrometer combines the high sensitivity and the selectivity of a mass spectrometer and the better accuracy of the Knudsen effusion technique for the high temperature thermodynamic study of alloys. Even then, only a limited number of reports exist in literature where multiple chambers Knudsen cell sources have been used for such studies. The problems encountered in the design and use of multiple chamber cells have been discussed by Chatillon et al. (5,6). The necessity of reliable and accurate positioning devices in such studies was treated there and one such positioning device also was reported (5). In the previous study, the design and use of a positioning device with about the same reliability but much simpler construction is described. In addition to the precise positioning of the twin-orifices alternately *in situ*, the crossover of the vapor species between the two effusing vapor beams is practically eliminated with the location of a pair of almost identical tungsten tubes (18).

## ACKNOWLEDGMENT

This work has been supported by the National Science Foundation under Grant CHE-8709916.

## REFERENCES

1. Kubaschewski, O. and Evans, L., *Metallurgical Thermochemistry*, 1958, Pergamon, NY, pp. 139–174.
2. Lewis, G. N. and Randall, M., *Thermodynamics*, 2d ed., Revised by Pitzer, K. and Brewer, L., McGraw Hill, NY, 1961.
3. Chatillon, C., Pattoret, A., and Drowart, J., *High Temp. High Press.* **7,** 119 (1975).
4. Drowart, J., *Proc. International School on Mass Spectrometry*, Ljubjana, Yugoslavia, 1969, pp. 187–242.
5. Chatillon, C., Senillou, C., Allibert, M., and Pattoret, A., *Rev. Sci. Instr.* **47,** 334 (1976).
6. Chatillon, C., Allibert, M., and Pattoret, A., *Adv. Mass Spect.* **74,** 615 (1978).
7. Buchler, A. and Stauffer, J. L., *Thermodynamics*, vol. 1, 1966, IAEA, Vienna, pp. 271–290.
8. Jones, R. W., Stafford, F. E., and Whitmore, D. H., *Met. Trans.* **1,** 403 (1970).
9. De Maria, G. and Piacente, V., *Bull. Soc. Belges.* **81,** 155 (1972).
10. Chatillon, C., Allibert, M., Moracchioli, R., and Pattoret, A., *J. Appl. Phys.* **47,** 1690 (1976).

11. Das, D., Dharwadkar, S. R., and Chandrasekharaiah, M. S., *J. Nucl. Mater.* **130,** 217 (1985).

12. Gingerich, K. A., *Proc. 10th Materials Research Symposium on Characterization of High Temperature Vapors and Gases*, Gaithersburg, MD, NBS Special Publ. 651, 1979, pp. 289–300.

13. Hultgren, R., Desai, P. D., Hawkins, D. P., Gleiser, M., Kelley, K. K., and Wagmann, D. D. *Selected Values of the Thermodynamic Properties of Elements*, American Society for Metals, Metals Park, OH, 1973, p. 78.

14. Chandrasekharaiah, M. S., Stickney, M. J., and Gingerich, K A., *J. Less Common Metals* **142,** 373 (1988).

15. Hultgren, R., Desai, P. D., Hawkins, D. P., Gleiser, M., and Kelley, K. K., *Selected Values of the Thermodynamic Properties of Binary Alloys*, American Society for Metals, Metals Park, OH, 1973, pp. 303, 386–390.

16. Gingerich, K. A., Kingcade, Jr., J. E., Stickney, M. J., and Chandrasekharaiah, M. S., *J. Less Common Met.* **143,** 373 (1988).

17. Cima, M. J., Lawerence Berkeley Laboratory Report, LBL-2195, 1986.

18. Stickney, M. J., Chandrasekharaiah, M. S. and Gingerich, K. A., *High Temp. High Press*, in print.

# Thermochemistry and Models

# Reference Books and Data Banks on the Thermodynamic Properties of Inorganic Substances

LEV V. GURVICH

*Thermocenter of the USSR Academy of Science, IVTAN, Izhorskaya 13/19, Moscow, USSR 127412*

## ABSTRACT

Data on the thermodynamic properties of substances published in the reference books or stored in computer memories should satisfy two requirements—the recommended data should be reliable within the frame of modern knowledge and these data should be given for a comprehensive set of elements and their compounds. Taking into consideration these requirements, some reference books and two data banks are compared and discussed.

**Index Entries:** Thermodynamic properties; data banks; inorganic substances; JANAF tables; NBS tables; IVTAN; VINITI; TPIS; thermodata.

## INTRODUCTION

During recent decades, the methods of thermodynamics have been finding more and more applications in investigating various processes and creating new technologies. The development of new methods of providing humankind with energy, the choice of the optimum paths for utilization of raw material resources and reprocessing of industrial wastes, precautionary measures for preventing environment pollution and forecasting the aftereffects of accidents, including those at nuclear power stations, could not have been achieved without preliminary analysis based on thermodynamic modeling. Therefore, it is essential to continue the study of thermodynamic properties of individual substances and the accumulation of the available data, their critical analysis, processing and systematization in a form acceptable to scientists and engineers working in different branches of science and technology. The resulting evaluated data on the thermodynamic properties must be issued in the

High Temperature Science, Vol. 26     © 1990 by the Humana Press Inc.

form of periodically updated and easily accessible reference books or stored in data centers in computer memories from whence these data may be obtained via communication lines or on magnetic media.

Reference data on the thermodynamic properties of substances, published in print or stored in databases, should satisfy the following requirements

1. Recommended data should be selected as the result of critical analysis of all data published in the literature, using correct methods of processing the primary information and calculating the thermodynamic properties.
2. The recommended values of thermodynamic properties should represent a system of mutually consistent quantities, including consistency of the thermochemical data and the thermal functions.
3. The recommended values should be based on fundamental constants, key thermochemical values, and atomic weights recommended by the appropriate international organizations.
4. For each recommended value, an estimate of its reliability should be given.
5. Brief texts should be available on the evaluation and calculation of the recommended quantities and the estimate of their uncertainties.
6. A bibliography of all papers used in the preparation of the reference data should be available.
7. The values of the recommended properties should be given for a wide and logically chosen set of substances.

The existing reference books can be divided into two types: first, the so-called critical reference books and, second, compilative, secondary (and even tertiary) handbooks. Critical reference books, the number of which is extremely limited in the world literature, are based on the critical analysis of primary literature and independent calculations of the recommended thermodynamic properties. They have to satisfy most of the requirements formulated above. The first such publications were *The International Critical Tables* and the reference book by Bichowsky and Rossini.

Compilative handbooks form the main part of the reference literature on thermodynamic properties of substances. They are based wholly or to a considerable extent on the reproduction of data from other reference publications (frequently combining data from several of them), or, they use uncritical data published in periodicals. In these handbooks, there is no internal consistency of recommended values and inconsistent values may be usually given.

In this review, we restrict the discussion to reference books and data banks dealing with such properties as heat capacity, heat content, entropy, and free energy function as well as enthalpy of formation, enthalpy of phase transitions, dissociation energy, ionization potential, and electron affinity of inorganic and simple organic substances.

The critical reference publications on thermodynamic properties can be divided into two groups, according to their content. Reference books, in which the thermochemical properties of substances at 298.15 K and the phase transition temperatures are presented, belong to the one type. Those containing the thermodynamic properties of individual substances over a wide range of temperatures belong to the other one.

The most well known reference books of the first type are those prepared at the National Bureau of Standards of the US and at the Academy of Sciences of the USSR. The book *The NBS Tables on Chemical Thermodynamic Properties* (1) contains data on thermodynamic properties at 298.15 K for about 9000 substances in the solid, liquid, gaseous states, and aqueous solutions as well. During the preparation of this book, the authors have thoroughly analyzed all the primary literature and also developed a method for obtaining a system of internally consistent values of thermodynamic properties for all compounds of the chosen element. The method, based on creating catalogs of the investigated chemical reactions, includes software for solution of the overdetermined system of the corresponding linear equations, enabling one to determine the required thermochemical quantities and their uncertainties.

Unfortunately, this reference book does not contain the data for phase transitions in contrast to its well-known predecessor *The NBS Circular 500*. The bibliography is absent as well. At present, reaction catalogs for all elements are being prepared in the Chemical Thermodynamics Data Center of NBS (NIST). The catalogs for rubidium and thorium compounds were published together with bibliographies. The catalogs for elements from oxygen through nitrogen (standard thermochemical order), alkali metals, and lanthanum are under preparation. Plans are to prepare catalogs on rare earth elements next.

The reference publication *Thermal Constants of Substances* (2), prepared under the direction of the late Academician V. P. Glushko and the late V. A. Medvedev in The Institute for High Temperatures of the USSR Academy of Science (IVTAN) and VINITI with the participation of many Soviet scientists, is the most complete publication of this type. In its coverage of substances (more than 25,000) and properties, it considerably exceeds other publications, and during its preparation all the requirements formulated above were satisfied, except for the third and fifth. The Hemisphere Publishing Corporation is now completing the preparation of the English edition of this book.

While preparing these two reference books, the authors from NBS and IVTAN have been working in constant contact, exchanging bibliographies and literature and jointly evaluating the most doubtful data. This has assisted in raising the level of both publications and eliminating some errors. Strictly speaking, the recommended values in these publications are incompatible, since they are based on different sets of key thermochemical values though in the majority of cases they are close to one another.

The books *Thermodynamic Properties of Individual Substances* (TPIS) and *The JANAF Thermochemical Tables*, prepared in the USSR and the US, are the critical reference publications of the second type. Both of them include the results of long term projects. Being designed first to assist development of rocket-space technology, they both soon expanded beyond this framework, providing specialists from different branches of science and engineering with necessary information. Rates of their development during the period of more than 20 years are shown in Table 1.

During the preparation of all three editions of the "TPIS" book (3–5), attention was paid to the development of methods of calculating the thermal functions of gases, methods of processing primary experimental data, and semi-empirical methods for estimating constants when the experimental data are incomplete or absent. In addition, a program was organized on measurements of molecular, thermochemical and thermodynamic constants, required for calculating the thermodynamic properties of substances for a wide temperature range. Many laboratories have participated in these investigations for more than 30 years.

In the 1956 edition, for the first time in the literature, the total internal consistency of all recommended values was achieved, including consistence of thermochemical quantities and thermal functions. In the 1962 edition, the uncertainties of all recommended values, including the tables of thermal functions were estimated for the first time. In fact, this edition satisfied all the requirements formulated above, however, only for a small set of compounds of 31 elements. During the preparation of the third edition (5), a number of new methods and programs were developed as the compounds of d- and f-elements were included in this edition. These methods allowed an increase in the accuracy of calculations of the thermal functions of gases, as well as correct estimation of the reliability of calculated quantities, even when the data on the molecular constants were absent. In Table 2, the values of entropy and enthalpy of formation of UO(g), calculated in TPIS (5) on the basis of estimated molecular constants as well as those calculated by Pedley and Marshall (16) on the basis of their own estimation, are compared with those calculated in IVTANTERMO Data bank using the results of UO spectroscopic investigations in the Molecular Spectroscopy Laboratory of THERMOCENTER in 1985–87. One can see in the Table that the values of S (UO, g), calculated in TPIS in the absence of the experimental data, are in satisfactory agreement with the new calculations within the whole temperature range.

The preparation of the fifth volume of this book is now coming to the end. The volume will include about 300 compounds of six more elements (Mn, Fe, Co, Ni, Cu, Zn), as well as additional compounds of Ge, Sn, Pb, Ti, Cr, and 40–50 updated tables from vols 1 and 2. The updating of all materials of previous editions is a peculiarity of TPIS. The Hemisphere Publishing Corporation has started to publish the English edition of this book (17). This edition will be a modified and updated translation of the third one and thermodynamic properties for many substances will be recalculated using experimental data published in 1978–87. It will contain

Table 1
The Growth of Information Presented in the TPIS and JANAF

| Number of | TPIS | | | JANAF | | |
|---|---|---|---|---|---|---|
| | 1st Ed 1956 | 2nd Ed 1962 | 3rd Ed 1978–82 | 1st Ed 1965 | 2nd Ed 1971 | 3rd Ed 1985 |
| Elements | 21 | 31 | 50 | 23 | 29 | 47 |
| Substances | | | | | | |
| Total | 206 | 336 | 1100 | 560 | 570 | 1050 |
| In gaseous state | 178 | 335 | 1060 | 460 | 625 | 905 |
| In condensed state | 29 | 45 | 278 | 220 | 275 | 400 |

Table 2
The $S°(T)/J \cdot (K \cdot mol)^{-1}$ and $\Delta_f H°(298.15\ K)/kJ \cdot mol^{-1}$ of UO(g)

| T/K | TPIS 1982 | Pedley 1983 | IVTANTERMO 1987 | THERMODATA 1986 |
|---|---|---|---|---|
| 298 | 248.8 ± 4 | 260.6 | 252.0 ± 0.1 | 241.4 |
| 1000 | 308.1 | 302.4 | 304.6 ± 0.1 | 282.7 |
| 2000 | 343.3 | 328.0 | 336.0 ± 1.5 | 308.2 |
| 3000 | 362.9 ± 8 | 343.1 | 356.3 ± 1.8 | 323.3 |
| 4000 | 376.5 | 353.9 | 372.8 ± 2 | 334.0 |
| 5000 | 387.1 | — | 386.9 ± 3 | — |
| 6000 | 396.0 ± 12 | — | 399.1 ± 4 | — |
| $\Delta_f H$ | 30.5 ± 17 | 25 ± 10 | 18 ± 15 | −49.4 ± 42 |

data for 56 elements and more than 1300 compounds. We hope that the final volume of the English edition will be published soon after publication of the fifth volume of the Russian edition.

The reference book *The JANAF Thermochemical Tables* arose from a number of projects carried out in the US starting from the 1950s. Although, strictly speaking, the recommended data have not been internally consistent in the JANAF Tables, there have not been significant internal contradictions between the data in the two last editions. Unfortunately, information on the uncertainties of tabulated thermal functions is absent in all three editions of the JANAF tables though in the last one they are given as a rule for the properties at 298.15 K. For calculating thermal functions of gases approximate methods and different approaches were used in many cases. The thermal functions of diatomic gases in most cases were calculated by the method of Mayer and Goeppert-Mayer. As a result, errors have arisen at low temperatures in the calculated values of free energy function and enthalpy content of molecules which have multiple ground states. Also for molecules which have dissociation energies below 250 kJ mol$^{-1}$ the calculated values of entropy, heat capacity and heat content have errors at elevated temperatures. Some inconsistency exists as a result of different approaches in calculation of electronic partition functions of monatomic, diatomic and polyatomic gases. In the second and third editions some tables were adopted from previous editions. Thus in the last edition the tables of properties calculated in 1962–7 are retained for some substances though they are based on incorrect data.

Due to the large amount of data in the TPIS and JANAF books, the probability of accidental errors in analysis and calculations is quite high, especially taking into account that the whole procedure for preparing the tables of thermodynamic properties has not been completely automatic and computerized until now. As a result in TPIS (5), for example, the thermal functions of $O_2(g)$ below and above 6000 K are in disagreement; one set of molecular constants was adopted for methyl alcohol in the text, whereas the calculations were done using another, less accurate set of constants. In the last edition of JANAF Tables (8) during the recalculations of the thermal functions of titanium dichloride from calories to joules the enthalpy of fusion was kept in the old units. Also, it is difficult to understand the source of errors in the JANAF tables in the thermal functions of gaseous monatomic zirconium over the whole temperature range, though they were correct in its previous edition.

International cooperation is a natural way to increase the speed of preparation of such reference books as TChT (1), TCS (2), TPIS (3–5), and JANAF (6–8) and to overcome their weak points. Experience in such cooperation has been accumulated in the last few years. Thus in the middle of the 1970s the IAEA started a project on the thermodynamic properties of actinides and their compounds (9). Scientists from a number of countries, engaged in research on thermodynamics of these substances, participated in this project. During 12 y, 9 of the 14 planned parts were published; however, the work remains incomplete. The principal shortcoming of the issued parts is the absence of a common approach to the analysis and processing of experimental data as well as to the calculation of thermal functions.

The experience of the CODATA task group on Chemical Thermodynamic Tables seems to be more successful. The group generalizes the experience accumulated in the course of preparation of the above mentioned reference books at NBS, IVTAN and JANAF Group as well as the experience accumulated, at the University in Grenoble, the National Physical Laboratory in Teddington and the Atomic Energy Research Establishment in Harwell in calculations of thermodynamic properties of alloys and solutions. The task group presented the worked out methods in the CODATA Bulletin. Then the group, including experts from US, UK, USSR, and France, prepared as a prototype of the future international tables, a book on the thermodynamic properties of calcium and some of its compounds, alloys and solutions. The book was published in 1987 (10). About 200 copies have been sent to scientists working in this field for comments. A considerable number of replies have been received. The Group is examining these notes and proposals and will take them into account in the future work. It is necessary to mention that though most of the corresponding data have been evaluated before in the reference books TPIS (5), JANAF (8), and TChTP (1), the Task Group still spent more than 3 years for analysis and unification of this data. At present, a book on the properties of iron, its compounds and alloys is being prepared. In spite of some difficulties, concerning the evaluation of thermodynamic properties of iron oxides, we hope that the work will be finished next spring. The preparation of a similar book on properties of cobalt and nickel has been started simultaneously. The group would be thankful to all scientists who are engaged in the investigations of thermodynamic properties of these elements and their compounds for information about the results of completed research as well as of measurements in progress.

As can be seen from this discussion, the preparation of a large set of evaluated reference data on thermodynamic properties in a printed form is an extremely long and tedious task. It seems reasonable to suppose that a computerized data system is a more effective way to create the comprehensive set of reliable data on thermodynamic properties of substances. Unfortunately, the majority of existent and emerging data banks on thermodynamic properties are analogous to secondary and tertiary handbooks. Their databases are being filled with the information from the published reference books without any critical analysis. These adoptions are frequently illegal, as the critical reference books are usually copyrighted. Furthermore, the books often contain old data, the recommendations of different books may disagree and can't form a set of consistent values; sometimes the data may be even incompatible. The thermodynamic data bank MALT demonstrated by the group of Japanese scientists at the Xth CODATA conference in Ottawa (11) in 1986 is an example of such a data bank. Its bases are filled with the information from reference books *The NBS TChTP, TPIS, JANAF*, and other publications. The scientific community should be careful since unpredictable errors may arise when using the information from such databanks. Another example is the STN International System, which collects in its databases independent sets of recommended values from JANAF Tables and TChT. In these two books different values were selected in many cases. As an example, their recommendations for $\Delta_f H(Al_2S_3$, cr, 298.15 K) in kJ.mol$^{-1}$ are $-724$ (TChT) and $-651.35 \pm 3.8$ (JANAF).

The ideal computerized system on the thermodynamic properties should not accumulate data from the published reference books, but generate evaluated thermochemical values and thermal functions on the basis of the previous accumulation, critical analysis, and processing of the primary data obtained during calorimetric, mass-spectral, spectroscopic, and other measurements. The data bank should also contain reliability estimates for both the primary and evaluated data. This is a large and labor-consuming task that can be performed only by highly trained experts. That is why there are only a few systems of such a kind now. These are THERMODATA and IVTANTERMO for inorganic substances and the system developing in the Texas A&M University for organic substances. As this paper concerns the properties of the inorganic substances, we shall dwell on the first two data banks only.

THERMODATA (12,13) is an integrated information system providing users via communication lines with the bibliography on thermodynamic properties of substances, with selected and evaluated thermodynamic data for inorganic substances, with data on phase diagrams and thermodynamics of binary and multicomponent systems. Though THER-MODATA used information from such reference books as JANAF, how-ever, these data are generally transformed and consistent with its own data. THERMODATA is the result of work of scientists from a number of West European countries. It is functioning under the leadership of the Scientific Group European THERMODATA including specialists of UK, France, FRG, and Sweden. The group supports the high scientific level of the information, and is capable of organizing the programs for measure-ments of numerical data in European laboratories. The important part of THERMODATA is the complex of software providing via communication lines the information stored in the databases as well as performing various thermodynamic calculations. The latter varies from calculations of thermodynamic parameters for some chemical reactions to calculations of phase diagrams for multicomponent systems.

The computerized system "IVTANTERMO" (14,15) contains numer-ical data on thermodynamic properties of individual substances over a wide temperature range. The data are internally consistent within the framework of the basic laws of thermodynamics and the reliability of all the recommended values is estimated. The recommended values are calculated using the constants chosen as a result of critical analysis and processing of all primary data from the literature. The corresponding processing and calculations are performed using a set of methods and programs that were created during the preparation of the reference book TPIS (5,17) and have been developed further by its authors.

The IVTANTERMO system consists of a set of databases and pro-grams. IVTANTERMO presently performs the following functions:

1. Storage and processing of primary numerical data to obtain the constants required for calculation of thermodynamic properties over a wide temperature range;
2. Calculation of thermodynamic properties of individual subs-tances in the solid, liquid, and gaseous states;
3. Assessment of reliability of the primary data and recom-mended values;
4. Provision of requested data to users.

Information from IVTANTERMO may be obtained on magnetic tapes for the subsequent input of data to the user's computer (on flop-pydiscs for PC compatible with IBM PC AT, since 1989). The information on the magnetic carriers consists of the database and the program sys-tem, and enables the users to get the catalog of substances, the tables of thermodynamic properties and the equations approximating thermal functions of substances over the wide temperature range and in different formats. They make it possible to calculate the thermodynamic parame-ters of chemical reactions as well as the composition and properties of arbitrary multielement systems.

The IVTANTERMO database is being regularly expanded and up-dated by including chemical elements and substances not considered

Table 3
The Growth of Information Presented
in the IVTANTERMO Database

| Number of | | 1983 | 1985 | 1987 | 1990 |
|---|---|---|---|---|---|
| Elements | | 56 | 61 | 78 | 90 |
| Substances | | | | | |
| | Total | 1350 | 1650 | 1950 | 2300 |
| In gaseous state | | 1300 | 1600 | 1900 | 2200 |
| In condensed state | | 350 | 425 | 530 | 600 |

before as well as by correcting and updating the accumulated data. The rate of development of IVTANTERMO is shown in Table 3.

One can see (and the same is true for THERMODATA) that the rate of development of computerized systems is appreciably higher than the rate of preparation of critical reference books, such as TPIS and JANAF. However, the computerized systems have one shortcoming. Users get the numerical data from their databases as from the "black box," practically without any accompanying information. THERMODATA notes only that the data are adopted from JANAF or any other source, or are calculated by THERMODATA. IVTANTERMO postulated that all its recommendations are based on the critical analysis and the calculations carried out by the experts of the THERMOCENTER of the USSR Academy of Science using IVTANTERMO software.

In contrast to the reference books JANAF and TPIS, the numerical data in THERMODATA and IVTANTERMO are not accompanied by text describing the analysis carried out and calculations. The information about the constants used for calculations of thermodynamic properties is absent as well.

The complete bibliography of papers dealing with the investigations of properties of a given substance may be received from THERMODATA separately, but as far as is known, without indication what information was used for evaluation procedure. In Table 2, the values of entropy and enthalpy of formation for $UO(g)$ calculated in THERMODATA in 1986 are given. It seems that they are incorrect but it is possible only to guess what is the source of errors. THERMOCENTER is going to supply the tables of thermodynamic properties, accumulated in IVTANTERMO bases with brief texts, values of the constants used in calculations and with a bibliography as well, but not earlier than 1991.

It is obvious that the number of substances presented in the databanks THERMODATA and IVTANTERMO is insufficient to satisfy scientists and engineers today, especially if we take into account that both data banks contain information mainly for the same substances. That is why it is desirable to have an international program for creating a computerized data system on evaluated thermodynamic properties of individual substances. Such a project may permit us to divide the work among many persons and to speed the task of producing a comprehensive and reliable data system using the experience accumulated by

CODATA TG on CTT and such databanks as THERMODATA, IVTANTERMO, and the databank at Texas A&M University.

## REFERENCES

1. Wagman, D. D., Evans, W., et al., *The NBS Tables of Chemical Thermodynamic Properties*, J. Ph. Chem. Ref. Data, Suppl. 2 (1982).
2. Medvedev, V. A., Bergman, G. A., et al., *The Thermal Constants of Substances*, in 10 parts, Glushko, V. P., et al., eds., VINITI (1965–82).
3. Gurvich, L. V., Yungman, V. S., et al., *Thermodynamic Properties of the Components of Combustions's Products*, in 3 volumes, Glushko, V. P., ed. USSR Ac. of Sci., M. (1956).
4. Gurvich, L. V., Khachkuruzov, G. A., et al., *Thermodynamic Properties of Individual Substances*, in 2 volumes, Glushko, V. P., ed., USSR Ac. of Sc., M. (1962).
5. Gurvich, L. V., Veitz, I. V., et al., *Thermodynamic Properties of Individual Substances*, in 4 volumes, Glushko, V. P., et al., eds., Nauka, M. (1978–82).
6. Stull, D. R., et al., *The JANAF Thermochemical Tables*, PB-168370 (1965).
7. Stull, D. R. and Prohet, H., *The JANAF Thermochemical Tables*, 2nd ed., NSRDS-NBS-37, W. (1971).
8. Chase, M. V., Davies, C. A., et al., *The JANAF Thermochemical Tables*, 3rd ed., J. Ph. Chem. Ref. Data, v. 14, Suppl. 1 (1985).
9. *The Chemical Thermodynamics of Actinide Elements and Compounds*, Oetting, E. L., et al., eds., IAEA, Vienna (1976).
10. *CODATA Thermodynamic Tables. A Prototype Set of Tables*, Garvin, D., et al., eds., Hemisphere Publ. Co., W., NY, L. (1987).
11. Yokokawa, H., Yamauchi, S., and Fujeda, S., *Computer Handling and Dissemination of Data*, Proc. of the Tenth CODATA Conf., Ottawa, Canada, Glaeser, Ph. S., ed., pp. 257–261, North Holland, NY (1987).
12. Cheynet, B., *THERMODATA, On-line Integral Information System*.
13. Ansara, I. and Sundman, B., *Computer Handling and Dissemination of Data*, Proc. of the Tenth CODATA Conf., Ottawa, Canada, Glaeser, Ph. S., ed. pp. 154–158, North Holland, NY (1987).
14. Gurvich, L. V., *Vestnik AN SSSR*, No. 3, pp. 54–63 (1983).
15. Gurvich, L. V., *Computer Handling and Dissemination of Data*, Proc. of the Tenth CODATA Conf., Ottawa, Canada, Glaeser, Ph. S., ed., pp. 252–256, North Holland, NY (1987).
16. Pedley, J. B. and Marshall E. M., *J. Ph. Chem. Ref. Data*, No. 4, pp. 967–1032 (1983).
17. Gurvich, L. V., Veitz, I. V., et al. Thermodynamic Properties of Individual Substances, in 5 volumes, Hemisphere, NY, L., vol 1 in 2 parts, 1989.

# Thermodynamic Properties
# of the Alkaline Earth Hydroxides

## A JANAF Case History

MALCOLM W. CHASE* AND RHODA D. LEVIN

*National Institute of Standards and Technology,
Gaithersburg, MD 20899*

## ABSTRACT

With the recent flurry of experimental and theoretical studies on many of the alkaline earth hydroxides, the time is appropriate for a thorough critique of the thermodynamic properties of these mono- and dihydroxides. We have started such a process, recently completed a thorough examination of the literature, and have generated annotated bibliographies on the thermodynamically-related information on the mono- and dihydroxides of the alkaline earth metals. In addition, the process of extracting all the pertinent data and constructing tabular and graphical summaries of the information has started. Until recently, the lack of sufficient experimental and theoretical data required that estimation schemes be used to produce the required thermochemical tables of all gas phase tables. Since 1983, considerable experimental data has been generated in the area of the spectroscopic properties of the monohydroxides. Four (Mg, Ca, Sr, and Ba) monohydroxides (and their corresponding deuterated species) are now well-characterized spectroscopically, and new experimental studies have confirmed the enthalpy of formation values for the condensed phase dihydroxides. In this presentation, the historical development of the thermodynamic properties of the monohydroxides are emphasized.

If you are interested in the thermodynamic information on the alkaline earth hydroxides, where would you go to find it? In the following, some of the readily available resources are discussed.

**Index Entries:** Thermochemical tables; monohydroxides; dihydroxides; spectroscopic properties; formation values; enthalpy of.

*Author to whom all correspondence and reprint orders should be addressed.

High Temperature Science, Vol. 26    © 1990 by the Humana Press Inc.

# INTRODUCTION

In 1952, the well-known NBS Circular-500 (*1*) was published. This critical evaluation of (ambient temperature) thermodynamic information does not have any reference to the gaseous alkaline earth monohydroxide or dihydroxide chemical species; only the condensed phase of the dihydroxides is covered. Enthalpies of formation at 298.15 K are given for all solid phase alkaline earth dihydroxides (except radium). Entropy information is given only for solid $Mg(OH)_2$ and $Ca(OH)_2$. Even in the update to the Circular 500 publication, the NBS Tables (*2*)—formerly referred to as Tech Note 270 and dated 1967–1970 for the Group IIa metals—complete information is not available. Enthalpies of formation at 298.15 K are given for the gaseous monohydroxides of Ca, Sr, and Ba, as well as for the gaseous dihydroxides of Be, Mg, Ca, Sr, and Ba. No entropy information is given for any of the monohydroxides and no information is given on the radium hydroxides. For the condensed phase dihydroxides, the enthalpy of formation at 298.15 K is given for five of the dihydroxides, but entropy information is only given for $Be(OH)_2$, $Mg(OH)_2$, and $Ca(OH)_2$. Low temperature heat capacity was available only for $Mg(OH)_2$ and $Ca(OH)_2$.

The 1960s reviews by Kelley (*3,4*) provide experimental high temperature enthalpy results for the condensed phase dihydroxides of Mg, Ca, and Ba, but only entropy values at 298.15 K (as derived from experimental low temperature heat capacity studies) for the Mg and Ca dihydroxides. The gaseous alkaline earth hydroxide species are not considered by these Kelley publications. The entropy at 298.15 K reported by Kelley (*4*) and Rossini et al. (*1*) agree for $Mg(OH)_2$, but differ for $Ca(OH)_2$. This difference is owing to the use of a low temperature heat capacity study by Kelley (*4*); the same study not being available to Rossini et al. (*1*) when they performed the data analysis. Kelley (*4*) and the more recent study by Wagman et al. (*2*) agree. This is expected since the recommended results are derived from the same experimental studies. In summary, with the exception of the $Mg(OH)_2$ and $Ca(OH)_2$ condensed phase species, little information on the alkaline earth hydroxides was available prior to the 1960's.

# DISCUSSION

The following is a discussion of the more recent reviews of thermochemical tables of the alkaline earth hydroxides. There have been three concerted efforts to characterize, as a family, these Group IIA hydroxide species. Critical evaluations for the temperature-dependent properties have been performed by Chase et al. (*5*) for the JANAF Thermochemical Tables project (JANAF), Gurvich et al. (*6*) for the Thermodynamics of Individual Substances project (TPIS), and Jackson (*7*) for a Lawrence Livermore Laboratory project. The JANAF and TPIS projects considered the condensed and gaseous phases of all the hydroxides of the Group IIA metals except radium. The study by Jackson in 1971 treats only the gas phase of the hydroxides, including the radium species. It should also be noted that the TPIS study was published in 1981, whereas the JANAF study, although published in 1985, actually contains critical evaluations of the alkaline earth hydroxides that were made in 1975.

Other publications containing temperature-dependent properties, e.g., Barin and Knacke (*8*), Robie et al. (*9*), and Kubachewski et al. (*10*) also include suggested values for the properties of some of these chemicals, primarily for the condensed phases. (The numerous US Bureau of Mines publications by Pankratz, which appeared in the 1980's do not include any hydroxides.) Robie et al. (*9*) only tabulates information for the condensed phases of the calcium and magnesium dihydroxides. For calcium dihydroxide, the authors state that their results are based on an earlier JANAF evaluation, although they specifically give an additional reference (also used by JANAF) for the basis of the entropy value; for magnesium dihydroxide, the authors provide original literature citations for the thermal functions and an earlier NBS publication (*2*) for the enthalpy of formation. Barin and Knacke (*8*) and Kubaschewski et al. (*10*) provide information on the condensed phases for the dihydroxides (with the exception of the strontium and radium) and the BeOH(g) species. These three publications typically refer to existing reviews, unless newer data are available. There also exists an International Atomic Energy Agency (IAEA) publication by Spencer et al. (*11*) for the beryllium species; this study uses the same references as the JANAF study. Of the various reviews that have been mentioned, all have relied on the same scant information for the calculation of the thermal functions, but have used slightly different sources for the formation properties.

More specific collections of data, on which the thermodynamic properties of these mono- and dihydroxides are based, are available. An example is the spectroscopic properties by Jacox (*12*), which contains information on CaOH, SrOH, and BaOH. The sources included here are the same as those that will be used for the upcoming critical review for the JANAF project, but were not available at the time of the previously mentioned reviews.

The rationale for the early study of the dihydroxides of Mg and Ca in the 1930's and 1950's is clear. These compounds, milk of magnesia and hydrated lime, respectively, and their corresponding oxides, were used extensively in day-to-day life. $Be(OH)_2$, through its possible role in propellant combustion, gained importance in the defense agency activities with the space program in the late 1950's.

Recent spectroscopic publications (*13–16*) provide a historical picture for the origins of study and importance of three alkaline earth monohydroxide species. CaOH, SrOH, and BaOH were thought to be important in the color of flames. Although these species were observed in 1823, the actual molecular species were not predicted until the mid-1950's. Most quantitative evidence was of the indirect variety (except for some IR and ESR studies) until the 1980's when the species were formally produced, observed, and characterized. Studies on BeOH, performed in support of several US government agencies, were initiated to explain propellant combustion problems in the late 1950's. Then, in support of the space program, BeOH and MgOH were studied because of their presumed importance in propellant combustion processes. The heavier metal compounds were of lesser importance to the Air Force at that time and were not added to the JANAF collection until the mid 1970's. At that time, these species had gained importance through studies of the stellar atmosphere and the monitoring of chemical species in the upper atmosphere.

The early studies of the monohydroxides, as included in the JANAF Thermochemical Tables and elsewhere, were based on estimates of the structure and the vibrational frequencies and experimentally-derived enthalpies of formation. The early formation measurements for BeOH were reported in a series of defense agency progress reports. Since that time in the 1960s, BeOH has not been studied experimentally except for an ESR study (19). All recent work has dealt with theoretical calculations for the structure, vibrational frequencies, and the enthalpy of formation.

Work in the 1980's by Harris and Bernath groups (13–16) on the vibrational frequencies and the geometry and Murad (17) on the dissociation energies has changed the thermodynamic picture from an estimated description to a picture based on solid experimental evidence. Having a more solid footing for four monohydroxides, we are now in a favorable position to estimate properties for the radium species and evaluate the theoretically-derived properties of the beryllium species.

Table 1 presents a summary of the calculated values for the entropy and enthalpy of formation of BeOH(g) and MgOH(g), as recommended by the JANAF staff since 1960. The numbers in parentheses are the uncertainties of the recommended values. It is important to recognize that the entropy values are all dependent on the estimation scheme used to determine the structure and vibrational frequencies. There was no experimental data on which to base these results. Although the early estimates invoked a bent structure, later estimation schemes were tied to the experimental studies of the alkali metal monohydroxides, which had been determined to be linear in the late 1960's. The uncertainty in these values is of the order of 10 or more J/K/mol, so that all values are essentially the same. The 4/86 value for MgOH is based on the experimental study of Harris, as reported by Kinsey-Nielson et al. (16); the uncertainty being an order of magnitude less than the estimated values. Additional changes that occurred in these evaluations are reflected in the changes of the enthalpy of formation for these two species. The changes in the enthalpy of formation values are based on preliminary progress report information and comparison-type information. Note that the values from one evaluation to the next often exhibit changes outside the estimated uncertainty.

The data on the remaining monohydroxides through the early 1980's also were based on estimated structures and vibrational frequencies. The enthalpies of formation were inferred from information dealing with analyses of flames. The differences in the reported recommended values

Table 1
Thermodynamic Properties at 298.15 K and 1 bar

| Entropy, J/K/mol | | Date | Enthalpy of Formation, kJ/mol | |
| --- | --- | --- | --- | --- |
| BeOH(g) | MgOH(g) | | BeOH(g) | MgOH(g) |
| 223.1(20) (bent) | 241.7(20) (bent) | 12/60 | −172(63) | −51(42) |
| 223.1(20) (bent) | 221.5(13) (linear) | 2/63 & 6/67 | −105(42) | −218(84) |
| 209.6(13) (linear) | 226.5(8) (linear) | 12/75 | −115(42) | −165(38) |
| | 230.7(2) (linear) | 4/86 | | |

for the monohydroxides all stem from differences in the estimation scheme used and the interpretation of the flame spectral information.

During the course of the three updates to the JANAF Thermochemical Tables, the changes that appear for the BeOH and MgOH species are all geared to changes in the preferred method for estimating the missing quantities or interpretation of sparse data. In addition, the early assumption that BeOH is bent is now felt to be correct. Unfortunately, in the intermediate years, we convinced ourselves that the molecule was linear. The recent experimental studies show that four of the monohydroxides— Mg, Ca, Sr, and Ba—are linear in their ground state. However, it appears that MgOH is quasilinear, suggesting that BeOH is bent. This conclusion is supported by some quantum mechanical calculations. In addition, the information would suggest that a reliable assumption is that RaOH is linear.

A comparison of the molecular and structural information for BaOH(g) is given in Table 2. This is the information needed for the calculation of a thermochemical table. Recall that all differences are because of the estimation scheme used. Calculations based on the experimental results of Kinsey-Nielson et al. (*16*) are included.

It is interesting to note the difference in the selected input by the various reviewers. The resulting entropy values for the three reviews are within each other's uncertainty, which is of the order of 10 J/K/mol. The experimentally based results (*16*) have the smaller uncertainty (approximately an order of magnitude less). This is owing to the fact that entropy calculation now is based on well-defined experimental data.

The calculated entropies (at 298.15 K and 1 bar) for four of the monohydroxides are given in Table 3. In all cases, the studies by the critical reviewers assumed the molecules to be linear. The experimental measurements confirmed the linear structure. Within the large uncertainty of all the values derived from estimation schemes, the listed values are the same.

The situation for the enthalpy of formation has improved greatly because of the use of a more direct experimental technique. The early studies dealt with the extraction of data from the analysis of flames. The results were indirectly obtained by the use of the hydroxide–water concentrations in the flames. Recent determinations were based on the production of the high temperature species, followed by a mass spec-

Table 2
Comparison of Molecular and Structural Information for BaOH (g)[a]

|  | Jackson (1971) | Janaf (1975) | Tpis (1981) | Bernath (1985) |
|---|---|---|---|---|
| $\nu_1$ (cm$^{-1}$) | 298 | 469 | 450 | 492.4 |
| $\nu_2$ (cm$^{-1}$) | 277 | 431 | 305 | 341.6 |
| $\nu_3$ (cm$^{-1}$) | 3700 | 3650 | 3700 | [b] |
| Ba–O(Å) | 2.40 | 2.17 | 2.20 | 2.201 |
| O–H(Å) | 0.96 | 0.96 | 0.96 | 0.923 |
| Angle | 180 | 180 | 180 | 180 |
| S° (J/K/mol) | 258.1 | 252.9 | 257.8 | 251.4 |

[a]At 298.15 K and 1 bar.
[b]Kinsey-Nielson, et al. (*16*) did not measure the $\nu_3$ frequency; in the calculation, an estimated value of 3650 cm$^{-1}$ was used.

trometric determination of the dissociation energy. At this time, there are reliable determinations of the dissociation energy of the MgOH, CaOH, SrOH, and BaOH (*17*). The dissociation energy value for BeOH rests on a theoretical calculation, whereas the corresponding value for RaOH is to be obtained from a study of the trends in the other four experimentally-determined results.

Assuming the necessary thermochemical information has been found in the publications just mentioned, the next concern is to appreciate the quality required for the given application. It is very important not to arbitrarily accept a thermochemical table without having an understanding of the effect of its uncertainty on the results of a calculation. Thus, one should carefully examine the uncertainties specified in the recommended values and assess whether such uncertainty can be tolerated in the application. Preferably, a sensitivity analysis should be used to better gauge the effect of any uncertainty owing to the thermochemical values and the need for additional experimental and/or theoretical study.

Time evolution knowledge in the characterization of the thermodynamic properties of the monohydroxide species provides an excellent backdrop for the understanding of the reliability of the thermochemical results. It also highlights the experimental and/or theoretical data that are needed for improved reliability in the recommended results. Such insight hopefully will permit the user to make more effective use of data from different sources. The selection of the set of values should be considered in light of the estimation scheme used and its purported reliability for the type of molecules under consideration. The choice of uncertainties in this case may be biased by the evaluator's experience with a particular technique. Thus, care must be taken of the value of the uncertainty and the potential bias owing to the estimation method used. Large changes in the estimated values are expected. As more related data become available, the foundations for the estimation schemes change. Subsequently, these changes are reflected in the adopted values. When experimental data eventually become available, the resulting thermodynamic information should be more firm with less uncertainty.

Table 3
Entropy at 298.15 K for Four Monohydroxides

|  | J/K/mol | | | |
|---|---|---|---|---|
|  | MgOH | CaOH | SrOH | BaOH |
| Jackson (1971) | 227.3 | 236.3 | 246.8 | 258.1 |
| JANAF (1975) | 226.5 | 235.5 | 246.5 | 252.9 |
| TPIS (1981) | 232.6 | 240.6 | 251.7 | 257.8 |
| Experimental | 230.7 | 233.9 | 243.9 | 251.4 |

Such historical information is being provided in the upcoming publications in support of the JANAF Thermochemical Tables. There will be a series of publications in which all reported information is to be summarized. Literature citations will be given (18), as well as tabular and graphic summaries of all available numerical data. The user will be able to observe the quantity and quality of the information. The proper use of the information for a given application should become easier as a better understanding of the values and the associated uncertainty becomes more obvious.

## REFERENCES

1. Rossini, F. D., Wagman, D. D., Evans, W. H., Levine, S., and Jaffe, I., *Natl. Bur. Stand. Circ.*, **500** (1952).
2. Wagman, D. D., Evans, W. H., Parker, V. B., Schumm, R. H., Halow, I., Bailey, S. M., Churney, K. L., and Nuttall, R. L., Thermodynamic Properties. *J. Phys. Chem. Ref. Data*, **11** (1982).
3. Kelley, K. K., *US Bur. Mines Bull.*, **584** (1960).
4. Kelley, K. K., and King, E. G., *US Bur. Mines Bull.*, **592** (1961).
5. Chase, M. W., Jr., Davies, C. A., Downey, J. R., Jr., Frurip, D. J., McDonald, R. A., and Syverud, A. N., *J. Phys. Chem. Ref. Data*, **14,** (1985).
6. Gurvich, L. V., Veits, I. V., Medvedev, V. A., et al., *Thermodynamic Properties of Individual Substances*, 3rd edition, vol. 1 (1978), vol. 2 (1979), vol. 3 (1981), vol. 4 (1982).
7. Jackson, D. D., *Thermodynamics of the Gaseous Hydroxides*, Lawrence Livermore Laboratory, UCRL-51137 (1971).
8. Barin, I., Knacke, O., and Kubaschewski, O., *Thermochemical Properties of Inorganic Substances*, Springer-Verlag, Berlin (1973); and Barin, I. and Knacke, O., *Thermochemical Properties of Inorganic Substances*, Supplement, Springer-Verlag, Berlin (1977).
9. Robie, R. A., Hemingway, B. S., and Fisher, J. R., U.S. Geological Survey Bulletin 1452, 456 pp (1978).
10. Kubaschewski, O. and Alcock, C. B., *Metallurgical Thermochemistry*, 5th edition, Pergammon, New York (1983).
11. Spencer, P. J., von Goldbeck, O., Ferro, R., Girgis, K., and Dragoo, A. L., *Atomic Energy Review* **4** (Special Issue) (1973).
12. Jacox, M. E., *J. Phys. Chem. Ref. Data* **17**, 269–512 (1988).
13. Nakagawa, J., Wormsbecher, R. F., and Harris, D. O., *J. Mol. Spectrosc.* **97**, 37–64 (1983).

14. Hilborn, R. C., Qingshi, Z., and Harris, D. O., *J. Mol. Spectrosc.* **97,** 73–91 (1983).
15. Brazier, C. R. and Bernath, P. F., *J. Mol. Spectrosc.* **114,** 163–173 (1985).
16. Kinsey-Nielsen, S., Brazier, C. R., and Bernath, P. F., *J. Chem. Phys.* **84,** 698–708 (1986).
17. Murad, E., *J. Chem. Phys.* **75,** 4080–4085 (1981).
18. Chase, M. W., *NBS Technical Note 1243* (1987).
19. Brom, J. M. and Weltner, W., *J. Chem. Phys.* **64,** 3894–5 (1976).

# Thermodynamic Modeling of Solution Phases and Phase Diagram Calculations

I. ANSARA

*Laboratoire de Thermodynamique et Physico-Chimie Métallurgiques, ENSEEG, BP 75, 38402 Saint Martin d'Hères Cédex, France*

## ABSTRACT

In recent years, the development of phase diagram calculations for multicomponent systems using a thermodynamic approach has proven to be very important in resolving industrial problems. This development is mainly a result of the progress in the description of the thermodynamic behavior of solution phases, particularly of alloys, and of advances in computer software. Realistic statistical models have been developed to take into account chemical ordering in solution phases.

Moreover, methods of predicting thermodynamic stabilities from fundamental physical calculations have been developed that provide a sound basis for the thermodynamic description of metastable phases.

The significant progress in the modeling of the thermodynamic properties of solution phases is presented, together with some examples of evaluated and assessed phase diagrams.

**Index Entries:** Thermodynamics; mixtures; modeling; phase diagram calculations.

## INTRODUCTION

The study of phase diagrams has long been an important tool in the development of science and technology. In metallurgy, it has made a major contribution to the design of new alloys, partly owing to considerable improvements in experimental techniques that have reached a high level of accuracy. However, with increasing demands of new high technological materials, which are generally complex, it is important that theoretical predictions guide the work of materials scientists. One such possibility is to apply the principles of physical chemistry in order to compute multicomponent and multiphase equilibria. Such an approach

High Temperature Science, Vol. 26 © 1990 by the Humana Press Inc.

can be very powerful in both planning and in considerably reducing the number of costly experiments.

After the pioneering work of Meijering (1), followed by the important contributions of Kaufman (2), great progress has been observed, since the *NBS Workshop on Phase Diagrams*, in the application of calculated phase diagrams generated from the underlying thermochemical data. This is owing mainly to the improvement of models for the solution phases, and to an increasing amount of reliable and consistent assessed data that is now being generated. The calculation of phase diagrams has been used successfully in a wide range of applications, including crystal growth of semiconductor materials by liquid phase epitaxy, prediction of phase equilibria for superalloys, light metal alloys or high speed steels, and chemical vapor deposition. This has been possible because of the development of sophisticated application software, which is now being used in numerous thermochemical computer databanks throughout the world.

A further important aspect of phase diagram calculation is its use in the teaching of thermodynamics and materials science in general. Very simple computer programs have been used to demonstrate graphically how changing interaction parameters or lattice stabilities affects the phase diagram, whereas the more powerful application programs used in their simplest modes can be used to illuminate the deceptive simplicity of concepts, such as, Gibbs phase rule.

The calculation of phase diagram requires the knowledge of the thermodynamic properties of the pure components, the compounds, and solution phases, in both stable or metastable physical states, and, clearly, consistency of such data between different phase systems is very important.

## ELEMENTS

### Lattice Stabilities

Let us first consider the elements. Since many elements can dissolve in a phase that is not stable for that element itself (e.g., Cr(bcc) dissolves extensively in Ni(fcc)), it is necessary to estimate the thermodynamic properties of elements in states that are metastable at all temperatures. Kaufman (2,3) presented sets of estimated parameters for the enthalpy and entropy differences between the bcc and hcp and the fcc and hcp structures of the transition metals applying Richard's rule for the entropy of melting to many of the elements. Later, it was shown that this rule is not satisfactory for refractory metals, whose entropy of fusion increases with increasing melting temperature. However, Kaufman's values of the Gibbs energy differences, also called lattice stabilities, have been extensively used throughout the world in conjunction with solution phase data to compute phase diagrams of metallic systems.

Very recently, Saunders et al. (4), evaluated the lattice stabilities for the metastable fcc(A1), and bcc(A2), and hcp (A3) forms of 43 elements.

Their results are based on assessed stable phase data, phase boundary extrapolations from binary alloys, extrapolations from pressure–temperature phase diagrams, relationships between the crystal structure, the entropy of fusion and the melting temperature, stacking fault energies, and first principle electronic energy calculations.

For the transition metals, the lattice stabilities are now much closer in magnitude to *ab initio* predictions. This is owing mainly to the recent reassessment of stable phase data and the assumption that the entropies of fusion of metastable forms behave in a similar fashion as the stable one, and to advances in electronic energy calculations.

However, most early expressions for lattice stabilities implied that the heat capacities of the two phases involved are the same at all temperatures. This, in general, is not true for two phases even at their transition temperature, and Anderson et al. (5) described a better, but not perfect, approach used by the Scientific Group Thermodata Europe (SGTE). The heat capacity of the liquid phase of an element is assumed to approach that of the stable form at a temperature of about $0.5\ T_{fus}$, and, similarly, the heat capacities of all solid phases above the melting temperature should approach that of the liquid at higher temperatures. In this assumption, the difference in heat capacities at the melting temperature is taken into account, its effect being less important as the temperature difference with respect to the melting temperature increases. A better estimate of the lattice stability below the melting temperature could certainly be made if a reliable model for the thermodynamics of the transition from liquid to glass were available.

## Magnetic Heat Capacity

The magnetic contributions to the Gibbs energy for magnetic materials were not, until recently, treated explicitly. It was merely included in the overall expression of the Gibbs energy of transformation with respect to temperature.

Inden (6) proposed an empirical and approximate analytical formula for the magnetic specific heat that represents the experimental data below and above the critical temperature $T_c$ by the following equations

$$C_p = K_1\ R\ \ln\ ((1 + t^3)/(1 - t^3)) \qquad \text{for } t = T/T_c < 1 \qquad (1)$$
$$C_p = K_2\ R\ \ln\ ((1 + t^{-5})/(1 - t^{-5})) \qquad \text{for } t > 1$$

$K_1$ and $K_2$ are constants for an element in its ferromagnetic and paramagnetic states. They can be determined from experimental data and they are related to the total magnetic entropy

$$\Delta S^{mag} = R\ \ln\ (\beta + 1) \qquad (2)$$

where $\beta$ is the mean atomic moment expressed in Bohr magnetons.

The expressions of the Gibbs energy obtained by integration are rather lengthy and complicated. Hillert and Jarl (7) preferred to expand the expression of $C_p$ in power series and suggested the following expressions

$$C_p = 2 K_1 R (t^3 + t^9/3 + t^{15}/5) \qquad \text{for } t < 1 \qquad (3)$$

$$C_p = 2 K_2 R (t^{-5} + t^{-15}/3 + t^{-25}/5) \qquad \text{for } t > 1$$

Recently, Chuang et al. (8) also suggested an expression for the heat capacity owing to magnetic ordering, which is given by the following equations

$$C_p = K'_1 t \exp (-4 (1 - t)) \qquad \text{for } t < 1 \qquad (4)$$

$$C_p = K'_2 t \exp (8 p (1 - t)) \qquad \text{for } t > 1$$

where $p = 1$ for bcc lattice and $p = 2$ for the fcc-lattice. The quantities $K'_1$ and $K'_2$ are related to the entropy of magnetic disordering. Figure 1 compares the magnetic heat capacity calculated from Eq. 2, 3, and 4 for bcc-iron. The results, in general, are similar except in the vicinity of the critical temperature.

### Pressure Dependence

Outside the geological field, rather little work has been undertaken to include the pressure dependence of the Gibbs energy of the pure elements. Fernandez Guillermet (9–12) has recently reassessed the thermodynamic properties of iron, molybdenum, cobalt, and zirconium, whereas Gustafson reevaluated those of tungsten (13). Figure 2 shows an assessed pressure–temperature diagram for iron.

## COMPOUNDS

In a series of papers that eventually led to the publication of a book, Miedema et al. (14), developed a semiempirical approach to treat the energy effects in metallic systems as interfacial energies generated at the contact interfaces between neighboring atomic cells. For binary alloys, the enthalpy of formation is proportional to the sum of two terms, a negative term proportional to the square of the chemical potential for electronic charge differences of the pure metals, X, and a positive repulsive contribution that is related to the difference of the electron densities of these elements at the Wigner-Seitz cell boundary $n_{ws}$, according to the following equation

$$\Delta_f H \ \alpha = P (\Delta X)^2 + Q (\Delta n_{ws}^{1/3})^2 \qquad (5)$$

P and Q are constants that are derived from fits to the experimental enthalpies of formation. This expression can be converted to a function of composition and can be applied not only to ordered compounds but also to liquid alloys.

Watson and Bennett (15) used a simple band theory model to predict the enthalpy of formation of 276 transition metal alloys at the equiatomic composition. Some of the input parameters, namely the bandwidth, the

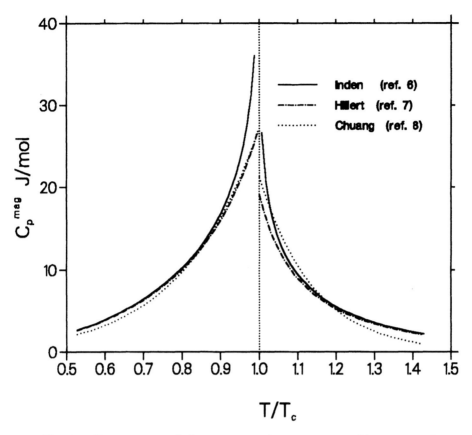

Fig. 1. Comparison of the magnetic heat capacity of bcc–iron calculated from Eqs. 1, 3, and 4.

Fermi level position, and the number of electrons in the band, are
allowed to vary within certain constraints in order to approximate known
values of the enthalpy of formation. More recently, Colinet et al. (16) also
developed a simple electron band theory model to predict enthalpies of
formation of transition metal alloys as a function of composition. They
used a tight-binding model considering the moments of the density of
states and calculated the enthalpies of formation for 210 binary alloys.
They also applied this approach to transition-rare earth metal alloys (17).

These approaches are very useful to predict the enthalpy of formation of metastable solid alloys of given composition. In the study of
multicomponent systems, the knowledge of the enthalpy of formation of
metastable as well as stable compounds is needed. For example, the
thermodynamic description of $(Cr,Fe)_{23}C_6$ or $(Cr,Fe)_7C_3$ carbides using a
sublattice model requires the enthalpy of formation of $Fe_{23}C_6$ and $Fe_7C_3$,
which are unstable in normal conditions, and they may be obtained by
such methods.

## SOLUTION MODELS FOR METALLIC SYSTEMS

Solution theories have also been the subject of great progress and
lead now to reasonable representations of the thermodynamic behavior

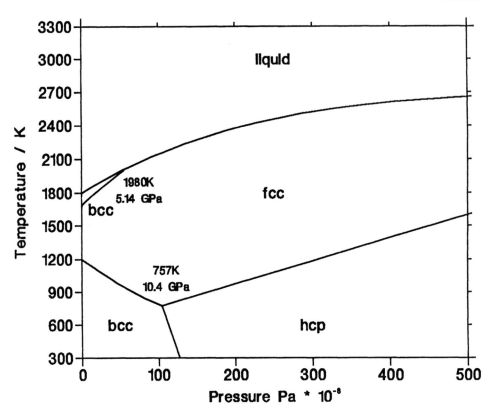

Fig. 2.   Pressure-temperature diagram for Fe (9).

of multicomponent phases. For many years, simple power series expansions have been used to describe the thermodynamic behavior of solution phases as well as the quasichemical approximation that, to a certain extent, introduced the concept of ordering in the solution phase. A great variety of empirical equations based on geometrical weighting have been used and have been reviewed by the present author (18). For most of the substitutional solutions, a simple power series expansion is now preferred, because it allows a very satisfactory representation of the thermodynamic properties of mixing with respect to composition.

## Magnetic Contributions

For solutions that are not ferromagnetic at a temperature below the Curie temperature of one of the constituents, the thermodynamic properties are referred to a hypothetical paramagnetic state of the phase where the magnetic moments are completely disordered. Hence, the Gibbs energy of mixing can be expressed by a sum of two terms, one describing the Gibbs energy of mixing and the second describing the ferromagnetic effect for the alloy.

The heat capacity for a solution phase describing the magnetic ordering contribution can be expressed, according to Hertzman et al. (19), by an equation identical to (3) but with the composition dependence of $T_c$

expressed by a power series with respect to the molar fraction, for example

$$T_C = x_A \, T_{C,A} + x_B \, T_{C,B} + x_A x_B \, (T_C^o + T_C^1 \, (x_A - x_B)) \qquad (6)$$

$$\beta = x_A \, \beta_A^{o} + x_B \, \beta_B^{o} + x_A x_B \, (\beta_{A,B})$$

The parameters $T_C^o$, $T_C^1$, $\beta_{A,B}$ are evaluated from experimental information. $T_{C,A}$, $T_{C,B}$, and $\beta_A^o$, $\beta_B^o$ are, respectively, the Curie temperature and the mean atomic moments expressed in Bohr magnetons of the pure elements. Figure 3 illustrates an isothermal section of the Fe–Co–Zn where the effect of magnetic ordering is taken into account for the calculation of the phase boundaries (20).

### Sublattice Model

Many binary systems exhibit intermediate phases with a fixed number of sites having a narrow range of nonstoichiometry. In phase diagram calculations, these compounds were generally assumed to be "line compounds." However, it may be necessary to describe the temperature dependence of the composition, when alloying elements are added to these compounds. The sublattice model developed by Hillert et al. (21) based on Temkin's model for ionic solution (22) and extended by Sundman et al. (23) to take into account more than two types of sites that can be represented by the general formula

$$(A_1, B_1, C_1, \ldots)_a (A_2, B_2, C_2, \ldots)_b \ldots (A_m, B_m, C_m, \ldots)_n$$

where a site fraction for each constituent in every sublattice is defined, their sum being equal to 1 for each lattice. a, b, . . . n represent the number of sites. For example, III-V compounds, reciprocal ionic or interstitial solutions, respectively, of the type $(Ga,In)(As,P)$, $(Na^+,K^+)(Cl^-,F^-)$, or $(Cr,Fe)(C,v)$ can easily be represented by two sublattices.

The number of sublattices for solid phases and the elements, including vacancies or ions that can occupy them, is generally obtained from structural information.

The Gibbs energy of formation of such phases requires the knowledge of the Gibbs energy of formation of the compounds formed by combining a constituent in one sublattice with a constituent in the other sublattices. It should be noted that to describe the thermodynamic behavior of the phase $(Cr,Fe)(C,v)$, the Gibbs energy of the hypothetical compound FeC and CrC are required. These Gibbs energies are generally obtained by means of optimization procedures that will be discussed further on.

Solid phases may have more complex structures, such as $\sigma$ or $\mu$-phases, and generally some simplifications are made by reducing the number of sublattices to avoid increasing the number of parameters.

The sublattice model has also been applied to describe order–disorder transformations in the Al–Ni (24) and Al–Ti (25) systems. The ordered phase, such as the fcc–$L1_2$ in the Al–Ni can be described by a model with two sublattices, both of which contain Al and Ni. Mathemati-

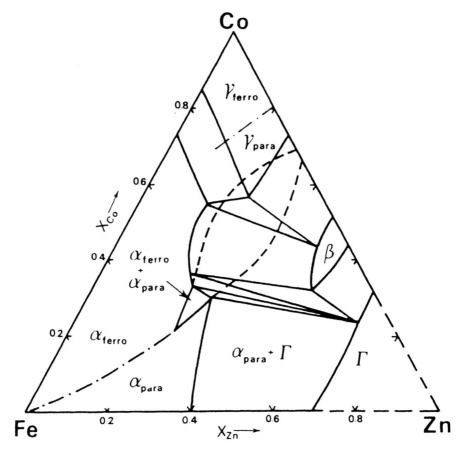

Fig. 3.  Isothermal section at 1036 K of the Fe–Co–Zn system (*20*).

cal constraints are introduced in order to represent both disordered fcc–A1 and ordered fcc–L1$_2$ phases with the same equation. The same model was applied to the Al$_3$Ni$_2$ phase, which has a D5$_{13}$ structure. The calculated diagram is shown in Fig. 4.

## Cluster Variation Method

The cluster variation method (CVM) developed by Kikuchi (*27*) is now being extensively used to calculate phase diagrams of systems exhibiting ordered phase regions. In this method, the distribution variable is a basic cluster of lattice points. For face-centered-cubic or body-centered-cubic phases, the basic cluster can be a four-point tetrahedron. The Gibbs energy of the system is expressed in terms of these basic distribution variables. If the formulation is done using the pair approximation, the entropy of the CVM is equivalent to that of the quasichemical model.

De Rooy et al. (*28*) used that model to calculate the fcc–part of the Cu–Ni–Zn system, the pairwise interaction potentials being obtained from a pseudopotential model. This method has also been applied to determine isothermal sections for coherent equilibria in Cu–Ag–Au by

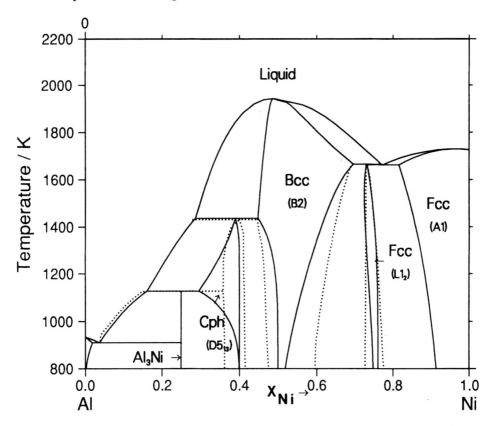

Fig. 4. Comparison between experimental (-----) (26) and calculated (—) Al–Ni phase diagram.

De Fontaine (29), where both clustering equilibria and ordering and their mutual interactions have been calculated from an analysis of the corresponding binary phase diagrams (Fig. 5).

## Monte Carlo Simulations

More sophisticated statistical methods are the Monte Carlo simulations, which are also being developed to describe order–disorder phenomena, usually based on the Ising model using pair interaction energies as phenomenological parameters. In a recent paper, Binder (30) reviewed the various methods of statistical mechanics that have been applied to describe atomic interactions for binary metallic alloys and compared them to each other, emphasizing the merits as well as the limitations of, for example, the cluster variation method and Monte Carlo simulation.

## Associate Model

An analysis of the experimental results shows that for certain metallic solutions, or even salt systems, the enthalpy of formation as a function of composition presents strongly asymmetrical or even triangular shaped curves. This was attributed to a tendency to form compounds in the melt

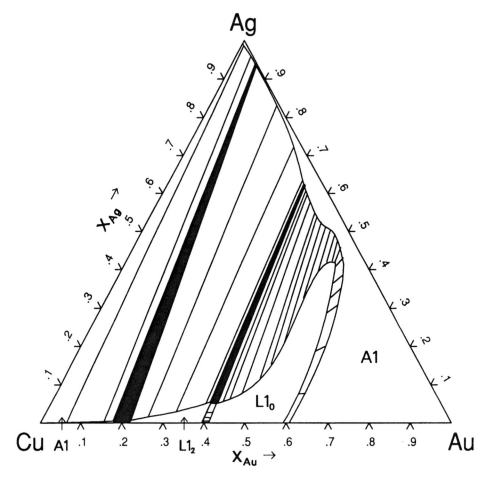

Fig. 5.   Isothermal section of Cu–Ag–Au coherent phase diagram at 513 K.

such as in the group II-VI or IV-VI systems. The associate model was basically developed by Dolezakek (*31*) followed by Prigogine et al. (*32*); more recently, Jordan (*33*) and Sommer (*34*) have both applied it to metallic melts. The model assumes that, in the melt, free atoms of the pure elements coexist with associates and has been very successful in representing the enthalpy of formation of such systems.

The derived expressions introduce a term that corresponds to the enthalpy of formation of the associate, whose value has to be adjusted to experimental results, and another that describes the mixing of the different particle types into an associated solution. Similarly, the entropy of mixing contains terms expressing ideal behavior between the free atoms and the associates, the entropy of formation of the associate, and the excess entropy of mixing. The stoichiometry of the associates is generally taken to be the same as a stable solid compound.

## Solution Models for Oxide Systems

In the last few years, an increased effort has been made in modeling the high temperature thermodynamics properties for silica-containing

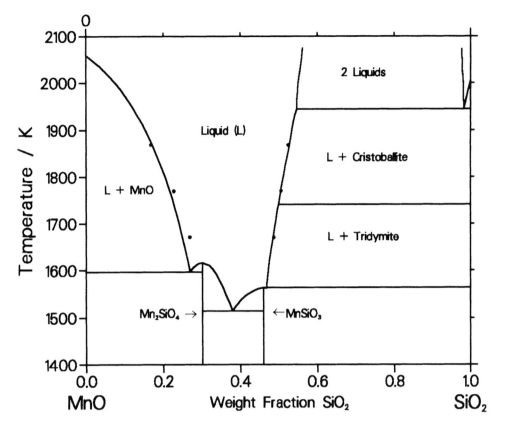

Fig. 6. Calculated MnO–SiO$_2$ phase diagram. (42).

oxide melts. These melts are characterized by a complex ionic structure, and little experimental information concerning them is available.

As for associated metallic solutions, the enthalpy of mixing tends to exhibit a negative triangular shaped peak near the composition of maximum ordering, whereas the entropy of mixing has an "m"-shape with a minimum near this composition. Among the numerous models that have been developed, one of the most recent is owing to Gaye et al. (35), who used a cellular model based on Kapoor's (36) description of the melt.

To account for the high ionization tendency in ionic solutions, Hillert et al. (37) extended the sublattice model to off-stoichiometric compositions by introducing neutral atoms and charged vacancies into the anion sublattice. This model has recently been used in assessments of the thermodynamic properties of the MgO–SiO$_2$ (38) and CaO–SiO$_2$ (39) systems.

It should be noted that for binary systems, their equation is identical to that of the associate solution model if the stoichiometric number of the anion is taken equal to 1. However, for higher order systems, the two models cannot be made identical.

Pelton et al. (40) used a modified quasichemical model, in order to describe molten silicate systems. Their formalism applies to both ordered and disordered systems. They calculated the MnO–SiO$_2$ (41) phase diagram over the entire composition range, as shown in Fig. 6. Michels et al.

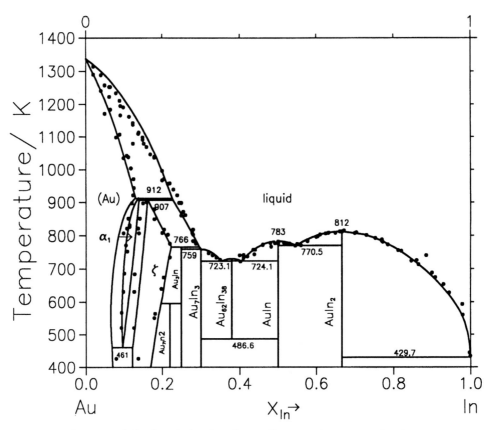

Fig. 7.   Calculated Au–In phase diagram (● exp. points).

(42) have developed a network model and described the variation of the interaction parameter with composition by assuming a functional dependence of the degree of polymerization of the silicate network. This model was used recently to describe the thermodynamic behavior of the CaO–SiO$_2$, MgO–SiO$_2$, and Al$_2$O$_3$–MgO–SiO$_2$ systems (43).

The conformal ionic solution theory, developed by Blander et al. (44–46) has proved to be very successful in calculating the thermodynamic properties for additive and for charge symmetric and asymmetric reciprocal salt ternary systems. Bale et al. (46) extended this theory to reciprocal quaternary systems of the type Li$^+$, Na$^+$, K$^+$ ‖CO$_3^{2-}$, So$_4^{2-}$, or Na$^+$, K$^+$ ‖CO$_3^{2-}$, OH$^-$, SO$_4^{2-}$. The excess Gibbs energy is expressed in terms of equivalent ionic fractions.

## OPTIMIZATION

The need to achieve consistency between phase diagram data and thermodynamic properties of various phases forming a system has led to developments of optimization procedures such as those developed by Lukas et al. (47), Pelton et al. (48), and Jansson (49). These procedures calculate optimized values of the interaction parameters of the solution models by taking into account all available experimental information,

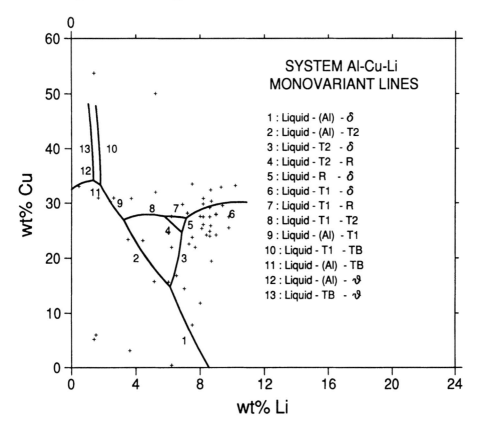

Fig. 8. Calculated monovariant lines in the Al–Cu–Li system (+ selected alloy compositions).

such as phase diagram data, enthalpy of mixing, partial Gibbs energies, and heat contents, various models for the solution phases being built into the program. In certain cases, with a limited knowledge of the thermodynamic behavior of certain phases, missing data may be evaluated when combined with phase boundary information. Figure 7 shows an example of a thermodynamically assessed binary phase diagram of the Au–In system (50). Figure 8 shows an optimized projection of the monovariant lines in the Cu–Li–Mg system (51), taking into consideration the liquidus temperatures obtained by differential thermal analysis, the enthalpies of formation of the different ternary compounds being derived from calorimetric measurements.

## CONCLUSION

Pertinent data are now generated in many countries, and the task of assembling, compiling, and collating data on thermochemical properties and phase diagram data is enormous. Much of this task is now being organized among different groups, on an international scale, for instance, by the CODATA Task Group of Thermochemical Tables (CTT), the Scientific Group Thermodata Europe (SGTE), the Alloy Phase Diagram International Commission (APDIC), and many other national endeavors. A better coordination of the work of critical evaluation and assessment still has to be organized in order to share this important task on an international scale.

## REFERENCES

1. Meijering, J. L., *Acta Metall.* **1**, 257 (1957).
2. Kaufman, L. and Bernstein, H., *Computer Calculation of Phase Diagrams*, 1970, Academic Press, NY.
3. Kaufman, L., *Phase Stability in Metals and Alloys*, Rudman, P. S., Stringer, J., and Jaffee, R. I., eds., McGraw Hill, NY, 1967, p. 125.
4. Saunders, N., Miodownik, A. P., and Dinsdale, A. T., *Calphad* **12**, 4, 351 (1988).
5. Andersson, J. O., Sundman, B., Fernandez-Guillermet, A., Gustafson, P., Hillert, M., Jansson, B., Jönsson, B., Sundman, B., and Agren, J., *Calphad* **11**, 1, 93 (1987).
6. Inden, G., *Z. Metallkde* **66**, 577 (1975).
7. Hillert, M. and Jarl. M., *Calphad* **2**, 3, 227 (1978).
8. Chuang, Y. Y., Schmid, R., and Chang, Y. A., *Metall. Trans. A.* **16A**, 153 (1985).
9. Fernandez Guillermet, A. and Gustafson, P., *High Temperature-High Pressures* **16**, 591, (1985).
10. Fernandez Guillermet, A., *J. of Thermophysics* **6**, 4, 367 (1985).
11. Fernandez Guillermet, A., *J. of Thermophysics* **8**, 4, 367 (1987).
12. Fernandez Guillermet, A., *High Temperatures-High Pressures* **19**, 119 (1987).
13. Gustafson, P., *Int. J. Thermophysics* **6**, 395 (1985).
14. de Boer, F. R., Boom, R., Mattens, W. C. M., Miedema, A. R., and Niessen, A. K., *Cohesion in Metals: Transition Metal Alloys*, vol. 1, de Boer, F. R. and Pettifor, D. G., North Holland (1988).
15. Watson, R. E. and Bennett, L. H., *Calphad* **5**, 1, 25 (1981).
16. Colinet, C., Pasturel, A., and Hicter, P., *Calphad* **9**, 1, 71 (1985).
17. Colinet, C. and Pasturel, A., *Calphad* **11**, 4, 323 (1987).
18. Ansara, I., *Int. Metal Rev.* **1**, 20 (1979).
19. Hertzman, S. and Sundman, B., *Calphad* **6**, 1, 67 (1982).
20. Nishizawa, T., Hasebe, M., and Ko, M., *Proc. Calphad VIII* Stockholm, 113 (1979).
21. Hillert, M. and Staffanson, L. I., *Acta Chem. Scand.* **24**, 3618 (1970).
22. Temkin, M. *Acta Phys. Chim.* **20**, 411 (1945).
23. Sundman, B., and Agren, J., *J. Phys. Chem. Solids* **42**, 297 (1981).
24. Ansara, I., Sundman, B., and Willemin, P., *Acta Metall.* **36**, 4, 977 (1988).
25. Gros, J. P., Sundman, B., and Ansara, I., *Scripta Met.* **22**, 1587 (1988).
26. Hansen, M. and Anderko, K., *Constitution of Binary Alloys*, McGraw-Hill, NY (1958).
27. Kikuchi, R., *Acta Metall.* **25**, 195 (1977).
28. de Rooy, A., Van Royen, E. W., Bronsveld, P. M., and de Hosson, J. Th. M., *Acta Metall.* **29**, 1339 (1980).
29. de Fontaine, D., *Physica* **103B**, 57 (1981).
30. Binder, K., *Adv. Solid State Phys.* **26**, 133 (1986).
31. Dolezalek, F., *Z. Phys. Chem.* **64**, 727 (1908).
32. Prigogine, I. and Defay, R., *Thermodynamique Chemique*, Dunod, Paris (1950).
33. Jordan, A. S., *Metall. Trans.* **1**, 239 (1970).
34. Sommer, F., *Z. Metallkde* **73**, 2, 72 (1982).

35. Gaye, H. and Welfringer, J., *Second International Symposium on Metallurgical Slags and Fluxes*, Fine, H. A., and Gaskell, D. R., eds. Lake Tahoe, 357 (1987).
36. Kapoor, M. L. and Frohberg, M. G., *Chemical Metallurgy of Iron and Steel*, The Iron and Steel Inst., UK (1971).
37. Hillert, M., Jansson, B., Sundman, B., and Agren, J., *Metall. Trans.* **16A,** 261 (1985).
38. Hillert, M., Sundman, B., and Wang, X., *Report TRITA-MAC-0381*, Mat. Res. Center, The Royal Inst. of Technology, Stockholm, Sweden (1988).
39. Hillert, M. and Wang, X., *Report TRITA-MAC-0392*, Mat. Res. Center, The Royal Inst. of Technology, Stockholm, Sweden (1989).
40. Pelton, A. D. and Blander, M., *Metall. Trans.* **17B,** 805 (1986).
41. Pelton, A. D. and Blander, M., *Calphad* **12,** 1, 97 (1988).
42. Rao, B. K. and Gaskell, D. R., *Met. Trans.* **12B,** 2, 311 (1981).
43. Michels, M. A. J. and Wesker, E., *Calphad,* **12,** 2, 111 (1988).
44. Blander, M. and Yosim, S. J., *J. Phys. Chem.* **39,** 2610 (1963).
45. Saboungi, M. L. and Blander, M., *J. Amer. Ceram. Soc.* **58,** 1-2, 1 (1975).
46. Saboungi, M. L. and Blander, M., *J. Chem. Phys.* **73,** 5800 (1980).
47. Bale, C. W. and Pelton A. D., *Metall. Trans.* **17B,** 805 (1986).
48. Lukas, H. L., Henig, E. Th. and Zimmermann, B., *Calphad* **1,** 225 (1977).
49. Bale, C. W., Pelton, A. D., and Thompson, W. T., *F*A*C*T, User's Guide Suppl.*, FITBIN, 1981, McGill Univ. Ecole Polytechnique, Montréal, Québec, Canada.
50. Jansson, B., Ph.D. thesis, 1984, Div. Phys. Met., Royal Institute of Technology, Stockholm, Sweden.
51. Ansara, I. and Nabot, J. P., unpublished work.
52. Dubost, B., Colinet, C., and Ansara, I., *Proc. Vth International Aluminium-Lithium Conf.*, Williamsburg, VA, March 1989, in press.

# F*A*C*T Thermochemical Database for Calculations in Materials Chemistry at High Temperatures

ARTHUR D. PELTON,*,1 WILLIAM T. THOMPSON,2 CHRISTOPHER W. BALE,1 AND GUNNAR ERIKSSON1

1Centre for Research in Computational Thermochemistry, Ecole Polytechnique, PO Box 6079, Station A, Montreal, Quebec, Canada H3R 3A7; and 2Royal Military College of Canada, Kingston, Ontario, Canada H7L 2W3

## ABSTRACT

This paper summarizes the software and thermodynamic data that are offered through the online F*A*C*T thermochemical database system. Particular emphasis is placed on the treatment of materials at high temperatures. In the F*A*C*T system, extensive databases are being prepared for the thermodynamic properties of a large variety of multicomponent solutions. Models are used to calculate the thermodynamic properties of multicomponent phases from evaluated data of the binary and ternary subsystems. A Unified Interaction Parameter Formalism has been developed to treat dilute solutions, such as steels. A modification to the Quasichemical Model is employed to represent structured liquids such as silicates. Ionic systems such as molten salts are treated by the Sublattice Model.

**Index Entries:** Thermodynamics; database; materials chemistry; software; pyrometallurgy.

## INTRODUCTION

Computerized thermochemical data and associated software for the calculation of chemical equilibria can be powerful tools in studying the chemistry of materials at high temperatures. Reliable thermodynamic data, which are retrieved and manipulated by computer programs, enable scientists and engineers to analyze chemical reactions and complex

*Author to whom all correspondence and reprint orders should be addressed.

equilibria of metals production, electronic and ceramic materials development, alloy design, coatings technology, process analysis, nuclear waste disposal, and so on.

Such calculations require

1. State-of-the-art computer software;
2. Self-consistent critically assessed thermodynamic data;
3. A vehicle to make the above accessible to the scientific and engineering community.

These conditions are satisfied by the F*A*C*T (Facility for the Analysis of Chemical Thermodynamics) online thermochemical database (1,2). Originally developed as a research tool for chemical metallurgists, F*A*C*T is now employed in many diverse fields of chemical thermodynamics by chemical engineers, corrosion engineers, organic chemists, geochemists, ceramists, electrochemists, and so on.

This paper outlines the software and databases of F*A*C*T, with particular emphasis placed on the programs and thermodynamic solution models employed in treating materials at high temperatures.

## F*A*C*T THERMOCHEMICAL DATABASE

F*A*C*T—Facility for the Analysis of Chemical Thermodynamics (1,2)—is a fully integrated thermochemical database which couples proven state-of-the-art software with self-consistent critically assessed thermodynamic data. International communications networks (Datapac, Tymnet, Telenet) make the on-line database readily accessible to the scientific and engineering community.

A summary of the F*A*C*T computer programs and thermodynamic databanks is given in Table 1. With the aid of these programs and thermodynamic data, one is able to: search for and display data on compounds and solutions at any temperature and composition; calculate thermodynamic property changes for any type of chemical reaction; perform heat balances; store and use private data; calculate vapor pressures; compute complex chemical equilibria; calculate and plot E-pH and predominance diagrams; calculate (or estimate) and plot binary and ternary phase diagrams; predict multicomponent multiphase complex chemical equilibria involving alloys, slags, molten salts; and so forth.

The benefits to a user of F*A*C*T are convenience, reliability of self-consistent calculations, and immediate access to the most recent thermodynamic data and programs.

## PURE SUBSTANCES AND IDEAL SOLUTIONS

In the literature, there are many algorithms that enable one to manipulate thermodynamic data in order to do such things as calculate phase diagrams, study chemical reactions, perform heat balances, calculate solubilities, and calculate solid-gas equilibria. However, the most

Table 1
F*A*C*T* Thermochemical Programs and Databanks

Programs
| | |
|---|---|
| MAIL | Messages to and from the manager |
| INSPECT | Search and list private or system pure substances database |
| DATAENTRY | Entry and storage of private compound data |
| REACTION | Extensive state property calculation for chemical reactions |
| GASMIX | Potential constrained gas mixture calculation |
| PREDOM | Calculate and plot isothermal predominance diagram for up to 7 elements |
| EPH | Calculate and plot E-pH Pourbaix diagram containing up to 6 elements |
| EQUILIB | SOLGASMIX-based Gibbs-energy minimization incorporating SAGE |
| POTCOMP | T or P vs composition phase diagram |
| FITBIN | Binary thermodynamic solution properties and phase diagram optimization |
| TERNFIG | Calculate and plot ternary phase diagram |
| RECIPFIG | Calculate and plot reciprocal ternary phase diagram |
| THERMDOC | Bibliographic database with 30,000 citations on thermodynamics and phase diagrams |
| SOLUTION | Entry and storage of private solution data and solution models |

Databanks
  Over 4,000 inorganic stoichiometric compounds (5,000 phases), including aqueous and gaseous ions
  Molten slag solution: $FeO$ - $MgO$ - $SiO_2$ - $MnO$ - $CaO$ - $Na_2O$ - $Al_2O_3$ - $K_2O$ - $ZrO_2$
  Liquid steel dilute solution database with 30 solutes
  Molten salt solution: $Li^+$, $Na^+$, $K^+$, and $F^-$, $Cl^-$, $CO_3^=$, $SO_4^=$, $NO_3^-$, $OH^-$ (9 components)
  Molten salt solution: $Li^+$, $Na^+$, $K^+$, $Rb^+$, $Cs^+$ and $F^-$, $Cl^-$, $Br^-$, $I^-$ (8 components)
  Several solid ceramic solutions
  THERMDOC bibliographic database with 30,000 citations
  Organic databases for several hundred binary organic systems

Under development
  Solid steel dilute solution database
  Liquid Cu, Al, and Ni dilute solution databases
  Liquid sulfides (mattes) solution databases
  More components added to the slag and salt solution databases
  Concentrated alloy databases

important ingredients in this type of calculation are reliable and self-consistent thermodynamic data.

The F*A*C*T pure substances databank has stored thermodynamic data for pure elements and compounds for over 5,000 species. Among other things, the following thermodynamic data are stored for each phase: the standard enthalpy of formation, $\Delta H°$ (298 K); the standard entropy, $S°$ (298 K); an expression for the heat capacity, $C_p$ (T); and the temperature range of $C_p$ (T). The data are taken primarily from standard compilations (3–8).

An example of the data stored in the pure substances databank is shown in Fig. 1, where the program INSPECT has listed the thermodynamic data for NiO. From Fig. 1, it is seen that NiO has three solid allotropes: s1, s2, and s3. The temperatures (and enthalpies) of transformation for NiO(s1) to NiO(s2) and for NiO(s2) to NiO(s3) are 525 K (42 J) and 565 K (42 J), respectively.

Figure 2 illustrates the use of the program REACTION (1,9), where the partial pressure of $O_2(g)$ in equilibrium with Ni(s)/NiO(s) mixtures is calculated. This mixture is often employed as the reference electrode in solid electrolyte oxygen sensors. The user has the entered the appropriate chemical reaction. The values in parentheses below each species correspond to "temperature (K), pressure (atm), and phase (allotrope)," respectively. REACTION automatically retrieves the thermodynamic data, including any private data, for all the species and performs the necessary thermodynamic calculations.

In the spreadsheet format of Fig. 2, among other things, the user has determined that the equilibrium partial pressure of oxygen above Ni(s)/NiO(s) mixtures is $P(O_2) = 10^{-9}$ atm at 1375.6 K.

The relative stabilities of materials in various chemical environments at high temperatures can be represented by predominance area diagrams. Figure 3 illustrates the use of the program PREDOM (1,10), where the stabilities of Ca-bearing materials in gas mixtures containing $SO_2$, $O_2$, CO, and $CO_2$ in the temperature range 1100–1200 K are plotted. CaO is often employed to remove undesirable $SO_2(g)$ emissions from fluidized-bed roasters and stack gases. Figure 3 is a two-dimensional isothermal slice of the Ca-C-O-S quaternary system.

The actual composition of the gas phase at any point on the diagram can be computed as shown in Fig. 4, which is a partial listing made by the program PREDOM (1,10) following the calculation of the diagram in Fig. 3. The algorithm which calculates the gas phase composition is based on the "Constrained Chemical Potential Method" (11).

The pure substances databank can be used for calculations involving ideal gas mixtures and pure condensed phases. For example, Figs. 5a, b illustrate a calculation by the Gibbs-energy minimization program EQUILIB (2), which is based on an extended up-to-date version of the SOLGASMIX algorithm (12). Gibbs-energy minimization calculations are powerful tools in treating complex chemical equilibria. The user specifies the reactants together with the conditions of the reaction (temperature, pressure, and so on). EQUILIB then computes the most stable (minimum

```
FORMULA: NI*O
NAME: NICKEL MONOXIDE
FORMULA WEIGHT:  74.709

PHASE    NAME         CP RANGE (K)
S1       SOLID-A      298.0 -  525.0   D H TRANS ( 525.00 K) =   0.042 (K J )
S2       SOLID-B      525.0 -  565.0   D H TRANS ( 565.00 K) =   0.042 (K J )
S3       SOLID-C      565.0 - 1800.0
G1       GAS          298.0 - 2000.0


         CP = A + 1.0E-3*B*T(K) + 1.0E5*C*T(K)**-2 + 1.0E-6*D*T(K)**2


*********************************************************************************
PHASE    DH(298)    S(298)   DENSITY       A          B          C         D
         (K J )     ( J /K)  (G/CM**3)   -----------( J /K)------------
*********************************************************************************
S1       -240.580   38.074   6.670       -20.878    157.235    16.276    0.0
S2       -241.409   35.354   6.670        58.074     0.0        0.0      0.0
S3       -239.326   40.394   6.670        46.777     8.452      0.0      0.0
G1        309.616  241.417   IDEAL        40.522     0.946     -6.184    0.0
```

REFERENCE:

```
"CONTRIBUTIONS TO THE DATA ON THEORETICAL METALLURGY",
XVI. THERMODYNAMIC PROPERTIES OF NICKEL AND ITS INORGANIC COMPOUNDS,
ALLA D. MAH AND L.B. PANKRATZ,
U.S. BUREAU OF MINES BULLETIN 668, WASHINGTON, 1976.
                    APPENDED TO:
"THERMOCHEMICAL PROPERTIES OF INORGANIC SUBSTANCES",
I. BARIN, O. KNACKE, AND O. KUBASCHEWSKI,
SPRINGER-VERLAG, BERLIN, 1977.
```

Fig. 1.  Thermodynamic data for NiO taken from the pure substances databank and listed by the program INSPECT.

```
            2 Ni    +    O2    =    2 Ni*O
          (T,1,S2)     (T,P,G)      (T,1,S3)

    CALCULATIONS ARE BASED ON THE INDICATED NUMBER OF GRAM MOLES

*********************************************************************************
  (T)       (P)       DELTA H     DELTA G     DELTA V     DELTA S
  (K)      (ATM)       ( J )       ( J )        (L)        ( J /K)
*********************************************************************************
?
 1000        1

  1000.0  0.100E+01   -471541.5   -300542.7   -0.820E+02  -170.999
?
 (1000 1500 100) *        *           0

  1000.0  0.200E-15   -471541.5      0.0      -0.411E+18  -471.542
  1100.0  0.344E-13   -470481.1      0.0      -0.262E+16  -427.710
  1200.0  0.249E-11   -469408.4      0.0      -0.395E+14  -391.174
  1300.0  0.926E-10   -468315.6      0.0      -0.115E+13  -360.243
  1400.0  0.204E-08   -467197.9      0.0      -0.563E+11  -333.713
  1500.0  0.295E-07   -466052.1      0.0      -0.417E+10  -310.701
?
    *        1.0E-9        *           0

  1375.6  0.100E-08   -467473.6      0.0      -0.113E+12  -339.836
```

Fig. 2.  Spreadsheet format of the program REACTION. Entries are made in free format on the lines following the "?". Unknown quantities are entered as asterisks "*"

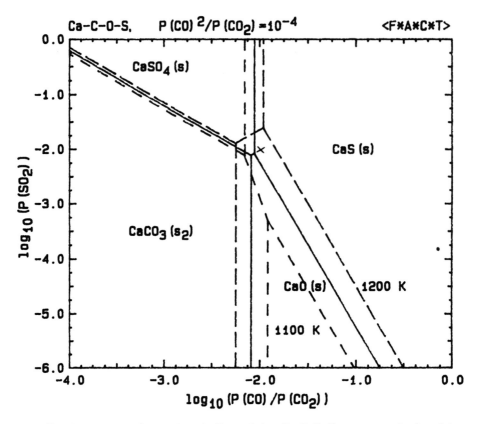

Fig. 3. A two-dimensional slice of the Ca-S-O-C system calculated by the program PREDOM.

Gibbs energy) combination of species and phases that satisfies the phase rule within the constraints specified by the user.

Figures 5a, b show an example of a calculation involving chemical vapor deposition (CVD) associated with vapor-phase epitaxial growth of GaAs from $GaCl_3$ and $AsH_3$. As reactants, the user has specified 1 mol $GaCl_3$, 1 mol $AsH_3$, and 100 mol $N_2$. As possible products, he has specified an ideal gas mixture (codes 1–9, 11–28) at 1 atm and 700 K, and the possible formation of pure condensed phases (codes 29, 30, 36–39, 42). The program calculates that 0.0393 mol GaAs(s) are deposited from the gas phase. The fourth value in parentheses after each phase corresponds to the activity of that species. It is noted that none of the other pure condensed species is formed ($a_{As(s)}$ = 0.6212, $a_{GaN}$ = 0.002). There are 101.32 mol of gas at equilibrium and the equilibrium partial pressures of $N_2$ = 0.987 atm, $H_2$ = 0.0051, and so on.

The relative molar Gibbs energy, $\Delta g$, of a solution phase may be represented by

$$\Delta g = \Delta g(\text{ideal}) + {}^E g \tag{1}$$

For simple nonelectrolytes, the ideal term for an $N$ component system is given as

CALCULATED ACTIVITIES AND PARTIAL PRESSURES AT 1200 K WHEN
log10(Y) =  -2.000, log10(X) =  -2.000
```
     1 CaSO4(s)              a = 4.42E-01
     5 CaS(s)                a = 3.21E-01
     6 CaS(g)                p = 2.80E-18 atm.
     7 CaCO3(s)              a = 3.02E-01
     8 CaCO3(s2)             a = 3.18E-01
*   10 CaO(s)                a = 1.00E+00
    13 CaC2(s2)              a = 1.14E-24
    14 Ca2(g)                p = 1.42E-40 atm.
    18 Ca(l)                 a = 3.20E-17
    20 COS(g)                p = 3.87E-05 atm.
    21 S2O(g)                p = 2.93E-07 atm.
    22 SO3(g)                p = 4.70E-09 atm.
    25 SO2(g)                p = 1.00E-02 atm.
    26 SO(g)                 p = 4.15E-06 atm.
    28 CS2(g)                p = 2.01E-10 atm.
    29 CS(g)                 p = 2.53E-11 atm.
    30 S8(g)                 p = 2.63E-26 atm.
    31 S7(g)                 p = 8.57E-23 atm.
    32 S6(g)                 p = 2.80E-19 atm.
    33 S5(g)                 p = 3.21E-16 atm.
    34 S4(g)                 p = 8.82E-14 atm.
    35 S3(g)                 p = 2.69E-09 atm.
    36 S2(g)                 p = 1.54E-05 atm.
    41 S(g)                  p = 1.61E-09 atm.
    42 C3O2(g)               p = 2.50E-22 atm.
    44 CO2(g)                p = 1.00E+00 atm.
    45 CO(g)                 p = 1.00E-02 atm.
    46 O3(g)                 p = 8.33E-28 atm.
    47 O2(g)                 p = 3.04E-12 atm.
    48 O(g)                  p = 4.33E-14 atm.
    50 C5(g)                 p = 1.32E-60 atm.
    52 C4(g)                 p = 5.49E-55 atm.
    54 C3(g)                 p = 3.04E-42 atm.
    56 C2(g)                 p = 2.67E-38 atm.
    58 C(s)                  a = 1.89E-06
    59 C(s2)                 a = 9.33E-07
    60 C(g)                  p = 1.76E-29 atm.
```

TOTAL PRESSURE IN GAS PHASE OF Ca   SPECIES =    2.80E-18 atm

TOTAL PRESSURE IN GAS PHASE OF ALL SPECIES =    1.02E+00 atm

Fig. 4.   Composition of the gas phase at $P_{SO_2} = 10^{-2}$ and $p_{O_2} = 10^{-2}$ atm (point x) in Fig. 3 calculated by the program PREDOM.

$$\Delta g(ideal) = RT(X_A \ln X_A + X_B \ln X_B + \ldots + X_N \ln X_N) \quad (2)$$

where $X_i$ is the mole fraction of component $i$. In ideal solutions, $^E g = 0$ and $\Delta g = \Delta g$ (ideal).

Figure 6 shows the results of a volume constrained gas-solid calculation with the program EQUILIB involving various Fe–Mn alloys. The user has specified that both phases are ideal solutions and that the volume of the gas is 0.010 liters. As in the previous cases, all the data are retrieved automatically from the pure substances databank.

```
****** ENTER REACTANTS ******
?
      0.6  Ga*Cl3    +    0.4  As*H3      +    100  N2    =

**************************************************************************
  T PROD    P PROD    V GAS    DELTA H   DELTA G   DELTA S   DELTA U   DELTA A
   (K)      (ATM)      (L)    <---- MAY SPECIFY IF SUBSCRIPTS PROVIDED ----->
**************************************************************************
?
   700        1

  - DATABASES BEING SEARCHED
  GASEOUS   SPECIES   1  -  28
  LIQUID    SPECIES  29  -  35
  SOLID     SPECIES  36  -  45

  ENTER CODE NUMBERS, ENTER "LIST" TO DISPLAY, OR ENTER "HELP"
?
  LIST

  POSSIBLE PRODUCT COMPOUNDS FOUND ("<---" WILL IDENTIFY YOUR PRIVATE DATA)

      1   GA*CL3                  <---   G1  GAS        200.0 K -   1500.0 K
      2   GA2CL6                  <---   G1  GAS        200.0 K -   1500.0 K
      3   AS                      <---   G1  GAS        298.0 K -   1500.0 K
      4   AS*CL3                         G1  GAS        308.0 K -   1000.0 K
      5   AS*N                           G1  GAS        295.0 K -    300.0 K
      6   AS*H3                          G1  ARSINE     298.0 K -   2000.0 K
      7   AS4                            G1  GAS        298.0 K -   1200.0 K
          ....
     27   H2                             G1  GAS        298.0 K -   3000.0 K
     28   H                              G1  GAS        298.0 K -   6000.0 K

     29   GA*CL3                  <---   L1  LIQUID     200.0 K -    800.0 K
     30   AS                      <---   L1  LIQUID    1090.0 K -   1500.0 K
     31   GA*AS                          L1  LIQUID    1238.0 K -   1400.0 K
     32   AS*CL3                         L1  LIQUID     298.0 K -    309.0 K
     33   GA*CL3                         L1  LIQUID     351.0 K -    575.0 K
     34   GA                             L1  LIQUID     303.0 K -   2478.0 K
     35   N2H4                           L1  LIQUID     295.0 K -    300.0 K

     36   N*H4CL                  <---   S1  SOLID      273.0 K -    800.0 K
     37   GA*CL3                  <---   S1  SOLID      200.0 K -    600.0 K
     38   AS                      <---   S1  ALPHA      298.0 K -   1090.0 K
     39   GA*AS                          S1  SOLID      298.0 K -   1238.0 K
     40   AS                             S1  SOLID      298.0 K -    800.0 K
     41   GA*CL3                         S1  SOLID      298.0 K -    351.0 K
     42   GA*N                           S1  SOLID      298.0 K -   1773.0 K
     43   GA                             S1  SOLID      298.0 K -    303.0 K

  ENTER CODE NUMBERS, ENTER "LIST" TO DISPLAY, OR ENTER "HELP"
?
  /1-9,11-28/29,30,36-39,42
```

Fig. 5a. Input for a chemical vapor deposition calculation by the program EQUILIB. All data are retrieved automatically from the pure substances databank. The gas is entered as an ideal mixture and the condensed species are pure.

According to Fig. 6, at 1450 K, the Fe–Mn alloy containing 80.0 atom % Mn is in equilibrium with $0.874 \times 10^{-8}$ mol of gas at $0.420 \times 10^{-3}$ atm, containing almost 100% Mn. It is important to note that the user has a prior knowledge of the thermodynamics and chemistry of the system. He knows that ideality for the solid Fe–Mn alloy solution is a reasonable assumption.

However, most condensed phases form solutions that are not ideal. In such cases, it is necessary to have access to proven solution models and reliable solution data.

```
0.6  GA*CL3    +    0.4  AS*H3      +    100  N2   =

    101.32                (   0.98694        N2
                          +   0.51043E-02    GaCl3                <---
                          +   0.50996E-02    H2
                          +   0.16337E-02    HCl
                          +   0.88945E-03    As4
                          +   0.23463E-03    GaCl
                          +   0.97273E-04    Ga2Cl6               <---
                          +   0.34487E-05    NH3
                          +   0.89195E-06    As2
                          +   0.52870E-09    AsCl3
                          +   0.39213E-09    AsH3
                          +   0.19518E-09    As3
                          +   0.84992E-10    AsN
                          +   0.35119E-14    As                   <---
                          +   0.17902E-14    H
                          ..............
                          +   0.55580E-28    NH
                          +   0.64628E-32    N
                          +   0.81673E-34    N3)
                          (   700.0, 1.00    ,G)

                          +   0.39329E-01    GaAs
                          (   700.0, 1.00    ,S1,   1.0000     )

                          +   0.00000E+00    As                   <---
                          (   700.0, 1.00    ,S1,   0.62120    )

                          +   0.00000E+00    As                   <---
                          (   700.0, 1.00    ,L1,   0.14711    )

                          +   0.00000E+00    GaN
                          (   700.0, 1.00    ,S1,   0.25638E-02)

                          +   0.00000E+00    GaCl3                <---
                          (   700.0, 1.00    ,L1,   0.81177E-04)

                          +   0.00000E+00    GaCl3                <---
                          (   700.0, 1.00    ,S1,   0.90988E-05)
```

GASEOUS IONIC SPECIES ARE SUPPRESSED BELOW 3000 K

DATA ON SPECIES IDENTIFIED WITH '<---' HAVE BEEN DRAWN FROM YOUR
PRIVATE DATA COLLECTION

Fig. 5b.  The output of the calculation from the input shown in Fig. 5a.

# NONIDEAL SOLUTIONS
## AND THERMODYNAMIC MODELS

Most condensed solution phases cannot be treated as ideal solutions and require special consideration. Various models have been proposed to represent the thermodynamic behavior of solutions, and these are well documented elsewhere (for example, *13,14*). It is important that the solution model provide a good representation of existing experimental data for the solution phase as well as be reliable for extrapolation and interpolation to compositions and temperatures outside the range of measurements.

Some of the most popular thermodynamic solution models used in metallurgical databases are: Margules-type polynomial expansions such as Redlich-Kister (*15*) or Legendre (*16*) expansions; dilute solution interaction parameter formalisms (*17–19*); Sublattice models (*20–22*); Conformal Ionic Solution Theory for molten salts (*23–25*); Quasichemical models

```
****** ENTER REACTANTS ******
?
<A>FE  +  <1-A> MN =
*********************************************************************
    <A>    T PROD  P PROD  V GAS  DELTA H  DELTA G  DELTA S  DELTA U  DELTA A
           (K)     (ATM)   (L)    <-- MAY SPECIFY IF SUBSCRIPTS PROVIDED --->
*********************************************************************
?
  (0.2,0.8,0.2) 1450   *      0.010

          1  FE                       G1   GAS          3135.0 K -  3600.0 K
          2  MN                       G1   GAS          2335.0 K -  2600.0 K
                                         .
          3  FE                       L1   LIQUID       1809.0 K -  3135.0 K
          4  MN                       L1   LIQUID       1517.0 K -  2335.0 K

          5  FE                       S1   ALPHA(MAG)    298.0 K -  1184.0 K
          6  FE                       S2   GAMMA        1184.0 K -  1665.0 K
          7  FE                       S3   DELTA        1665.0 K -  1809.0 K
          8  MN                       S1   SOLID-A       298.0 K -   980.0 K
          9  MN                       S2   SOLID-B       980.0 K -  1360.0 K
         10  MN                       S3   SOLID-C      1360.0 K -  1410.0 K
         11  MN                       S4   SOLID-D      1410.0 K -  1517.0 K

   ENTER CODE NUMBERS, ENTER "LIST" TO DISPLAY, OR ENTER "HELP"
?
  /1,2/6,10/

   <A> FE  +  <1-A> MN  =

         0.87427E-08        (  0.99997         Mn
                            +  0.26704E-04      Fe)
                            (  1450.0,0.420E-03,G)

     +   1.0000             (  0.80000         Mn
                            +  0.20000         Fe)
                            (  1450.0,0.420E-03,SOLN 2)

   WHERE 'A' ON THE REACTANT SIDE IS 0.2000

   <A> FE  +  <1-A> MN  =

         0.68356E-08        (  0.99993         Mn
                            +  0.71208E-04      Fe)
                            (  1450.0,0.315E-03,G)

     +   1.0000             (  0.60000         Mn
                            +  0.40000         Fe)
                            (  1450.0,0.315E-03,SOLN 2)

   WHERE 'A' ON THE REACTANT SIDE IS 0.4000
```

Fig. 6.   Example of volume constrained gas–solid equilibrium between ideal solutions calculated by the program EQUILIB. All data are retrieved automatically from the pure substances databank.

(26–29); the IRSID Cell Model (30). The models and their applications are well documented elsewhere (13,14,31–35).

## Margules-Type Polynomial Series

In binary solutions, the Margules-type polynomial expansion takes the general form

$$^Eg/RT = \Sigma \; \phi_{nm} \; X_A^n X_B^m \tag{3}$$

where $\phi_{nm}$ are temperature dependent coefficients. Differentiation of Eq. [3] leads to the corresponding expressions for the partial properties (35,36)

$$^Eg_A/RT = \ln \gamma_A = \Sigma \; \phi_{nm} \; [nX_A^{n-1} + (1-n)X_A^n] \; X_B^m \tag{4}$$

$$^E g_B / RT = \ln \gamma_B = \sum \phi_{nm} (m X_B^{m-1} + (1-n) X_B^m) X_A^n \qquad (5)$$

The terms in Eqs. [3–5] may be regrouped to form Redlich-Kister (15) or Legendre (16) polynomials.

Many nonionic solid and liquid solution phases are well represented by Margules-type polynomials (Eqs. [3–5]). For example, Fig. 7 shows the assessed Pb-Sn binary system (37) calculated from the thermodynamic and molar volume data listed in Table 2. The phase diagram was computed with the aid of the F*A*C*T program POTCOMP (1).

In order to obtain the coefficients listed in Table 2, it is necessary to take into account all possible thermodynamic and phase diagram data for each phase, including activity measurements, EMF data, solubilities, vapor pressures, and enthalpies of mixing. By performing a coupled phase diagram and thermodynamic analysis, one is able to obtain an optimized thermodynamic description of each phase that agrees with the thermodynamic data and generates the phase diagram. Details of how this may be done with the aid of the F*A*C*T program FITBIN are published elsewhere (36).

Once the phase diagram and thermodynamic properties have been assessed, it is possible to calculate the phase equilibria for conditions where experimental data are not available. For example, Fig. 8 shows the formation of a Pb–Sn intermetallic phase when the phase diagram is calculated (37) at high pressures from the data listed in Table 2.

It often arises that one wishes to calculate the thermodynamic properties and phase diagram for a multicomponent system for which there are no data. For those cases where the binary subsystems have been assessed, the Kohler (38) and Toop (39) interpolational models are employed by F*A*C*T to estimate the properties of the multicomponent systems solely from the binary data. Examples of the use of these models are presented elsewhere (35).

For metallurgical calculations involving dilute metallic solutions (liquid or solid solvent with several solutes), F*A*C*T employs activity coefficients of the solute at infinite dilution coupled with first-order interaction parameters (17), and second-order interaction parameters (18). These parameters are substituted into the "Unified Interaction Parameter Formalism" (19,40), which is thermodynamically self-consistent at all compositions and is a special case of the Margules-type polynomial expansion (Eq. [3]).

In the case of first-order interaction parameters, the Unified Interaction Parameter Formalism takes the following form:

$$\ln \gamma_i = \ln \gamma_i^\circ + \ln \gamma_{solvent} + \epsilon_{i1} X_1 + \epsilon_{i2} X_2 + \epsilon_{i3} X_3 + \\ + \epsilon_{iN} X_N \qquad (6)$$

where the activity coefficient of the solvent is given by:

$$\ln \gamma_{solvent} = -\tfrac{1}{2} \sum_{j=1}^{N} \sum_{k=1}^{N} \epsilon_{jk} X_j X_k \qquad (7)$$

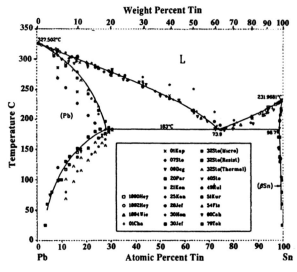

Weight Percent Tin

I. Karakaya and W.T. Thompson, 1988.

Fig. 7. Assessed Pb-Sn binary phase diagram taken from (37). Solid lines are calculated from the thermodynamic data listed in Table 2.

Table 2
Pb-Sn Thermodynamic Properties[a]

Thermodynamic properties of transformation with respect to the liquid
$G^0(Pb, L) = 0$
$G^0(Sn, L) = 0$
$G^0(Pb, fcc) = -4700 + 7.99\,T$
$G^0(Sn, bct) = -11\,131.8 + 22.874\,T - 0.5975 \times 10^{-2}T^2$
$\qquad\qquad -0.055T\ln T + 644\,170/T$
$G^0(Pb, G) = 192\,976 - 140.011T + 0.2715 \times 10^{-2}T$
$\qquad\qquad -0.4645 \times 10^{-6}T^3 + 5.426T\ln T - 616\,100/T$
$G^0(Sn, G) = 301\,610 - 121.062T + 20.330T\ln T$

Thermodynamic properties of hypothetical transformations
$Pb(fcc) \leftrightarrow Pb(bct)\qquad \Delta G^0 = 213\,580$
$Sn(bct) \leftrightarrow Sn(fcc)\qquad \Delta G^0 = 5\,510 - 9.49\,T$
$Pb(fcc) \leftrightarrow Pb(cph)\qquad \Delta G^0 = 3\,650 - 3.50\,T$
$Sn(bct) \leftrightarrow Sn(cph)\qquad \Delta G^0 = 2\,400 - 3.10\,T$

Thermodynamic properties of solutions
$\Delta H(Pb) = 6\,700\,X_{Pb}X_{Sn}$
$S^{ex}(Pb) = -0.365\,X_{Pb}X_{Sn}$
$\Delta H(\beta Sn) = -203\,068\,X_{Pb}X_{Sn}$
$S^{ex}(\beta Sn) = 0$
$\Delta H(L) = (4\,800 + 200\,X_{Sn})X_{Pb}X_{Sn}$
$S^{ex}(L) = (-2.815 + 1.215\,X_{Sn})X_{Pb}X_{Sn}$
$\Delta H(\varepsilon) = 7\,800\,X_{Pb}X_{Sn}$
$S^{ex}(\varepsilon) = 7.6\,X_{Pb}X_{Sn}$

[a]Values are in J/mol and J/mol K.

Pb-Sn Molar Volume Data[b]

| Component | Composition, at.%Sn | (Pb) | (βSn) | ε | L |
|---|---|---|---|---|---|
| Pb | 0 | 18.7(a) | 18.4(b) | 18.873 | 19.38(c) |
| Sn | 100 | 16.855(b) | 16.555(a) | 16.548 | 16.98(c) |

[b]Values are in cm$^{-3}$/mol.

242

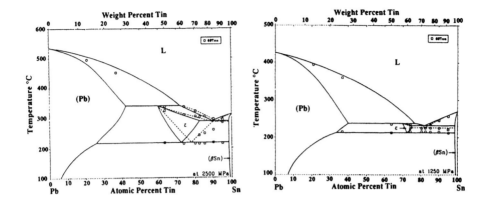

Fig. 8.   Effect of pressure on the Pb-Sn binary phase diagram calculated (37) from the thermodynamic and molar volume data listed in Table 2.

It has been shown (19,40), that Eqs. [6,7] are completely thermo-dynamically consistent in both dilute and nondilute regions.

### Quasichemical Model

The Margules-type polynomial expansion (Eqs. [3–5]) cannot ade-quately represent the excess thermodynamic properties of solutions which exhibit structural "ordering" such as is found, for example, in molten silicates containing CaO, MgO, or FeO. The quasichemical theo-ry, first proposed by Guggenheim (26), has been modified (27–29) in order to resolve this problem.

In the Modified Quasichemical Model, the partial Gibbs energies and activities take the following form in a binary system ($i = 1, 2$)

$$\Delta g_i = RTlnX_i + b_i \, RTln \, X_{ij}/Y_i^2$$
$$+ \, b_i \, (X_{12}/2) \, (1 - Y_i)\partial(\omega - \eta T)/\partial Y_i \qquad (8)$$

where $X_{ij}$ are fractions of i–j bonds, the term $(\omega - \eta T)$ is an energy change associated with the formation of i–j bonds from i–i and j–j bonds, and $Y_i$ are "equivalent fractions."

$$Y_1 = b_1X_1/(b_1X_1 + b_2X_2) \qquad (9)$$

$$Y_2 = b_2X_2/(b_1X_1 + b_2X_2) \qquad (10)$$

where $b_1$ and $b_2$ are charge parameters. The model may be easily extend-ed to ternary and multicomponent systems (27–29).

Figure 9a illustrates the use of the Modified Quasichemical Model in calculating the $CaO–Al_2O_3–SiO_2$ ternary phase diagram with the aid of the program TERNFIG (41). Details of the analysis for this particular system are given in (42). Figure 9b is the experimental phase diagram according to (43). Agreement everywhere is within the experimental error limits.

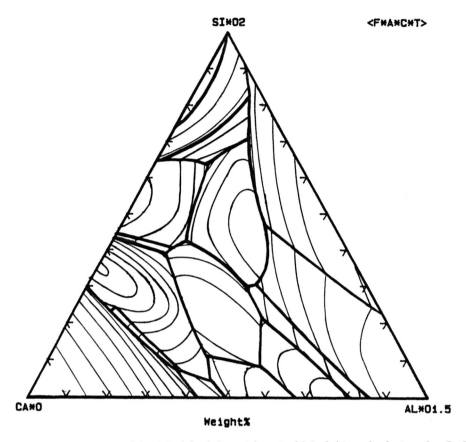

SI*O2                                                        <F*A*C*T>

CA*O                                                         AL*O1.5
                        Weight%

Fig. 9a.   Use of the Modified Quasichemical Model in calculating the CaO-Al$_2$O$_3$-SiO$_2$ ternary phase diagram with the aid of the program TERNFIG.

Figure 10a shows the input to the EQUILIB program to calculate metal and slag compositions during silico-manganese ladle deoxidation at 1873 K. Solutions are identified by entering the code numbers of the species between slashes, "/". The following models are employed for the solution phases: gas phase, ideal; liquid steel, Unified Interaction Parameter Formalism; liquid slag phase, Modified Quasichemical Model; MnO–FeO solid solution, regular solution. All pure condensed species are also included in the calculation by entering the code numbes, 33–79, after the solution phases.

In Fig. 10b, the output from the calculation gives the compositions for the stable solution phases (steel phase and slag phase) in weight per cent. The solid FeO–MnO solution does not form, nor do any other solid phases. The calculated equilibrium composition is close to silica saturation (activity of SiO$_2$(s) = 0.961).

### Sublattice Model

For ionic solutions, such as molten salts, it is necessary to consider the distribution of anions and cations on sublattices. Sublattice models for molten salts are well documented (20–25, 44, 45) and a general

CaO–Al₂O₃–SiO₂

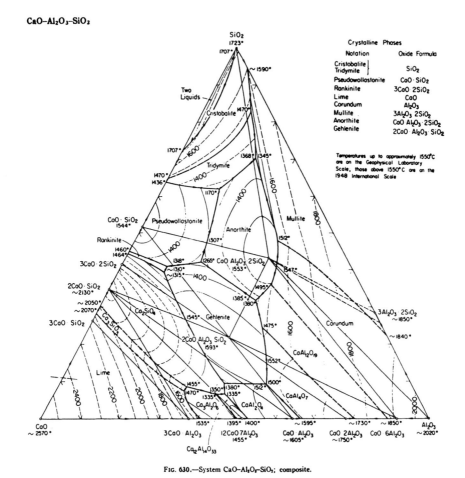

Fig. 630.—System CaO-Al₂O₃-SiO₂; composite.

Fig. 9b. Experimental CaO-Al₂O₃-SiO₂ ternary phase diagram taken from (43).

extension to multicomponent systems containing any number of components has recently been proposed by one of the authors (22).

In the Sublattice Model, the integral Gibbs energy per equivalent is given by

$$g = \sum_c \sum_a Y_c Y_a g^\circ_{c/a} + \Delta g(\text{ideal}) + {}^E g \qquad (11)$$

where $a$ = anion and $c$ = cation, and where $Y_i$ are equivalent cationic or anionic site fractions represented by

$$Y_A = q_A n_A / (q_A n_A + q_B n_B + q_c n_c + \ldots) \qquad (12)$$

where $q_i$ is the absolute charge of the ion $i$. The ideal term takes the form

$$\Delta g(\text{ideal})/RT = (\Sigma q_c X_c)^{-1} (\Sigma X_c \ln X_c) + (\Sigma q_a X_a)^{-1} (\Sigma X_a \ln X_a) \qquad (13)$$

Figure 11a shows the phase diagram of the ternary reciprocal salt system Li,K/CO₃, OH calculated by the program RECIPFIG (1). The Sublattice Model was used to calculate the thermodynamic properties of

```
****** ENTER REACTANTS ******
?
 100. Fe  + 0.16 O + 0.4 Fe + 0.4 Mn  + 0.3 Si + 0.08 Ar =

**************************************************************
   T PROD    P PROD    DELTA H    DELTA G    DELTA V    DELTA S
    (K)       (ATM)      (J)        (J)        (L)        (J/K)
**************************************************************
?
 1873          1

 - DATABASES BEING SEARCHED
GASEOUS  SPECIES   1 -  15
LIQUID   SPECIES  16 -  32
SOLID    SPECIES  33 -  79

Modified Quasichemical Model for molten oxides.
   80   SI*O2
   81   FE*O
   82   MN*O

Fe liquid dilute solution database (solutes - Ag,Al,B,C,Ca,Ce,Co,
Cr,Cu,H,La,Mn,Mo,N,Nb,Ni,O,P,Pb,Pd,S,Si,Sn,Ta,U,V,W & Zr).
   83   FE
   84   MN
   85   O2
   86   SI

Solid Oxide Solution
   87   FE*O
   88   MN*O

ENTER CODE NUMBERS, ENTER "LIST" TO DISPLAY, OR ENTER "HELP"
?
 /1-15/80-82/83-86/87,88/33-79
```

Fig. 10a.  Input to compute metal and slag compositions during silico-manganese ladle deoxidation calculated by the program EQUILIB.

the solution phase solely from the properties of the pure components and the four assessed binary subsystems. Details of the analysis for this system are given in (*46*). The calculated diagram can be compared with the experimental diagram that is shown in Fig. 11b (*43*).

## USER-SUPPLIED THERMODYNAMIC DATA

Although F*A*C*T has extensive stored thermodynamic data on pure substances and on binary and multicomponent solutions, there are many cases where the user wishes to employ one's own data. The program DATAENTRY (*1*) enables one to enter and store private thermodynamic data on pure substances.

A recent addition to the F*A*C*T system is the program SOLUTION. This software enables one to store private binary and multicomponent solution data that can be subsequently retrieved automatically by EQUILIB. The following models are supported by SOLUTION.

1. Margules-type polynomials, including Kohler and Toop interpolation models;

```
100. Fe  + 0.08 O + 0.4 Fe + 0.4 Mn  + 0.3 Si + 0.08 Ar =

         0.30793     litre (    99.943      vol% Ar
                           +   0.24987E-01 vol% Mn
                           +   0.24069E-01 vol% SiO
                           +   0.82057E-02 vol% Fe
                           +   0.79044E-07 vol% O
                           +   0.60192E-08 vol% Si
                           +   0.11200E-08 vol% O2
                           +   0.35385E-15 vol% Si2)
                             ( 1873.0, 1.00    ,G)

     + 0.18501     gram  (    49.948      wt.% SiO2
                           +   42.104      wt.% MnO
                           +   7.9478      wt.% FeO)
                             ( 1873.0, 1.00    ,SOLN 2)

     + 100.99      gram  (    99.400      wt.% Fe
                           +   0.33630     wt.% Mn
                           +   0.25426     wt.% Si
                           +   0.98375E-02 wt.% O2)
                             ( 1873.0, 1.00    ,SOLN 3)

     + 0.00000E+00       (    0.91726E-01      MnO
                           +   0.26259E-01      FeO)
                             ( 1873.0, 1.00    ,SOLN 4,0.118    )

                       +   0.00000E+00 gram SiO2
                             ( 1873.0, 1.00    ,S4, 0.96059    )

                       +   ....

                       +   0.00000E+00 gram (MnO)(SiO2)
                             ( 1873.0, 1.00    ,S1, 0.35128    )

                       +   ....
```

        GASEOUS IONIC SPECIES ARE SUPPRESSED BELOW 3000 K

Fig. 10b.   Output for the calculation of Fig. 10a. The gas phase (G) is ideal; the FeO(s)-MnO(s) solution phase (SOLN 4) is treated as a regular solution; the molten steel (SOLN 2) is represented by the Unified Interaction Parameter Formalism; the molten slag phase (SOLN 3) is described by the Quasichemical Model. All data are retrieved automatically from the various databases.

2. Unified Interaction Parameter Formalism for dilute solutions;
3. Modified Quasichemical Model for structured solutions (slags, highly complexed liquids); and
4. Sublattice Model for molten salts.

In all calculations involving pure substances and solution data, the user has the option of using the F\*A\*C\*T databanks and/or his private data stored by the programs DATAENTRY and SOLUTION.

## SUMMARY

F\*A\*C\*T (Facility for the Analysis of Chemical Thermodynamics) is an on-line thermodynamic database computing system. With the aid of

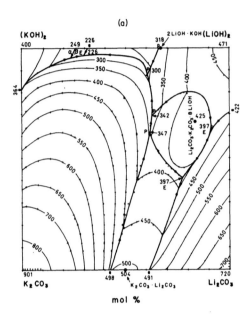

Fig. 11a.   Phase diagram of the ternary reciprocal salt system Li, K/CO₃, OH calculated by the program RECIPFIG mainly from the properties of the pure components and the assessments of the 4 binary subsystems. The Sublattice Model was employed to represent the thermodynamic data of the liquid.

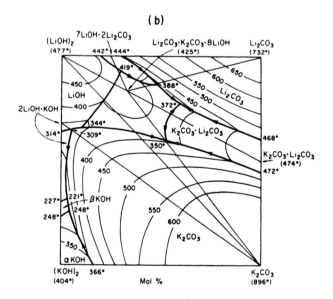

Fig. 11b.   Experimental Li ,K/CO₃, OH phase diagram according to (43).

F*A*C*T, a user is able to: search and display data on compounds and solutions at any temperature and composition; calculate thermodynamic property changes for any type of chemical reaction; perform heat balances; store and use private data; calculate vapor pressures; compute complex chemical equilibria; calculate and plot predominance diagrams; calculate (or estimate) and plot binary and ternary phase diagrams; and predict multicomponent multiphase complex chemical equilibria.

The Unified Interaction Parameter Formalism, the Modified Quasi-chemical Model, and the Sublattice Model, are the principal models employed to treat thermodynamic solution data. The solution data, along with a large pure substances databank for over 5,000 phases, are accessible by the F*A*C*T system.

## ACKNOWLEDGMENTS

Financial assistance from the Natural Sciences and Engineering Research Council of Canada is gratefully acknowledged.

## REFERENCES

1. Thompson, W. T., Pelton, A. D., and Bale, C. W. *F\*A\*C\*T—Facility for the Analysis of Chemical Thermodynamics, Guide to Operations*, McGill University Computing Centre, Montreal, pp. 500 (1988).
2. Thompson, W. T., Eriksson, G., Pelton, A. D., and Bale, C. W. *Int'l Sympos. on Computer Software in Chem. & Extract. Metal.*, Pergamon, vol. 11, *Proc. Metall. Soc. of CIM*, Montreal '88, pp. 87–106 (1988).
3. Barin, I. and Knacke, O. *Thermochemical Properties of Inorganic Substances*, Springer-Verlag, Berlin (1973).
4. Barin, I., Knacke, O., and Kubaschewski, O. *Thermochemical Properties of Inorganic Substances, Supplement*, Springer-Verlag, Berlin (1977).
5. *JANAF Thermochemical Tables, 2nd ed.* (and supplements), US Dept. of Commerce, N.B.S. (NIST) (1971).
6. Mills, K. C. *Thermodynamic Data for Inorganic Sulfides, Selenides and Tellurides*, Butterworths, London (1974).
7. Robie, R. A., Hemingway, B. S., and Fisher, J. R. *US Geol. Survey Bull. 1452*, US Government Printing Office, Washington DC (1979).
8. Stull, D. R. and Prophet, H. *JANAF Thermochemical Tables, 2nd ed.* (and supplements), Nat. Stand. Ref. Data, Nat. Bur. Stand., Washington, pp. 1141 (1971).
9. Thompson, W. T., Pelton, A. D., and Bale, C. W. *Engineering Education*, pp. 201–204 (1979).
10. Bale, C. W., Pelton, A. D., and Thompson, W. T. *Can. Met. Quart.* **25**, 107–112 (1986).
11. Bale, C. W., Melançon, J., and Pinho, A. *Can. Metall. Quart.* **19**, 363–371 (1981).
12. Eriksson, G., *Chemica Scripta* **8**, 100–103 (1975).
13. Ansara, I., *Int'l Metall. Rev.* Review 238, No. 1, pp. 20–53 (1979).
14. Hillert, M., *CALPHAD* **12**, 257–259 (1988).
15. Redlich, O. and Kister, A. T., *Ind. Chem. Eng.* **40**, 345 (1948).
16. Bale, C. W. and Pelton, A. D., *Metall. Trans.* **5**, 2323–2337 (1974).
17. Wagner, C., *Thermodynamics of Alloys*, Addison-Wesley, Reading, MA, p. 51 (1962).

18. Lupis, C. H. P. and Elliott, J. F., *Acta Metall.* **14,** 529–38 (1966).
19. Pelton, A. D. and Bale, C. W., *Metall. Trans.* **17A,** 1211–15 (1986).
20. Hillert, M. and Staffansson, L. I., *Acta Chem. Scand.* **24,** 3618 (1970).
21. Hillert, M., Jansson, B., Sundman, B., and Agren, J., *Metall. Trans.* **16A,** 261–266 (1985).
22. Pelton, A. D., *CALPHAD* **12,** 127–142 (1988).
23. Blander, M. and Yosim, S. J., *J. Chem. Phys.* **39,** 2610 (1963).
24. Blander, M. and Topol, L. E., *Inorganic Chem.* **5,** 1641 (1966).
25. Blander, M., *Int'l Sympos. on Computer Software in Chem. & Extract. Metal.,* Pergamon, vol. 11, *Proc. Metall. Soc. of CIM, Montreal '88,* pp. 3–14 (1988).
26. Guggenheim, E. A., *Proc. Royal Soc.* **A14B,** 304 (1935).
27. Pelton, A. D. and Blander, M., *Proc. AIME Sympos. Molten Salts and Slags,* 281–299, TMS-AIME, Warrendale, PA (1984).
28. Blander, M. and Pelton, A. D., *Geochim. and Cosmochim. Acta* **51,** 85–95 (1987).
29. Pelton, A. D. and Blander, M., *Metall. Trans.* **17B,** 805–815 (1986).
30. Kampoor, M. L. and Frohberg, M. G., *Archiv. Eisenhuttenw.* **45,** 663 (1974).
31. Kaufmann, L. R. and Bernstein, H., *Computer Calculation of Phase Diagrams,* Academic Press, New York (1970).
32. *Computer Coupling of Phase Diagrams and Thermochemistry (1977–1989), CALPHAD J,* Pergamon, vols. 1–13.
33. *Bull. Alloy Phase Diagrams (1980–89),* ASM International, Metals Park, OH 44073.
34. Kubaschewski O., Evans, E. L., and Alcock, C. B., *Metallurgical Thermochemistry,* Pergamon, NY, (1967).
35. Pelton, A. D. and Bale, C. W., *CALPHAD,* **1,** 253–273 (1977).
36. Bale, C. W. and Pelton, A. D., *Metall. Trans.* **14B,** 77–83 (1983).
37. Karakaya, I. and Thompson, W. T., *Bull. Alloy Phase Diag.* **9,** 144–152 (1988).
38. Kohler, F., *Mh. Chem.* **91,** 738 (1960).
39. Toop, G. W., *TMS-AIME* **233,** 850 (1965).
40. Bale, C. W. and Pelton, A. D., *Metall. Trans B* submitted
41. Lin, P.-L., Pelton, A. D., and Bale, C. W., *CALPHAD* **4,** 47–60 (1980).
42. G. Eriksson and A. D. Pelton, private communication (1989).
43. Levin, E. M., Robbins, C. R., and McMurdie, H. F., *Phase Diagrams for Ceramists,* compiled by NBS. (NIST), publ. The American Ceramic Soc. Inc., 1964, and supplements 1969, 1975.
44. Temkin, M., *Acta Phys. Chim. USSR* **20,** 411 (1945).
45. Saboungi, M.-L. and Blander, M., *J. Am. Ceram. Soc.* **58,** 1 (1975).
46. Pelton, A. D., Bale, C. W., and Lin, P.-L., *Canad. J. Chem.* **62,** 457–474 (1984).

# Application of MTDATA to the Modeling of Multicomponent Equilibria

R. H. DAVIES,*,1 A. T. DINSDALE,1 T. G. CHART,1
T. I. BARRY,1 AND M. H. RAND2

1Division of Materials Applications, National Physical Laboratory,
Teddington, Middlesex TW11 OLW, UK;
and 2Materials Development Division, Harwell Laboratory,
Didcot, Oxfordshire, OX11 ORA, UK

## ABSTRACT

MTDATA embodies the principle that equilibria in multicomponent, multiphase systems can be calculated from a knowledge of the thermodynamic data for the subsystems. A number of modules are incorporated for manipulating and retrieving the data, making various types of calculation and plotting binary, ternary, multicomponent, and predominance area diagrams. The principles and operation of MTDATA are illustrated by reference to systems of practical importance.

**Index Entries:** MTDATA; multicomponent equilibria; multiphase systems.

## AN OVERVIEW OF MTDATA

The main function of MTDATA is the calculation of chemical equilibria in multicomponent systems. For this purpose, it includes software modules for manipulation of the data, evaluation of thermodynamic functions, equilibrium calculation, and the plotting of results, as well as some utilities. The primary database is that of SGTE, Scientific Group

*Author to whom all correspondence and reprint requests should be addressed.

Thermodata Europe. MTDATA as a whole and most of the modules are fully documented (1). It is normally used interactively, but can also be driven in batch mode and in addition, macros can be used to customize the program or replace commonly entered sequences of commands.

Data are retrieved from one or more databases by means of the Access module. The system is chosen by a definition of its components, which may be elements or compounds. A typical command allowing data to be recovered for the $CaO-FeO-Al_2O_3-SiO_2$ system from the oxid1 database might be

define system = 'CaO,FeO,Al2O3,SiO2,O/ − 2' source
= oxid1 out = 'CFAS' !

Here, the command "define" is used to name the system, the source database, and the datafile that will be generated when the command "save" is issued. Any unique abbreviation of the keywords will suffice and four levels of user experience are catered for. The normal command sequence has the following syntax

command {parameter = value} !

The types of substance and phase/model for which data can be stored and retrieved include stoichiometric compounds, gases, aqueous species, other associates, nonideal solutions, and phases with solution on fixed or variable sublattices. The inclusion of the oxide ion in the above example allows phases bearing charged species to be part of the system. This is necessary because sublattice models, in particular, use charge to regulate the thermodynamic properties of liquid or solid solutions.

The datafiles generated by the command "save" have a standard format and provide the input data used by G-plot, Binary, Ternary, Multiphase, Transition, Liquidus, and Coplot, in fact, all the calculation modules in MTDATA except Thermotab, which deals with chemical equations rather than systems.

G-plot is used for tabulating and plotting thermodynamic properties for purposes of data assessment, mainly, but not exclusively, for binary systems. The assessments are based on all the available thermodynamic and phase equilibria data and, where these are insufficient or of doubtful validity, on well-established methods of estimation. Figure 1 illustrates the comparison of part of the experimental and calculated partial Gibbs energy data for the Mn–Si system (4). The phase diagram for this system calculated using the Binary module is shown in Fig. 2.

Ternary is used to generate isothermal sections for ternary systems. The process of calculating interlinked two and three phase regions is automatic. Figure 3 shows a diagram for the Fe–Mn–Si system, which is based on assessments for the three binary subsystems and some experimental studies of the ternary system.

The workhorse, if that is an appropriate word for such a powerful program, of all the calculation modules except Thermotab and Coplot, is Multiphase, which also gives its name to the most general of these

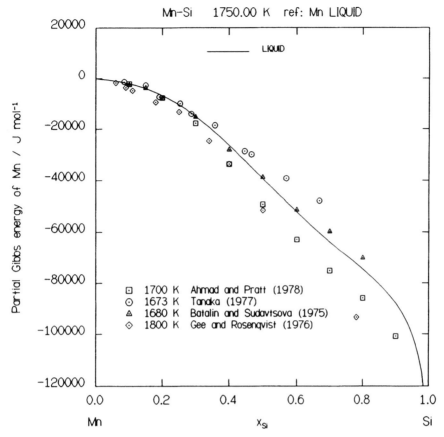

Fig. 1.   Comparison of experimentally derived and calculated partial
Gibbs energy values for Mn in liquid Mn–Si at 1750 K.

modules. Multiphase, which was developed at the National Physical
Laboratory by Susan Hodson, is described in the MTDATA handbook (2)
and more briefly by Dinsdale, Hodson, Barry, and Taylor (3). The command protocol of Multiphase allows composition, temperature, pressure
or volume, partial pressures, and activities to be set and one variable
stepped. Amounts can be specified in terms of mass or moles and can be
stepped in multidimensional component space. Depending on the nature of the problem, either or both of two calculation stages can be used.
The first employs a particularly reliable method of determining equilibrium when all component and phase amounts are greater than $10^{-6}$ mol.
The second stage is used in other cases. Means are provided to make the
calculations more efficient by using results or the program variables from
previous calculations as a starting point for the next calculation. A method of entering an initial guess is provided, but is never needed in
practice.

The calculation of equilibria in a system of many components generates an enormous quantity of information that can rarely be displayed on
a single graph and would be difficult to assimilate in tabular form. For
this reason Multiphase incorporates flexible procedures to control what is
to be plotted from the graph files calculated previously. Amounts of

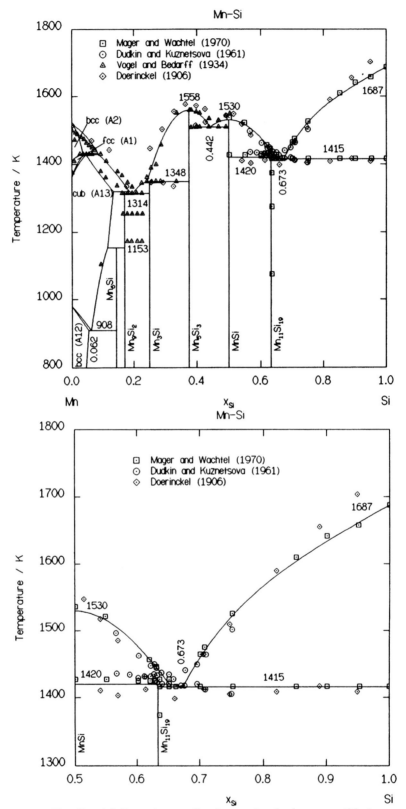

Fig. 2. (a) Experimentally determined phase equilibria and phase diagram calculated from an assessment of all relevant data for the Mn–Si system. (b) An expanded region of Fig. 2a.

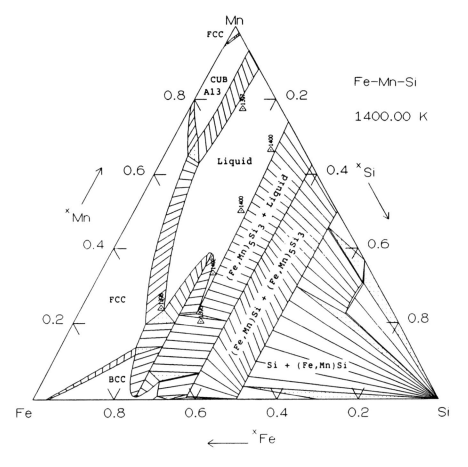

Fig. 3. Phase equilibria in the ternary system Fe–Mn–Si at 1400 K. Experimental studies in the ternary are confined to compositions having less than 0.5 mol fraction of silicon.

phases or substances (individually, in a phase, compounds of a particular component, or all), component distribution between phases or within a phase, component and compound activities, partial pressure, and Gibbs energy can all be plotted. The axis scales can be varied and made logarithmic, and amounts can be expressed in moles or mass terms irrespective of the way the problem was formulated. Moreover, the abscissa of the graph can be changed from the originally stepped variable to one of the calculated variables.

Transition is a new module used to search for phase boundaries with either composition or temperature as a variable. It has proven particularly useful in the determination of liquidus surfaces in slags used in pyrometallurgical processing. Figure 4 shows the results of a scan through the $CaO$–$FeO$–$Al_2O_3$–$SiO_2$ system in which only the liquidus temperature has been requested. The phase coexisting with the liquid is indicated by the symbol. In this mode of operation, Transition steps across composition space in a straight line unless it finds a phase with a short range of existence when it backtracks in order to determine more detail.

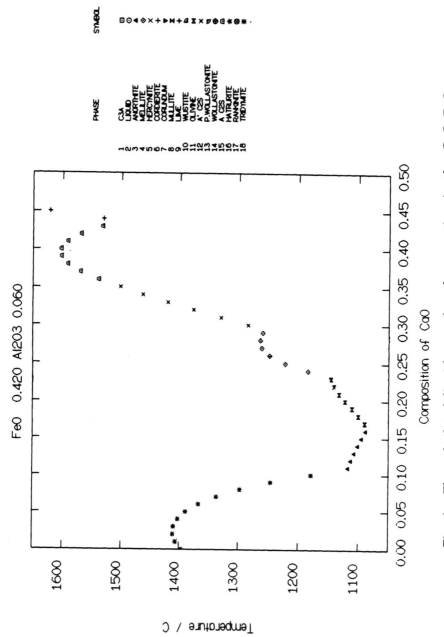

Fig. 4. The calculated liquidus surface for a section in the CaO–FeO–Al₂O₃–SiO₂ system. FeO = 40 wt%, Al₂O₃ = 10 wt%. The data for the liquid employ associated species interacting nonideally.

Coplot is used for drawing predominance area or Pourbaix diagrams for systems involving stoichiometric substances, gases, and ideal aqueous solutions. The chemistries of more than one element can be explored at the same time.

Thermotab is used for determining thermodynamic functions for chemical equations that can be balanced automatically. The data may be amended interactively and then saved into a specified database.

## OXIDE SYSTEMS

Oxide systems are of great importance in geochemistry and pyrometallurgy. Most ceramics are predominantly oxides, and the resistance of alloys and nonoxide ceramics to corrosion depends critically on the nature of the oxide formed on the surface. Oxides are also used in many types of electronic and magnetic devices. Coal contains silicates that form a sticky slag if the coal is overheated during combustion or gasification. Thus, knowledge of phase equilibria in oxide systems is of far-reaching economic importance.

Although several papers have been written indicating varying degrees of success in the calculation of equilibria involving liquid and solid oxides (5–8), none of the models used for the liquid phase has been well validated.* Part of the problem is that the provision of data for the many crystalline phases presents formidable problems despite the substantial amount of data and number of assessments by geochemists. For example, the thermodynamics of mixing in the melilite phase, between gehlenite, akermanite, and iron-bearing melilites, requires data for systems of four component oxides, whereas in alloy systems, it is often sufficient to use data from binary systems.

In the data for the $CaO–FeO–Al_2O_3–SiO_2$ system used for the calculation of Fig. 4 the liquid was represented using associated species in a nonideal solution: a total of 11 associates in all, including the components, six binary and one ternary associate. The Redlich-Kister equation was used to describe the interactions between associates and, hence, a total of 55 binary and 165 ternary terms, each of which could have temperature and composition parameters, were available to fit the experimental data. Even though the majority of the coefficients could be set to zero and none of the binary terms for excess Gibbs energy were large compared with RT, it is not desirable to have a large number of adjustable parameters in ternary and higher order systems. Cooperation between laboratories is desirable to devise and validate models and data for this important class of systems. To date, MTDATA has been used to examine the cellular model of Kapoor and Frohberg (5) using data of Gaye and Welfringer (8) supplemented by our own assessments, and the variable-site-sublattice model in cooperation with the group at the Royal Institute of Technology in Stockholm. Some of the data used in the work reported here were converted from assessments undertaken at Stockholm (9).

---

*Ed's Note: See other papers in this volume.

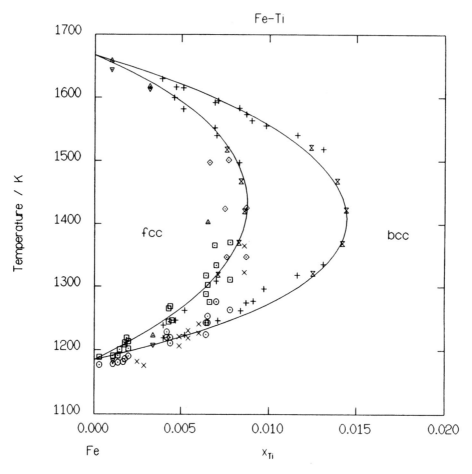

Fig. 5. The calculated gamma loop for the Fe–Ti system with experimental data superimposed.

Despite the large number of possible coefficients, an advantage of the associate model is that it is easy to gain a feel for the data by determining, for example, which associates make the greatest contribution to the Gibbs energy. This allows appropriate adjustments to be made to interaction terms. MTDATA is highly suited to this task.

## NITRIDING STEELS

Creep resistant alloys are needed that also resist void swelling when used in nuclear reactors. For this purpose, it was decided to investigate titanium nitriding steels. Titanium markedly reduces the composition range for the austenite phase, and the aim was to compensate by adding nickel so that nitriding could be performed in the austenitic region to impart the desired properties when the alloy was cooled to produce a ferrite steel.

The very narrow gamma loop (austenite–ferrite two-phase field) for the Fe–Ti system is shown in Fig. 5. The loop for the Fe–Cr system is

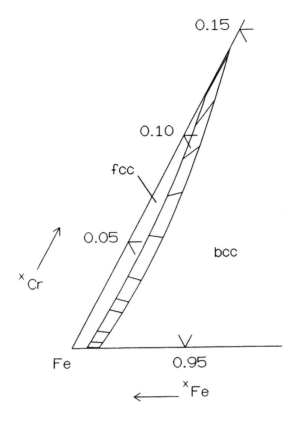

Fig. 6. The iron rich corner of the Cr–Fe–Ti systems showing the gamma loop calculated for 1300 K.

wider. This is implied by Fig. 6, which shows a corner of the phase diagram for the Cr–Fe–Ti system at 1300 K prepared using the zoom facility incorporated in the Ternary module of MTDATA. The diagram was calculated on the basis of critical assessments for the binary systems and shows close agreement with limited experimental data for the ternary system.

Calculations were made for the real steel that also contains nickel, silicon, and carbon. Fig. 7 shows the proportions of the austenite, ferrite, and carbide phases as a function of temperature in one of the compositions explored, and Fig. 8 shows the composition of the austenite. A fuller description of this project is given by Barry and Chart (*10*).

Calculations can be conducted much more rapidly than experiment and on the basis of the results of experimental trials, reported by Wilson, Gohil, and Laing (*11*), were made of two prototype steels. The measured phase equilibria have been shown to be in almost exact agreement with calculation.

## MOLYBDENUM SILICIDES

Predominance area diagrams usually show the chemistry of one component as a function of the potentials of two others. They are rela-

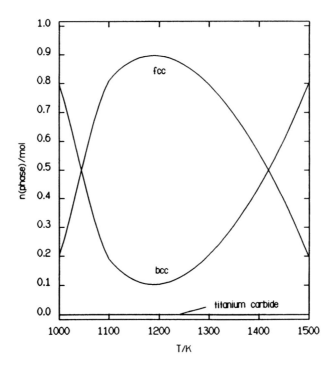

Fig. 7. Calculated amounts of the fcc, bcc, and carbide phases present in a C–Cr–Fe–Ni–Si–Ti alloy of composition in mol% 0.05 C, 12 Cr, 2 Ni, 0.5 Si, 1 Ti, and balance Fe as a function of temperature.

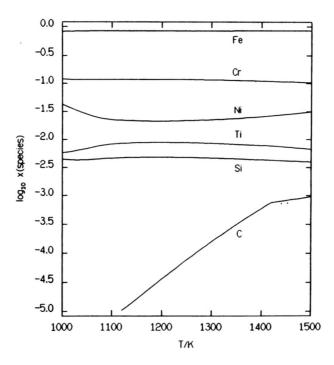

Fig. 8. The composition of the fcc, austenitic phase of Fig. 7.

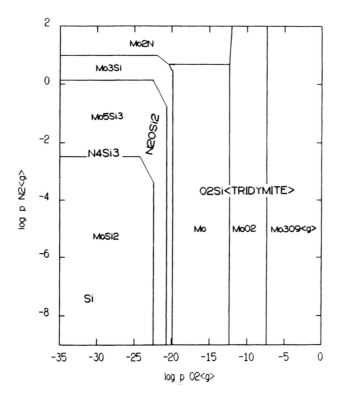

Fig. 9. Predominance area diagram for silicon and molybdenum as a function of oxygen and nitrogen potentials for 1423 K. The gas volume is the ideal gas volume, 0.1168 m$^3$, and the amounts of silicon and molybdenum are 1 and 0.001 mol, respectively. The diagram for silicon (dotted lines, large labels) is calculated first. The molybdenum diagram (solid lines, smaller labels) is then calculated for each area of the silicon diagram, the component for silicon being the compound of silicon oxygen and nitrogen that is stable in that area. The individual diagrams are then automatically combined.

tively simple to produce and are useful in gaining a broad understanding of the chemistry of aqueous systems of interest in corrosion and hydro-metallurgy, as well as high-temperature, gas-solid reactions.

The Coplot module in MTDATA incorporates a number of special features. In particular, it allows the chemistry of up to five elements in the environment of others to be explored simultaneously. For example, molybdenum disilicide is used for high temperature furnace elements. In Fig. 9, the behavior of molybdenum and silicon are examined as a function of oxygen and nitrogen pressure. Silica, which tends to form on the surface of these furnace elements, is taken to be present in greater amounts and, therefore, the chemistry of the silicon is largely independent of the molybdenum. However, the chemistry of the silicon contributes to the environment that decides the behavior of the molybdenum. The diagram shows how the formation of the various molybdenum silicides depends on which are the stable compounds of silicon with nitrogen and oxygen.

Because the axes of predominance area diagrams define potentials rather than compositions, it is possible for regions of the diagrams to be either overconstrained or experimentally inaccessible. For example, if the axes were log $p(O_2)$ and log $p(S_2)$, regions where both of these were even moderately large would correspond to enormous pressures of $SO_2(g)$. MTDATA allows such areas to be identified and excluded from the diagram.

## AVAILABILITY

MTDATA is currently available for mounting on Vax ® and Prime ® computers and in restricted form on IBM ® mainframe computers and personal computers using the MS-DOS ® operating system.

## ACKNOWLEDGMENTS

The authors gratefully acknowledge the work of many contributors to MTDATA and, in particular, their indebtedness to S. M. Hodson, J. A. Gisby, and N. J. Pugh. Users of MTDATA are also thanked for their comments, particularly B. B. Argent and I. Duckenfield of Sheffield University, who have helped in identifying and isolating system dependent features and in the provision of the version of Prime computers. The support of Mineral Industry Research Organization for the work on oxides and British Steel and UKAEA for the projects on steels systems is also acknowledged.

## REFERENCES

1. Barry, T. I., Davies, R. H., Dinsdale, A. T., Gisby, J. A., Hodson, S. M., and Pugh, N. J., "MTDATA Handbook: Documentation for the NPL Metallurgical and Thermochemical Databank," National Physical Laboratory, Teddington UK, 1989.
2. Hodson, S. M. "MULTIPHASE Theory," Reference 1, 1989.
3. Dinsdale, A. T., Hodson, S. M., Barry, T. I., and Taylor, J. R., *Proc. Annual Conf. Metall. CIM*, C. W. Bale, ed., Montreal, 1988.
4. Gisby, J. A. and Dinsdale, A. T., to be published.
5. Kapoor, M. L. and Frohberg, G. M., *Proc. Symp. Chemical Metallurgy of Iron and Steel*, The Iron and Steel Institute, London, 17–22, 1971.
6. Hillert, M., Jansson, B., Sundman, B., and Agren, J., *Metall. Trans.* **16A,** 261–266 (1985).
7. Pelton, A. D. and Blander, M., *Metall. Trans.* **17B,** 805–815 (1986).
8. Gaye, H. and Welfringer, J., *Proc. 2nd Internat. Symp. Metal Slags and Fluxes*, Lake Tahoe, NV, 357–375, 1984.
9. Hillert, M. and Sundman, B., Xizhen Wang, *Report*: Royal Inst. Tech., Stockholm, TRITA-MAC-0355, 1985 and private communication.
10. Barry, T. I. and Chart, T. G., *High Temperature Materials R&D for Industry*, E. Bullock, ed., Commission of the European Communities, 1989 (in press).
11. Wilson, A. M., Gohil, D. D., and Laing, K., *CALPHAD XV Conference Poster Session*, London, July 1986.

# Thermochemical Applications of Thermo-Calc

BJÖRN JÖNSSON* AND BO SUNDMAN

*Division of Physical Metallurgy, Royal Institute of Technology, S-100 44 Stockholm, Sweden*

## ABSTRACT

Thermo-Calc is a databank for thermochemistry and metallurgy, It consists of sophisticated and generalized software for calculation of equilibria and phase diagrams together with several databases for inorganic chemistry and metallurgy. Initially, its main application was for alloy design and the development of steels. Recently, a number of new databases have extended the applicability of Thermo-Calc considerably. A substance database from SGTE (Scientific Group Thermodata Europe) and a slag database assessed by IRSID have been incorporated into Thermo-Calc. Combined with a dilute solution database assessed by Sigworth and Elliott, the IRSID database forms a very useful tool for understanding the metal/slag equilibrium in practical applications. A new geological database for solid oxides and silicates assessed by Saxena and coworkers also includes high pressure properties. Finally, at KTH, new assessments of oxide systems have been started using a more generalized model for the liquid phase. Thermo-Calc is available on line from Europe or can be obtained on a license basis.

**Index Entries:** Thermodynamics; databases; slags; metallurgy; inorganic chemistry; software.

## INTRODUCTION

Thermo-Calc is a databank (1) for thermochemistry and metallurgy. It consists of sophisticated and generalized software for the calculation of chemical equilibria and phase diagrams in very general kinds of thermochemical systems. It was originally developed for calculation of equilibria in steels, and its main application was for alloy design and development.

---

*Author to whom all correspondence and reprint orders should be addressed.

High Temperature Science, Vol. 26    © 1990 by the Humana Press Inc.

It has been used for this purpose for almost ten years. In the last few years, there has been considerable interest to use Thermo-Calc also for other types of chemical systems, such as metallurgical slag systems or molten salts. For this purpose, a number of new databases have been incorporated into the Thermo-Calc system, thereby considerably extending its applicability. Examples of new databases are a substance database from SGTE (Scientific Group Thermodata Europe) (2) and a metallurgical slag database assessed by IRSID (Institut de Recherches de la Siderugie Francaise), a French steel research institute (3), which consists of assessed data for the liquid and condensed compounds in the system $Al_2O_3$–$CaO$–$FeO$–$Fe_2O_3$–$MgO$–$MnO$–$SiO_2$. For the liquid oxide, IRSID uses the "cell model" of Kapoor and Frohberg (3), and this has been implemented into the Thermo-Calc system. An enhancement made at KTH (The Royal Institute of Technology) was to merge the IRSID database, which only covers the oxides, with a database for dilute solutions of about 20 elements in liquid iron assessed by Sigworth and Elliott (4) in order to make it possible to use the Thermo-Calc software for realistic calculations of equilibria between the slag and the liquid steel. In addition to this, the Thermo-Calc system has been extended by a new geological database for solid oxides and silicates assessed by Saxena and co-workers that also includes high pressure properties using the Murnaghan model. Finally, at KTH, new assessments of alloys and oxide systems have been started using a generalized and consistent model for the liquid phase.

## SOFTWARE

The Thermo-Calc system consists of different modules, as shown in Fig. 1, in order to make it easier to handle for the user. All the modules have a uniform user interface with a menu of commands, and when the user has typed a command the program may ask questions. This kind of "conversation" between the user and the program makes the program easy to learn and for the novice there is extensive on-line help by just pressing the "?" whenever the program asks a question. The commands are very elementary and by combining these in almost any order an advanced user can control the action of the program in a very flexible way.

Each module in Thermo-Calc has a specific purpose. In the database module, the user selects database and defines his system, i.e., which components he is interested in. There are two equilibrium calculation modules, the older POLY-1 is dedicated to alloy systems, whereas the new, POLY-3, can handle more general equilibrium calculations, e.g., gas equilibria and equilibria with slags. Each of these modules has a postprocessor for tabular or graphical output of the results. In addition to these modules, there is one module for interactive manipulation of thermodynamic model parameters, and, finally, one module for tabulation of data or chemical reactions.

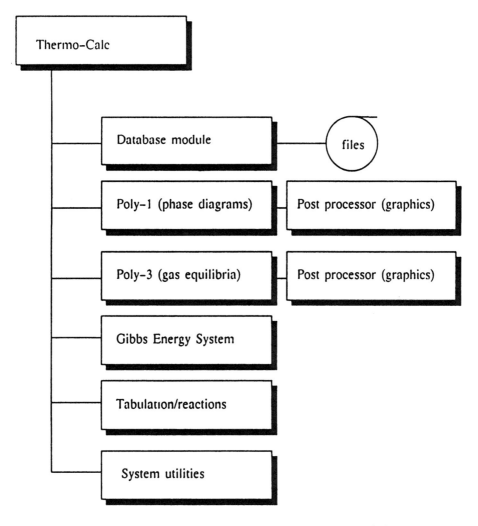

Fig. 1.  The modular structure of the Thermo-Calc.

## DATABASES

Once one has access to a general thermodynamic tool like the Thermo-Calc, the important question is, how good are the databases available for handling a particular problem? A thermochemical database is a unique type of database in the sense that one may have just a few assessed systems in it, e.g., the IRSID database has just twenty binaries, but it can be used to calculate and predict properties in multicomponent systems where there is no experimental information available. Currently, the Thermo-Calc has access to the following databases

- The SGTE solution database with about 150 assessed binary, ternary and higher systems;
- The SGTE substance database with about 2000 substances and 1000 gaseous species;

- The SGTE aqueous database with 500 species in dilute aqueous solutions;
- The Fe-base database with about 25 assessed ternary systems in the iron rich corner;
- The IRSID/Sigworth-Elliott database for oxide slag/liquid iron equilibria; and
- The SAXENA database for solid oxides and silicates up to high temperatures and pressures.

SGTE stands for Scientific Group Thermodata Europe (2), and is a consortium of seven research organizations and universities in England, France, Germany, and Sweden. SGTE has a long term project to provide a general purpose thermochemical database with assessed data.

The SGTE databases are extended each year. The other databases are updated only occasionally. Within the SGTE group, there is currently a great interest to develop a new consistent oxide database for slags, ceramics, and geological systems.

## ASSESSMENT PROCEDURE AND MODELS

It is a task requiring high qualifications to develop databases for thermochemical calculations. One cannot compare it with traditional bibliographical or other similar databases, because it is not sufficient just to collect or select the available data for a particular system. One also has to interpret experimental data found in terms of thermochemical models, and a number of model parameters must be determined. Both the selection of model and the decision which data to trust if there are inconsistencies can be a very difficult task.

An advantage of assessing the data in terms of thermochemical models is that one gets a convenient and more reliable way of extrapolating the properties of the system under consideration. Thus, one can easily calculate the thermochemical properties of a multicomponent system, although one only has access to assessed data from binary or ternary systems. In multicomponent systems, the amount of experimental data is usually very small and the possibility of predicting their properties is a very important feature of a system like Thermo-Calc.

In the IRSID database, the solid oxides are all treated with a fixed stoichiometry. For many oxide systems, this treatment is not satisfactory. Therefore, at KTH, we have started a project on assessing some oxide systems using more sophisticated models, i.e., the compound energy model (5) and a new ionic model for the liquid (6). One advantage of the new model for the liquid is that it can have more than one anion, whereas the cell model used by IRSID cannot easily handle more than one anion, i.e., oxygen. With the new assessments, we hope to incorporate sulfur or any other anionic species. Another advantage is that the new model gives a uniform description of the liquid, treating the oxide and metallic liquids as a single phase with a miscibility gap. This project has just recently started, and, so far, Fe–O, CaO–SiO$_2$, Al$_2$O$_3$–CaO, and

Ca–Fe–O are in various stages of completion. The data for the solid oxides are assessed from 298.15 K up to their melting points. This makes it possible to compute their enthalpies and heat capacities. The current project at KTH is to assess most ternary oxide systems in the Al–Ca–Fe–Mg–Si–O system within two years. Parallel projects will lead to an assessment of some sulfur systems starting with Cu–Fe–S.

Other groups are also working on databases for oxide systems, for example, the FACT (Facility for the Analysis of Chemical Thermodynamics) group of Montreal. They have developed a new model for the liquid using a quasichemical entropy expression. As already mentioned, we have the database of Prof. Saxena at Brooklyn College, Brooklyn, NY, who works with geological systems and has a special interest for high pressure properties. Part of his assessed data is already included into the Thermo-Calc. Figure 2 gives a simple example of what can be done with this database.

## MODELING THE EFFECT OF PRESSURE

In Thermo-Calc, one may use any type of explicit function to describe the pressure and temperature dependency of a substance. However, a rather simple but useful model developed by Murnaghan (7) has been adopted for most condensed compounds. The Murnaghan model assumes that the bulk modulus is a linear function of pressure

$$B = B_o(T) + nP \qquad (1)$$

where $B_o$ is the bulk modulus at zero, or atmospheric, pressure. It may depend on temperature. The pressure coefficient, $n$, is usually about 4. Unfortunately, there are very few experimental data for high temperatures and high pressures that are needed to determine if $n$ is temperature dependent or not. Since Thermo-Calc requires an explicit expression for the Gibbs energy of each substance (or each phase), we have to do an integration. The expression for G(T,P) that incorporates the Murnaghan equation and the expansivity is developed below.

The compressibility at constant temperature is equal to the inverse of the bulk modulus

$$K = 1/B \qquad (2)$$

and assuming isotropic behavior the compressibility is equal to the pressure derivative of the volume divided by the volume

$$1/V \; \delta V/\delta P = - K \qquad (3)$$

After inserting the Murnaghan expression for B(T,P), we can integrate and obtain

$$V = V_o(T) \, (1 + nP/B_o(T))^{-1/n} \qquad (4)$$

Since the volume is the pressure derivative of the Gibbs energy we formally obtain

THERMO-CALC (88.12.21:10.34) :

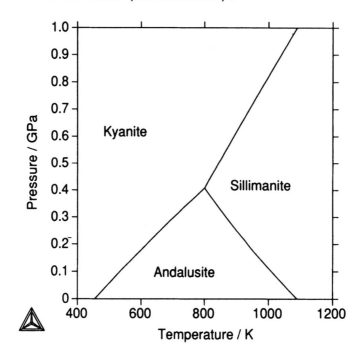

Fig. 2.   Calculated P-T diagram for Sillimanite, Andalusite, and Kyanite. Thermochemical data are from the Saxena database.

$$G_V = \int_O^P V \, dP = V_o(T) \left[ \frac{B_o(T)}{n-1} \right] [(1 + nP/B_o(T))^{1-1/n} - 1] \tag{5}$$

The thermal expansivity of the volume

$$\alpha(T) = 1/V_o \, \delta V_o/\delta T \tag{6}$$

is usually expressed as a polynomial in temperature. By integration we obtain

$$V_o(T) = V_o(T_o) \exp( \int_{T_o}^T \alpha(\tau) \, d\tau) \tag{7}$$

where $T_o$ is a reference temperature, usually 298.15K. Altogether this gives

$$G_V = V_o(T_o) \exp( \int_{T_o}^T \alpha(\tau) \, d\tau) \left[ \frac{B_o(T)}{n-1} \right] [(1 + nP/B_o(T))^{1-1/n} - 1] \tag{8}$$

which is the contribution to the Gibbs energy of the substance owing to the pressure dependence expressed by the Murnaghan model.

This expression can readily be included in a database if the coefficients $\alpha(T)$, $B_o(T)$, and $n$ are known or can be estimated. With Thermo-Calc one can also assess such coefficients from experimental data on the volume, thermal expansivity, and bulk modulus of a substance. For pure elements a number of such assessments have been made (Fe, Zr, W, Co, and so on) in addition to the oxide systems.

A special advantage of the Murnaghan model is that one may invert the function, and, instead of expressing the volume as a function of pressure, one may express the pressure as a function of the volume. In this way one may use the same data for a database based on Helmholtz energy rather than the Gibbs energy. For a more extensive discussion of the Murnaghan model see e.g. (8,9).

## SOME DISCUSSION
### WITH REFERENCE TO APPLICATIONS
### OF THERMO-CALC

In this section, we will discuss some of the different models available in Thermo-Calc and then show some examples. In principle these models may be regarded as extensions of the well known regular-solution model that is based on the concept of random mixing of the constituents. In order to take into account deviations from random mixing, polynomial corrections are introduced, if the deviations are moderate, and multi-sublattice occupancy (the compound-energy model (5) for solids and the ionic-liquid model (6) for liquids) and/or the introduction of molecular like species (the associate-solution model (10) if there are strong deviations.

One may use any type of explicit function to describe the pressure and temperature dependency of the Gibbs energy of a substance. A simple and widely used model is a straight forward polynomial in T and P. However, sometimes there are strong variations in the thermodynamic properties over a narrow range. In such a case, it is not so convenient to use a polynomial description because many terms are needed. It is then preferable to treat such an effect separately with a model based on the physical effect giving rise to the strong variation of the properties. One such model incorporated in Thermo-Calc is the Hillert-Jarl-Inden (11) model for the effect of a magnetic transition. As an example, we would like to give reference to a paper by Fernández Guillermet (12) entitled *A Thermodynamic Treatment of the Volume Effects in Ferromagnetic Metals*, which incorporates the use of Eq. 8 as well as the Hillert-Jarl-Inden model.

As some specific examples on applications of the compound energy model (5), we would like to give reference to two papers. Andersson and Sundman (13) applied it to the sigma phase in the Cr–Fe system and Hillert, Jansson and Sundman (14) applied it to non-stoichiometric $CeO_{2-x}$, and an orthopyroxene solid solution, and non-stoichiometric magnetit, and an olivin solid solution.

We give reference to three different papers on applications of the ionic-liquid model (6) in order to show its power to model the gradual change in character of the liquid as one moves from one side to another in a phase diagram where the pure elements are of different character.

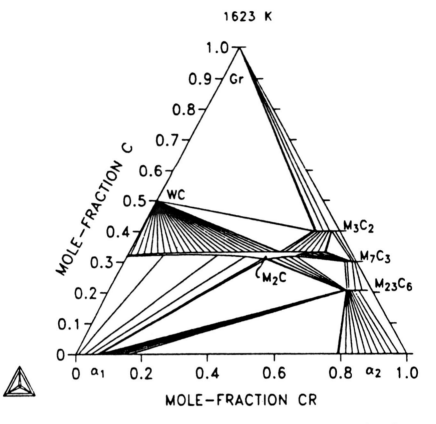

Fig. 3.    From Ref. *19*. Calculated isothermal section of the C–Cr–W system at 1623 K.

Fernández Guillermet et al. (*15*) applied it to the Fe–S system, and Jönsson and Ågren (*16*) to the Mg–Sb system. As an example on an application to an oxide system we refer to Hillert, Sundman, and Wang (*17*), who have assessed the CaO–SiO$_2$ system.

As indicated previously, Thermo-Calc can also handle gas-phase equilibria. As an example, we refer to Sproge and Ågren (*18*), who modeled gas consumption in a gas carburizing process.

Let us now consider the Poly modules that are used for equilibria calculations. The simplest way of using Poly is to do a single calculation in order to find the equilibrium state (it is also possible to calculate constrained equilibria) for a given pressure, temperature, and overall composition. However, one is often interested in calculating how some property varies with temperature (or pressure or composition or chemical potential or combinations thereof). For this purpose, Poly gives one the possibility of specifying stepping variables (chosen from the above mentioned) in order to do such mapping automatically in one, two, or three dimensions, with a few commands. We will give some illustrative examples below, all of them are outputs from the Poly postprocessor.

The first example is the phase diagram previously presented in Fig. 2. Stepping variables in this case were temperature and pressure. The next example is an isothermal section of the ternary system C–Cr–W, *see*

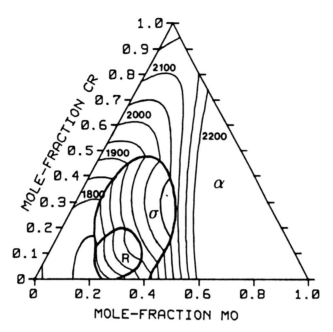

Fig. 4. From Ref. *20*. Calculated liquidus projection of the Cr–Fe–Mo system.

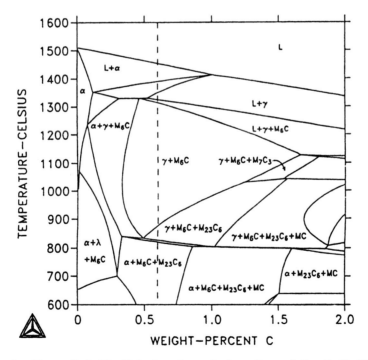

Fig. 5. From Ref. *21*. Calculated vertical section of the C–Cr–Fe–Mo–W system at 4 w/o Cr, 6 w/o Mo, and 6 w/o W.

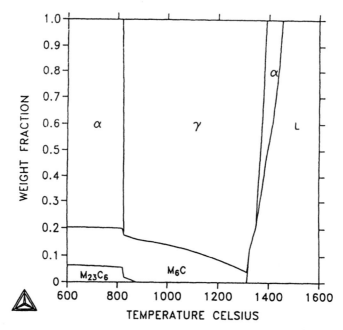

Fig. 6. From Ref. *21.* Calculated weight fractions of the phases present in the alloy indicated with the dashed line in Fig. 5.

Fig. 3. Stepping variables were fraction C and fraction Cr. Still other examples are Figs. 4–6. Figure 4 is a calculated liquidus projection with superimposed isothermal lines, stepping variables were the fractions of Mo and Cr, and the temperature. Figure 5 is a calculated vertical section in a quinternary system. Stepping variables were fraction C and temperature. Finally, Fig. 6 shows the calculated phase amounts along the dashed line in Fig. 5. The temperature was used as stepping variable.

We hope that these examples have given some flavor of what is possible to achieve with the aid of Thermo-Calc. It is, of course, possible to calculate many other types of diagrams, and, in addition, the Poly postprocessor allows the user to define and plot quite arbitrary functions of the thermodynamic quantities.

## ACKNOWLEDGMENTS

This work has been sponsored by The Swedish Board for Technical Development (STU).

## REFERENCES

1. Sundman, B., Jansson, B., and Andersson, J-O., *Calphad,* **9,** 153 (1985).
2. Ansara, I. and Sundman, B., *The Scientific Group Thermodata Europe,* Proc. Conf. CODATA, Ottawa, July, 1986.
3. Gaye, H. and Columbet, D., Report CECA no 7210-CF/301 (1984).
4. Sigworth, G. K. and Elliott, J. F., *Metal. Sci.* **8,** 298 (1974).

5. Andersson, J.-O., Fernández Guillermet, A., Hillert, M., Jansson, B., and Sundman, B., *Acta Metall.* **34,** 437 (1986).
6. Hillert, M., Jansson, B., Sundman, B., and Ågren, J., *Metall. Trans.* **16A,** 261 (1985).
7. Murnaghan, F. D., *Proc. Natn. Acad. Sci. USA* **30,** 244 (1944).
8. Fernández Guillermet, A., *J. Phys. Chem. Solids* **47,** 605 (1986).
9. Fernández Guillermet, A., *J. Phys. Chem. Solids* **48,** 819 (1987).
10. *See,* for instance: Sommer, F., *Z. Metallkde* **73,** 72 (1982).
11. Hillert, M. and Jarl, M., *Calphad* **2,** 227 (1978).
12. Fernández Guillermet, A., *High Temp. High Press.* **19,** 639 (1987).
13. Andersson, J.-O. and Sundman, B., *Calphad* **11,** 83 (1987).
14. Hillert, M., Jansson, B., and Sundman, B., *Z. Metallkde* **79,** 81 (1988).
15. Fernández Guillermet, A., Hillert, M., Jansson, B., and Sundman, B., *Metall. Trans. B.* **12B,** 745 (1981).
16. Jönsson, B. and Ågren, J., *Metall. Trans. A.* **18A,** 1395 (1987).
17. Hillert, M., Sundman, B., and Wang, X., *Trita-Mac-381*, Internal Report, Div. Phys. Met., Royal Institute of Technology, Stockholm, Sweden, Nov. 1988.
18. Sproge, L. and Ågren, J., *J. Heat Treating* **6,** 9 (1988).
19. Gustafson, P., *Metall. Trans. A.* **19A,** 2547 (1988).
20. Andersson, J.-O. and Lange, N., *Metall. Trans. A* **19A,** 1385 (1988).
21. Gustafson, P., *Trita-Mac-354*, Internal Report, Div. Phys. Met., Royal Institute of Technology, Stockholm, Sweden, Nov. 1987.

# Thermodynamic Modeling
# of High Temperature Melts
# and Phase Diagram Calculations

M. Gaune-Escard* and G. Hatem

SETT, UA 1168, Université de Provence, Centre de St Jérôme, Av.
Escadrille Normandie Niemen, 13397 Marseille Cédex 13, France

## ABSTRACT

Thermodynamic modeling was reviewed for different classes of
high temperature melts: symmetrical and asymmetrical, binary and
multicomponent, ionic mixtures, and also those systems that contain
at least one covalent component. Applications to phase diagram
calculation are presented.

## INTRODUCTION

There are several kinds of molten salt solutions according to the
nature and the number of involved components.

First, the solutions can be either fully ionic (obtained from components dissociated into ions when molten), or either partially or nonionic
(obtained from at least one covalent salt).

Classifications usually adopted for other classes of mixtures (alloys,
molecular solutions, and so on) do not adequately describe ionic melts.
These liquids consist of ions, and the mixtures can better be described
considering those ions rather than the number of components. Two-
component mixtures are unambiguously defined as "binary" for other
classes of solution (e.g., the Au-Ge binary alloy), but a two-component
molten salt mixture can be not a "true" binary: LiCl-NaBr, for instance, is
a ternary reciprocal system, whereas LiCl-NaCl is a common-ion (additive) binary system.

Ternary molten salt mixtures, also, can be either additive (common-
ion; e.g., LiCl-NaCl-KCl) or reciprocal (e.g., LiCl-NaCl-NaBr), both hav-

*Author to whom all correspondence and reprint orders should be addressed.

ing three independent components and four ions distributed either as 3 cations (anions)/1 anion (cation) or as 2 cations (anions)/2 anions (cations).

A schematic classification is given in Table 1 for charge symmetrical systems and, obviously, also holds for charge unsymmetrical mixtures. Additive systems and reciprocal systems not only represent two distinct types of molten salt solutions, but are parallel to two distinct types of metallic systems. The first type belongs to the class of additive systems in which either the types of positive ions differ and there is one type of anion (e.g., $A^+$, $B^+//X^-$) or the types of anions differ and there is one type of cation (e.g., $A^+//X^-$, $Y^-$). This class parallels substitutional alloys (e.g., Na + K). This can be readily seen if one considers that electrons, at least in a formal sense, are like anions, e.g., Na + K $\approx$ $Na^+$, $K^+//e^-$.

The second type is a member of the class of reciprocal systems that are systems containing at least two types of cations and two types of anions (e.g., $A^+$, $B^+//X^-$, $Y^-$). The simplest member of this class are ternary systems. Reciprocal systems are equivalent to interstitial alloys (e.g., $Fe^{2+}$, $Cr^{2+}//O^{2-}$, $e^-$).

Taking into account the nature of these different mixtures, appropriate relationships have been proposed for the ideal entropy of mixing. These are of primary importance to measure the deviation from ideality of the other functions of mixing. We give in the following the relationships expressing the ideal entropy in some typical ionic mixtures.

Another distinction among ionic melts lies in the possibly different valencies of the ions involved; in the previously quoted systems, all cations, and also anions, had the same charge, whereas a mixture like $NaCl$-$CaCl_2$ includes at the same time single-charged and double-charged cations. Also, those solutions obtained from at least one covalent salt have different features, generally characterized by complex species (e.g., $AlCl_4^-$, $Al_2Cl_7^{2-}$ in the aluminum chloride-based mixtures). A considerable amount of experimental thermodynamic investigations made evident that those different classes of melts do not behave identically and, therefore, also evident that specific thermodynamic models should be developed and used to describe their features (1).

## IONIC MELTS

### Symmetrical Mixtures ($AX + BX$ or $AX_2 + BX_2$)

These mixtures, e.g., $NaCl + KCl$ or $BaCl_2 + CaCl_2$, have been described as constituted by two interlocking lattices, one cationic and one anionic; for an ideal solution, the cations mix randomly on the cation sublattice and the anions on the anion sublattice. If the energetic interactions of $A^+$ and $B^+$ with their environments are the same, then the molar free energy of mixing is $-T\Delta S_{mix}$ and

$$\Delta S_{id}^{mix} = -R\left[x_A \ln x_A + x_B \ln x_B\right] \qquad (1)$$

Table 1
Schematic Classification of Ionic Melts

| Ions | System | Basic components |
|---|---|---|
| 3 ions<br>$A^+$, $B^+$, $X^-$<br>or $A^+$, $X^-$, $Y^-$ | Binary<br>Common ion | 2 components<br>AX, BX<br>or AX, AY |
| 4 ions<br>$A^+$, $B^+$, $C^+$, $X^-$<br>or $A^+$, $X^-$, $Y^-$, $Z^-$ | Ternary<br>Additive | 3 components<br>AX, BX, CX<br>or AX, AY, AZ |
| $A^+$, $B^+$, $X^-$, $Y^-$ | Ternary<br>Reciprocal | Any 3 among AX,<br>BX, AY, BY |
| 5 ions<br>$A^+$, $B^+$, $C^+$, $D^+$, $X^-$ | Quaternary<br>Additive | 4 components<br>AX, BX, CX, DX |
| $A^+$, $B^+$, $C^+$, $X^-$, $Y^-$ | Quaternary<br>Reciprocal | Any 4 among<br>AX, BX, CX, AY, BY, CY<br>or AX, AY, AZ, BX, BY, BZ |

where $n_A$, $n_B$ are the number of moles of ions in the mixture and the x's are the ionic fractions defined as

$$x_A = n_A/(n_A + n_B) \quad \text{and} \quad x_B = n_B/(n_A + n_B) \quad (2)$$

This relationship was proposed by Temkin (2) for the ideal entropy of mixing and the $x_i$'s are the so-called (Temkin) ionic fractions. The Surrounded Ion Model (SIM) (3–5), which is a statistical model of ionic mixtures, gives a more realistic description of a melt since it takes into account all the possible energetic interactions of $A^+$ and $B^+$, depending on the local environment of each ion: each "surrounded ion" has nearest neighbors of the opposite charge on its first coordination shell ($X^-$) and next-nearest neighbors of the same charge on the second coordination shell ($A^+$ and $B^+$). The same result is found for the ideal entropy of mixing that is given by a relationship identical to (1); the most interesting features of this model lie in the fact that it is able to take into account the asymmetry of the thermodynamic excess functions with only two energetic parameters with a physical meaning. The previous models, which were based on a pair-wise interaction concept, failed in the description of such asymmetries that are observed experimentally. At this stage, most authors arbitrarily assumed a linear dependence of the "interaction parameter" against composition (Hardy's (6) quasi-regular model) or arbitrarily used polynomial expansions with parameters without any physical meaning. Also, the more elaborated Guggenheim's "quasichemical model" (7), though able to account for the temperature dependence of the thermodynamic excess functions, was unable to account for experimental asymmetries. The SIM, which is a most general model and was also found successful in describing the thermodynamics of other kinds of molten salt mixtures, allows the previously quoted models to be deduced as particular cases.

Deviations from ideality ($\Delta G^{ex} \neq 0$, $g_i \neq 1$) arise from a lack of balance in the interionic forces between the different species. If the salts are similar in chemical nature (i.e., sodium chloride + potassium chloride), the mixture is nearly ideal. If they differ, the forces are usually greater between ions of the same species.

### Binary Common-Ion Mixtures

For a great many simple binary molten salt solutions, simply polyomial expansions in the ionic fractions provide a good representation of the excess Gibbs energy. Least-squares techniques have been developed (8), mainly for alloys, which permit coefficients of such empirical expansions to be determined by a simultaneous optimization of all available thermodynamic and phase diagram data. But they can be used confidently only for those melts which can be assimilated to alloys, e.g., additive symmetrical molten salt mixtures.

### Ternary Mixtures

Models have been developed to permit the properties of ternary common-ion (9,10) and reciprocal (11,12) molten salt solutions to be calculated from the binary coefficients. Empirical ternary coefficients may also be included in these equations.

For years, the chemical industry has recognized the importance of the thermodynamic and physical properties of solutions in design calculations involving chemical separations, fluid flow, and heat transfer. The development of calorimetry, mass spectrometry, and potentiometry has enabled the experimental determination of excess enthalpies and heat capacities of molten salt mixtures with convenience and accuracy. But even with modern instrumentation, experimental measurements of thermodynamic properties have become progressively more expensive and time consuming with each additional component beyond binary mixtures. In the chemical literature, properties for binary systems are relatively abundant, properties for ternary systems are scarce, and properties for higher order multicomponent systems are virtually nonexistent. Naturally, one of the primary goals of research in the area of solution thermodynamics has been the development of expressions for predicting the thermodynamic properties of multicomponent mixtures.

As was discussed earlier, a binary system contains three kinds of ions, e.g., $A^+$, $B^+//X^-$. Of the three ionic components, only two are independently variable because of the constraint imposed by electroneutrality $n_A + n_B = n_X$. Ternary ionic systems contain four ions which can be constituted in three different ways $A^+$, $B^+$, $C^+//X^-$ or $A^+//X^-$, $Y^-$, $Z^-$, or $A^+$, $B^+//X^-$, $Y^-$.

The first two are additive ternary systems and the last is a ternary reciprocal molten salt mixture. There are only three independent component salts from which one can make up the solution. Ternary, quaternary, . . ., $n$th order additive molten salt mixtures are ionic systems containing 3, 4, . . ., $n$ ions of the same species (cations or anions) and 1 ion of the other species (anion or cation). Of the 4, 5, . . ., $(n + 1)$ ionic constituents, only 3, 4, . . ., $n$ are independent because of the constraint imposed by electroneutrality:

$$n_A + n_B + n_C + \ldots = n_X + n_Y + \ldots$$

The binary models provide reasonable estimates for several systems of practical importance and they also address the problem of how to report experimental data. Enthalpies of mixing, vapor pressures, and other solution properties are worthless unless they are transmitted from the experimentalist to the design engineer. Mathematical representation provides a convenient method to reduce extensive tables of experimental data into a few equations, and this is of crucial importance for multicomponent systems.

This paragraph will be devoted to several of the empirical equations or models that have been suggested for parametrizing and predicting mixture data.

Redlich and Kister (13) proposed an expression for the excess Gibbs free energy of mixing of a ternary mixture

$$\Delta G^{ex} = x_1 x_2 \Sigma (G_n^{ex})_{12} (x_1 - x_2)^n + x_1 x_3 \Sigma (G_n^{ex})_{13} (x_1 - x_3)^n$$
$$+ x_2 x_3 \Sigma (G_n^{ex})_{23} (x_2 - x_3)^n \qquad (3)$$

that provisions for additional ternary parameters. The initial popularity of the Redlich-Kister equation arose because the first parameter $(G_0^{ex})_{12}$ could be determined conveniently from the experimental data at $x_1 = 0.50$ as $4 \Delta G^{ex}{}_{12}$. Remember that the computers were not available during the 1940s and the majority of experimental data were graphically presented in the literature. The Redlich-Kister equation provided a means to transmit data from the experimentalists to the chemical engineer designing distillation columns.

It is common to predict the properties of a ternary solution phase by a simple summation of the binary expressions, when they obey the regular solution model. Several "geometric" models have proposed that differ in the geometric weighting factors of the binary contributions. Figure 1 summarizes those derived by Kohler (14), Jacob (15), Colinet (16), and Muggianu (17).

Kohler (14) proposed an equation for the excess Gibbs free energy of mixing of a ternary solution

$$\Delta G^{ex} = (x_1 + x_2)^2 \Delta G_{12}^{ex} + (x_1 + x_3)^2 \Delta G_{13}^{ex} + (x_2 + x_3)^2 \Delta G_{23}^{ex} \qquad (4)$$

in which $\Delta G_{ij}{}^{ex}$ refers to the excess Gibbs free energies of the binary mixtures at a composition $(x_i^0, x_j^0)$ such that $x_i^0, = 1 - x_j^0, = x_i/(x_i + x_j)$.

Kohler's equation is symmetrical in that all three binary systems are treated identically. Its numerical predictions do not depend on the arbitrary designation of component numbering.

Colinet (16) established a slightly more complex relationship for expressing the thermodynamic excess properties of multicomponent systems

$$\Delta G^{ex} = \Sigma\Sigma x_i (1 - x_j)^{-1} \{\Delta G_{ij}^{ex}\} x_j \qquad (5)$$

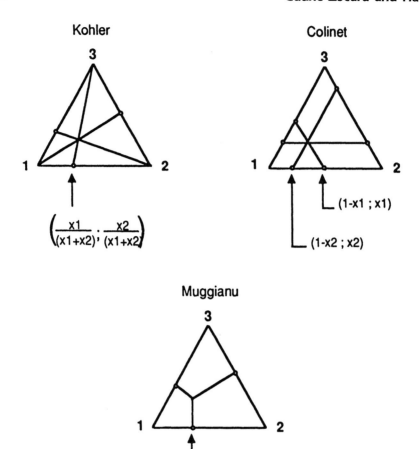

Fig. 1.   Geometric models for symmetrical systems.

in which $\{\Delta G_{ij}^{ex}\}x_j$'s are calculated from the binary data at constant mole fraction $x_j$. This equation, although perfectly symmetrical, requires the addition of the thermodynamic properties at six different binary compositions for a ternary mixture.

Muggianu (*16*) also developed a geometric model with different weightings of the binary contributions (*see* Fig. 1).

$$\Delta G^{ex} = \Sigma\Sigma x_i (1 - x_j)^{-1} \{\Delta G_{ij}^{ex}\}_{x1;x2}$$

So far, all the three methods that have been discussed treat the components in the same way and may thus be characterized as symmetric methods. However, sometimes there may be a physical reason to divide the component elements into different groups. For instance, if components 2 and 3 are similar to each other and differ markedly from component 1, then one should expect that the binary systems 1-2 and 1-3 to be similar and it may be advantageous to describe the ternary 1-2-3 system in such a way that the expression would reduce to the binary expressions if one could make 2 and 3 identical. A numerical method has been proposed by Toop (*17*), that has this asymmetric property. It is illustrated in Fig. 2, and it yields the following equation for the excess Gibbs free energy of mixing

$$\Delta G^{ex} = x_2(1 - x_1)^{-1} \Delta G_{12}^{ex} + x_3(1 - x_1)^{-1} \Delta G_{13}^{ex} \\ + (1 - x_1)^2 \Delta G_{23}^{ex} \tag{6}$$

in which the $\Delta G_{ij}^{ex}$ refer to the excess Gibbs free energies for the binary mixtures at compositions $(x_i^0, x_j^0)$ such that $x_i^0 = x_1$ for the 1-2 and 1-3 binary systems and $x_2^0 = x_2/(x_2 + x_3)$ for the 2-3 binary system.

Equation (6) possesses a desirable mathematical form since it is independent of binary parameterization. Other asymmetric "geometric" numerical methods have been proposed by Hillert (*18*) and are also illustrated in Fig. 2.

Differing from the previous empirical equations in that they are based on a physical description of the melt and on statistical mechanics principles, some models also provide expressions able to predict multi-component properties in terms of lower-order interactions. The Surrounded Ion Model (SIM) (*9,10*) yields an equation for the excess enthalpy of mixing of a ternary mixture

$$\Delta H = x_A x_B \{x_A \Delta H_{B(A)} + (1 - x_A) \Delta H_{A(B)}\} \\ + x_A x_C \{x_A \Delta H_{C(A)} + (1 - x_A) \Delta H_{A(C)}\} \\ + x_B x_C \{x_C \Delta H_{B(C)} + (1 - x_C) \Delta H_{C(B)}\} \tag{7}$$

where $\Delta H_{A(B)}$ is the limiting partial enthalpy of AX in the AX-BX binary mixture and the ideal Gibbs energy of mixing is always

$$\Delta G_{id}^{mix} = RT \{x_A \ln x_A + x_B \ln x_B + x_C \ln x_C\}$$

The Surrounded Ion MODEL (SIM) was also applied to quaternary reciprocal systems (5 ions, 4 independent components) (*19*). The Conformal Ionic Solution model (*20*) also gives a relationship for the excess enthalpy of mixing of a ternary solution

$$\Delta H = \Sigma \Sigma a_{ij} x_i x_j + \Sigma \Sigma b_{ij} x_i^2 x_j + \Sigma\Sigma c_{ij} x_i^2 x_j^2 \\ + P x_A x_B x_C + \Sigma \Sigma Q_i x_i^2 x_j x_k \tag{8}$$

The coefficients $a_{ij}$, $b_{ij}$, and $c_{ij}$ are evaluated from data on the three binary subsystems. The coefficients P and $Q_i$ of the "ternary" terms being calculated from the binary data.

Very recently, Hoch and Arpshofen developed a model which is applicable to binary, ternary and larger systems (*21, 22*). It was derived originally by looking at complexes in the solution, and the A-B bond strength (between species A and B) made dependent on the presence of other atoms in the complex. The model was applied to metal-salt (*23*), metal-metaloxide (*24,25*), silicate (*26,27*), and metallic systems (*28*). For a binary system the basic equation for the enthalpy of mixing is

$$\Delta H = Wn(x - x^n) \tag{9}$$

The term *n* is an integer (2, 3, 4, etc.), W is the interaction parameter, and *x* is the mole fraction of the component so that the maximum of $\Delta H$ (either positive or negative) is at $x > 0.5$. The quantity *n* is chosen such

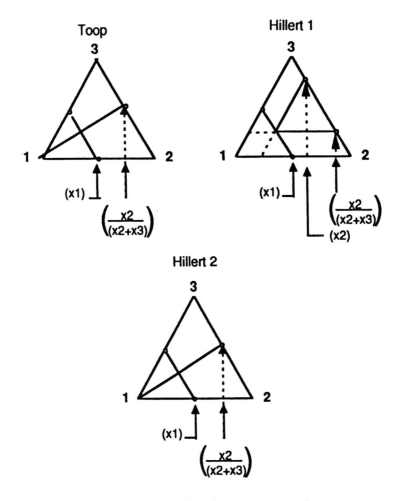

Fig. 2.   Geometric models for asymmetrical systems.

that $W$, determined from thermodynamic data, is independent of the composition.

The signs of the limiting partial enthalpies obtained from Eq. (9) are determined by $W$. In other words, one interaction parameter describes one type of interaction or reaction in a system. If two reactions exist in the binary system such as the $MgO$-$SiO_2$ (29), where a strong compound-forming tendency exists on the magnesium-rich side ($Mg_2$-$SiO_4$) and a miscibility gap exists on the $SiO_2$-rich side, then two interaction parameters, an attractive $W$ and repulsive $M$, are needed. Each has its major effect at different compositions with different dependency on composition ($n$ and $x$ in one case, $m$ and $y$ and the other). Thus

$$\Delta H = Wn(x - x^n) + Mm(y - y^m)$$

$W$ and $M$ are determined by least square analysis; confidence limit and error in $W$ and $M$ are also calculated.

This model was extended to ternary or larger system by evaluating the effect of each limiting binary system in the multicomponent mixture.

In an A-B-C mixture constituted of 3 components with the mole fractions $x$, $y$, and $z$, it comes for the effect of the binary A-B system in the ternary.

$$\Delta H = W4xy\{1 + (1-y) + (1-y)^2\} \tag{10}$$

This proliferation of similar expressions becomes confusing especially when the equations are encountered for the first time. There is no preferred way of knowing which method will provide the best predictions for a given system. The fact that so many empirical equations have been developed suggests that no single equation can describe all types of systems encountered.

To illustrate several of the predictive methods, let us assume one wishes to use the Redlich-Kister, Kohler, and SIM equations to predict the excess enthalpies of KF (1) + LiF (2) + NaF (3) mixtures at $x_1 = 0.50$, $x_2 = 0.25$.

The binary data (Holm and Kleppa (30)) are parameterized as

$$\Delta H_{12} = x_{1\times2}\{-19251 - 120x_1 + 4732x_2^2\} \qquad \text{J/mol}$$
$$\Delta H_{13} = x_{1\times3}\{-335\} \qquad \text{J/mol}$$
$$\Delta H_{23} = x_{23}\{-7565 + 368x_3\} \qquad \text{J/mol}$$

*Redlich-Kister equation*

$$\begin{aligned}
\Delta H_{123} = &\ (0.50)(0.25)\{-19251 - 120(0.5) + 4732(0.5)^2\} \\
&+ (0.5)(0.25)\{-335\} \\
&+ (0.25)(0.25)\{-7565 + 368(0.25)\} \\
= &\ -2774 \text{ J/mol}
\end{aligned}$$

*Kohler equation*

The excess enthalpies of the binary systems at $x_i^0$, $x_j^0$ are first calculated:

$$\Delta H_{12} = (0.67)(0.33)\{-19251 - 120(0.67) + 4732(0.67)^2\} = -3822 \text{ J/mol}$$
$$\Delta H_{13} = (0.67)(0.33)(-335) = -74 \text{ J/mol}$$
$$\Delta H_{23} = (0.50)(0.50)\{-7565 + 368(0.5)\} = -1845 \text{ J/mol}$$

and these quantities are then added in accordance to Eq. (4).

$$\begin{aligned}
\Delta H_{123} = &\ (0.50 + 0.25)^2(-3822) + (0.50 + 0.25)^2(-74) + \\
&\ (0.25 + 0.25)^2(-1845) = -2642 \text{ J/mol}
\end{aligned}$$

*SIM equation*

$$\begin{aligned}
\Delta H_{123} = &\ (0.5)(0.25)\{-19251(0.5) - 14639(0.5)\} \\
&+ (0.5)(0.25)\{-335(0.5) - 335(0.5)\} \\
&+ (0.25)(0.25)\{-7197(0.25) - 7565(0.75)\} \\
= &\ -2627 \text{ J/mol}
\end{aligned}$$

Up to this point, the primary emphasis has been on predicting multicomponent properties from binary data; several of the empirical expressions proposed during the past 40 years were summarized. These expressions also served as the point-of-departure for the mathematical representation of multicomponent excess properties. Differences between the predicted values and experimentally determined values are expressed as

$$(\Delta Z_{123}^{ex})_{exp} - (\Delta Z_{123}^{ex})_{cal} = x_1 x_2 x_3 Q_{123}$$

with $Q$-functions of varying complexity. The abbreviations *exp* and *cal* indicate experimental and calculated, respectively. For most systems commonly encountered, the experimental data can be adequately represented by the power series expansion

$$Q_{123} = A_{123} + \Sigma B_{12}^i (x_1 - x_2)^i + \Sigma B_{13}^i (x_1 - x_3)^i + \Sigma B_{23}^k (x_2 - x_3)^k \tag{11}$$

though it is unlikely data for multicomponent systems will be obtained with sufficient precision to warrant more than a few parameters.

### Reciprocal Mixtures

The $A^+$, $B^+/X^-$, $Y^-$ reciprocal mixture is also a ternary mixture; there are four constituent salts (AX, BX, AY, BY), but only three of them can be chosen as independent components: AX, BX, AY or AX, BX, BY or BX, AY, BY or AY, BX, BY. Therefore, a solution that contains a given amount of $A^+$, $B^+$, $X^-$ and $Y^-$ ions can be made in four different ways; for instance, a mole of the mixture containing $x_{A+} = 0.3$, $x_{B+} = 0.7$, and $x_{X-} = 0.5$ can be made up from four different combinations that lead to the same final solution.

|     | 1    | 2    | 3   | 4   |
|-----|------|------|-----|-----|
| AX  | -0.2 | 0.5  | 0.3 | —   |
| AY  | 0.5  | -0.2 | —   | 0.3 |
| BX  | 0.7  | —    | 0.2 | 0.5 |
| BY  | —    | 0.7  | 0.5 | 0.2 |

Though each of these ways leads to the same final solution, the free energy for the mixing process must differ in each of these. The negative quantities of constituents in the adjacent table can be understood from a composition diagram (*see* Fig. 3). Reciprocal systems must be represented by a square, whereas the composition of usual ternary systems is obtained from a triangle representation.

The addition of any component to a solution varies the composition linearly toward the corner of that component. Consider a mixture inside the two triangles AX, BX, BY and AY-BX-BY (Fig. 3): It can be made up of positive quantities of each of these three. On the other hand, to make up this composition from AX, AY, and BX and to get out of the triangle AX-AY-BX one must subtract AX from an AY + BX mixture. However, the activity and chemical potential of all four salts are defined and are the same no matter how the three components were chosen.

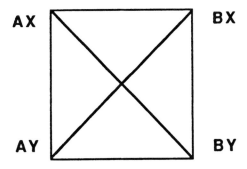

Fig. 3.   Composition square for a ternary reciprocal mixture.

For an ideal solution in our example $a_{AX} = 0.15$, $a_{AY} = 0.15$, $a_{BX} = 0.35$, and $a_{BY} = 0.35$. The free energy of mixing of such systems was first deduced by Flood, Førland, and Grjotheim ($31$).

$$\Delta G^{mix} = RT\{x_A \ln x_A + x_B \ln x_B + x_X \ln X_X + x_Y \ln x_Y\}$$

$$\pm x_i x_j \Delta G^0 \qquad (12)$$

where $\Delta G^0$ is the standard free energy change for the metathetical reaction

$$AX + BY = AY + BX \qquad (13)$$

and where $ij$ is the salt that is not a component. This last term expresses the idea that the three components, in effect, "react" to form the last constituent. The + sign is for the case when $ij$ is AY or BX and the − sign is for AX or BY. Everything beyond the last term given in Eq. (12) is the same no matter how the solution was made up.

Førland ($32$) improved the original FFG theory by including four binary interaction terms in (12).

$$\Delta G^{mix} = RT\{x_A \ln x_A + x_B \ln x_B + x_X \ln x_X + x_Y \ln x_Y\} \pm x_i x_j \Delta G^0$$

$$+ x_A x_B x_X \lambda_X + x_A x_B x_Y \lambda_Y + B\, x_X x_Y x_B \lambda_B \qquad (14)$$

with $\lambda_i$ the interaction parameter in the $i$ common ion binary mixture.

Later, Blander and Yosim ($33$) generalized the Conformal Ionic Solution theory ($34$), ($35$), which has been applied to binary molten salt mixtures by Reiss, Katz, and Kleppa ($36$). They found for the free energy of mixing a relationship identical to ($14$) and added a nonrandom term to this equation; this correction term was only taken by analogy with the one obtained from the quasi-lattice theory ($37$) developed for binary mixtures.

$$\Delta G^{mix} = RT\{x_A \ln x_A + x_B \ln x_B + x_X \ln x_X + x_Y \ln x_Y\} \pm x_i x_j \Delta G^0$$

$$+ x_A x_B x_X \lambda_X + x_A x_B x_Y \lambda_Y + x_X x_Y x_A \lambda_A + x_X x_Y x_B \lambda_B$$

$$+ x_A x_B x_X x_Y \lambda \qquad (15)$$

with $\lambda = -(-\Delta G^0)^2/2z\, RT$ and $z$ the cation-anion coordination number. The Surrounded Ion Model (SIM) was also applied to ternary reciprocal mixtures (11,12). The equation found for the Gibbs free energy of mixing is

$$\Delta G^{mix} = RT\{x_A \ln x_A + x_B \ln x_B + x_X \ln x_X + x_Y \ln x_Y\} \pm x_i x_j \Delta G^0$$
$$+ x_A \Delta G_A^{ex} + x_B \Delta G_B^{ex} + x_X \Delta G_X^{ex} + x_Y \Delta G_Y^{ex}$$
$$+ x_A x_B x_X x_Y \lambda \qquad (16)$$

with $\Delta G_i^{ex}$ the excess free energy of mixing in the i common ion binary mixture, $\lambda = -(-\Delta G^0)^2/2 \ zRT$ and z and the cation-anion coordination number.

It should be stressed that Eq. (16) is able to take into account experimental asymmetries in binary systems and contains the nonrandom term L that has been calculated and not estimated, by formal analogy, as previously. The Surrounded Ion Model has also been extended to quaternary reciprocal molten salt solutions (19).

## Asymmetrical Mixtures

### Binary Common-Ion-Mixtures ($AX_2$ + BX)

The same lattice description of the melt implies that the substitution to the $A^{++}$ divalent cation of the $B^+$ monovalent cation creates one vacancy on the corresponding sublattice. Several relationships have been proposed for the ideal entropy; they differ in the assumptions made on the vacancies. If the number of vacancies is assumed to be negligible, e.g., changing a divalent cation with a monovalent cation does not induce a local disorganisation of the corresponding sublattice, the same result (1) is found for the ideal entropy of mixing. If the number of vacancies is not negligible and if no particular association exists between the divalent ion and the vacancy, Førland (39) found the following relationship for the ideal entropy of mixing.

$$\Delta S_{id}^{mix} = -R\{2x_A \ln x_A' + (1 - x_A)\ln(1 - x_A')\} \qquad (17)$$

where

$$x_A = n_A/(n_A + n_B) \text{ and } x_A' = 2n_A/(2n_A + n_B) \qquad (18)$$

The $x_i$'s have been defined previously and the $ix_i$'s are the so-called equivalent ionic fractions. If the number of vacancies is not negligible and if the divalent ion and the vacancy are assumed to constitute a dimer, Flory (40) proposed for the ideal entropy of mixing.

$$\Delta S_{id}^{mix} = -R\{x_A \ln x_A' + (1 - x_A)\ln(1 - x_A')\} \qquad (19)$$

The SIM (41,42) provides, from different theoretical grounds, a relationship for the ideal entropy which is identical to the one (17) found by Førland.

It should be stressed that many authors do not take into account this charge-dependent expression for the ideal entropy of mixing but arbitrarily use equivalent ionic fractions (18) in the relationships expressing the other functions of mixing and Temkin's ideal entropy (1). It has been shown (43) that this may lead to some inconsistencies.

The "surrounded ion" model, already mentioned for melts including ions of same valency, was extended to asymmetrical mixtures of the kind $AX_2$-$BX_2$ or $AX_2$-$BX$. For these binary common-ion mixtures $AX_2 + BX$, the Surrounded Ion Model enabled some formal analogies to be evidenced between symmetrical and asymmetrical binary common ion mixtures; they are reported in Table 2.

*Additive Ternary Mixtures ($AX_2 + BX + CX$)*

Taking into account the analogies shown in Table 2, Eq. (7) obtained for symmetrical additive ternary mixtures becomes (44)

$$\Delta H/\lambda 1 + x_A\lambda = x'_A x'_B \{(\tfrac{1}{2})x'_B\Delta H_{A(B)} + (1 - x'_B)\,\Delta H_{B(A)}\}$$
$$x'_B x'_C \{x'_C\Delta H_{B(C)} + (1 - x'_C)\,\Delta H_{C(B)}\}$$
$$x'_C x'_A \{x'_A\Delta H_{C(A)} + (1 - x'_A)\,\Delta H_{A(C)}\} \tag{20}$$

with

$$x'_A = 2x_A/1 + x_A;\ x'_B = x_B/1 + x_A;\ x'_C = x_C/1 + x_A$$

*Reciprocal Mixtures*

The Surrounded Ion Model has been applied to asymmetrical ternary reciprocal mixtures ($A^{2+}$, $B^+/X^-$, $Y^-$) (45). By using the same analogies aforementioned the equation found for the molar heat of mixing is

$$\Delta H = x'_B x'_X(1 + x_A)\Delta H_0 + x'_x\Delta H_x$$
$$+ x'_Y\,\Delta H_Y + x_B\Delta H_B + x_A\Delta H_A$$
$$+ x'_A x'_B\, x'_x x'_Y\,(1 + x_A)\Lambda \tag{21}$$

where $\Delta H_0$ is the enthalpy change for the metathetical reaction

$$(\tfrac{1}{2})A_2X + BY = (\tfrac{1}{2})B_2Y + AX$$

$\Delta H_i$ the excess enthalpy in the $i$ common ion binary mixture, and $\Lambda = -(\Delta H^0)^2/2zRT$

## PARTIALLY IONIC MELTS

Very few results exist on the thermodynamics of such systems. The reason is twofold, since, owing to very different physical and chemical properties between the components of the mixtures, experimental investigations are not easy, and theoretical approaches are not simple. Some experimental investigations and some theoretical approaches were recently performed or are in progress for systems containing alkali halides and halides of aluminum, rare earth, niobium, tantalum, bismuth, zirconium, and also for the very novel low melting systems containing aluminum chloride and organic chlorides. Most of these investigations are related to industrial interests and are parts of national research programs in France, China, Germany, India, Norway, Poland and the USA. The applications fields are mainly the production of the metals from electrolysis of the melts, fabrication of glasses in relation to optical fibers, elaboration of new materials, and so on.

Table 2
Formal Analogies Between Symmetrical
and Asymmetrical Mixtures

| Symmetrical mixtures AX + BX | | Asymmetrical mixtures $AX_2$ + BX |
|---|---|---|
| $x_A = n_{AX}/\lambda n_{AX} + n_{BX}\lambda$ | $\rightarrow$ | $x'_A = 2n_{AX_2}/\lambda 2n_{AX_2} + n_{BX}\lambda$ |
| $\Delta H$ | $\rightarrow$ | $\Delta H' = \Delta H/\lambda 1 + x_A\lambda$ |

A proper description of the features of such melts can be given by models based on "associated solution" concepts. Very similar models have been developed recently and independently in Marseille (46) and in Trondheim (47) that are equally successful in describing the thermodynamics of $BiCl_3$ and $AlCl_3$-based mixtures.

Melts including at least one covalent salt may be either partially or non ionic. Mixtures obtained from alkali halides AX (e.g., fully ionic salts) and transition, rare-earth, actinide, metal halides $M_iX_j$ (e.g., covalent salts, generally, $p$ (> 0) and $q$ (< 0) are the electrovalencies of M and X respectively), are examples of such melts, generally characterized by the formation of heterogeneous ionic complexes such as $M_iX_{(j + 1)q}$, $M_{(i + 1)}X_{(j + 1)}^{p + q}$, and so on. The existence of complexes in molten salt mixtures has been the subject of lively discussions in the molten salt community (48). Development of investigation methods of structure brought new arguments to the controversy. These complexes contribute to the entropy of the mixture and should be taken into account to modelize the solution.

We have recently proposed an ideal associated model to describe the mixtures of alkali chlorides with bismuth chloride or mixtures of alkali fluorides with zirconium fluoride. The excess thermodynamic quantities of the melt are assumed to arise only from the formation of complex ionic species. For the ZrF4-based mixtures, for instance, the enthalpies and the constants of formation refer to the reactions

$$(M^+)(F^-) + ZrF_4 = (M^+)(ZrF_5^-) \qquad :(h_1, K_1)$$
$$3(M^+)\,3(F^-) + ZrF_4 = (M^+)_3\,(ZrF_7^{2-}) \qquad :(h_2, K_2)$$

and the enthalpy of mixing is obtained as

$$\Delta H = n_4/(n_1 + n_3)h_1 + n_5/(n_1 + n_3)h_2$$
$$\Delta H = x^*(1 - x^*)\{[K_1h_1 + K_2h_2x^{*2}]\}1 + K_1x^*(2 - x^*) + K_2x^{*3}(4 - 3x^*) \tag{23}$$

with $x = x(MF)$

$$x = x^*\{1 + K_1 + K_2x^{*2}(3 - 2x^*)\}/\{1 + K_1x^*(2 - x_3^*) + K_2x^{*3}(4 - 3x^*)\} \tag{24}$$

The ($h_i$'s $K_1$'s) solutions are obtained numerically.

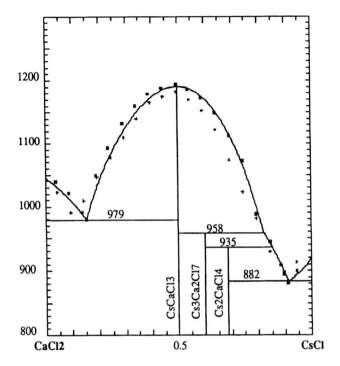

Fig. 4. Calculated phase diagram of the CaCl₂-CsCl system (temperature in K).

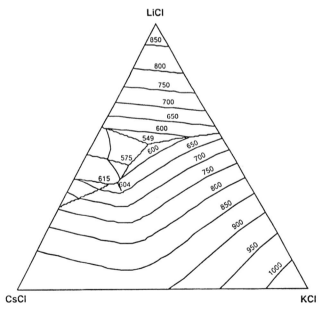

Fig. 5. Calculated liquidus of the CsCl-KCl-LiCl system (temperature in °C).

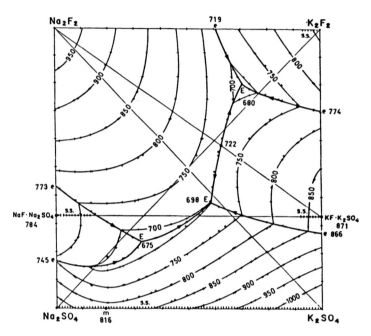

Fig. 6. Calculated liquidus of the Na, K/F, SO₄ system (temperature in °C).

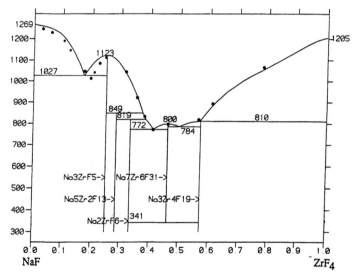

Fig. 7. Phase diagram of the NaF-ZrF₄ system (temperature in K). • Ref. (55). —Calculated.

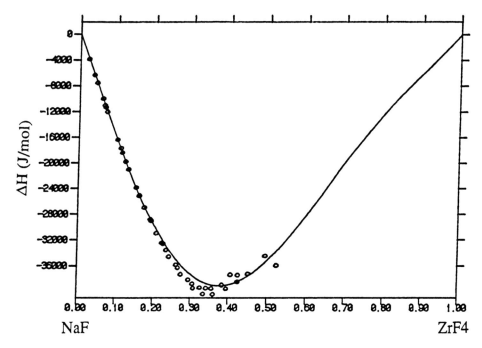

Fig. 8. Enthalpy of mixing of the NaF-ZrF$_4$ system. • Ref. (56). — Calculated (Redlich-Kister).

The four parameters $K_1$, $K_2$, $h_1$, and $h_2$ are obtained by numerical calculation. From the experimental enthalpy data it is possible to evaluate two approximate values for $h_1$ and $h_2$; they are used as the initial parameters $h_{10}$ and $h_{20}$ in the iterative procedure. Two arbitrary values are set for the initial parameters $K_{10}$ and $K_{20}$. For each experimental data set $\{x, \Delta H(x)\}$, the program solves Eq. (24) and calculates the solutions $x^*$.

A nonlinear regression program, applied to Eq. (23), yields optimized values of the parameters $h_{11}$, $h_{21}$, $K_{11}$, and $K_{21}$. If these values differ from the initial values, the program iterates with $h_{11}$, $h_{21}$, $K_{11}$, and $K_{21}$ as initial values. Convergence is attained after a few iterations.

## PHASE DIAGRAM CALCULATIONS

The principle of phase diagram calculation is minimization of the Gibbs free energy of the mixture under investigation. Several numerical calculation programs exist so far in Europe and North-America. These programs, developed by specialized research groups, differ in the method of description of the envisaged system (thermodynamic functions of mixing are described either by empirical polynomial expansions or by equations deduced from theoretically based models); and the mathematical methods worked out in the search of a minimum.

For metallic systems, for instance, a very efficient program has been elaborated, and successively improved, by H. L. Lukas (49) (Max-Planck-

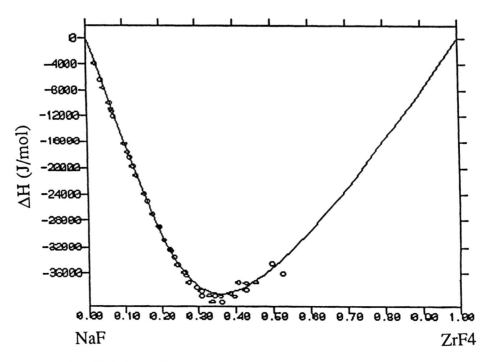

Fig. 9.   Enthalpy of mixing of the NaF-ZrF$_4$ system. • Ref. (56). —
Calculated (associated model).

Institut, Stuttgart). Least-square methods were developed that enable all
the thermodynamic and phase diagram data to be simultaneously optimized and thus to provide optimized coefficients in polynomial expansions of thermodynamic functions of mixing.

Using this program, it has been possible to calculate satisfactorily the
phase diagrams of common-ion binary (50) or multicomponent (51), (52)
molten salt mixtures. Modifications revealed indispensable in order to
perform such calculations with asymmetrical ionic systems. These were
made according to the previously quoted theoretical developments for
the Surrounded Ion Model.

The results obtained for the CaCl$_2$-CsCl melt are given as an
example. Figure 4 shows the phase diagram of this mixture. Other examples of SIM phase diagram calculations are depicted in Figs. 5 and 6 for
the ternary additive and reciprocal systems CsCl-LiCl-KCl (52) and
Na,K/F,SO$_4$ (53), (54), respectively.

For the system NaF-ZrF$_4$, Fig. 7 shows the optimized phase diagram
calculated from a Redlisch-Kister description of the thermodynamic functions. It is evident that it does not provide a good representation of the
experimental phase diagram data (55) and only a moderately acceptable
representation of the enthalpy of mixing (56) (Fig. 8). When the associated solution model is used a quite satisfactory agreement is obtained for
ΔH (Fig. 9). Therefore, the optimization programs are being modified
accordingly.

# REFERENCES

1. Gaune-Escard, M., *Pure Appl. Chem.* **55**, 105 (1983).
2. Temkin, M., *Acta Physicochim. URSS* **20**, 411 (1945).
3. Gaune-Escard, M., Mathieu, J. C., Desré, P., and Doucet, Y., *J. Chim. Phys.* **9**, 1390 (1972).
4. Gaune-Escard, M., Mathieu, J. C., Desré, P., and Doucet, Y., *J. Chim. Phys.* **9**, 1397 (1972).
5. Gaune-Escard, M., Mathieu, J. C., Desré, P., and Doucet, Y., *J. Chim. Phys.* **11–12**, 1666 (1973).
6. Hardy, H. K., *Acta Metallurgica* **1**, 202 (1953).
7. Guggenheim, E. A., *Mixtures*, Oxford, Clarendon Press, 1952.
8. Lukas, H. L., Henig, E. Th., Zimmermann, B., *CALPHAD* **1,3**, 225 (1977).
9. Gaune-Escard, M., *J. Chim. Phys.* **9**, 1167 (1974).
10. Gaune-Escard, M., *J. Chim. Phys.* **9**, 1175 (1974).
11. Gaune-Escard, M., Mathieu, J. C., Desré, P., and Doucet, Y., *J. Chim. Phys.* **6**, 1003 (1973).
12. Gaune-Escard, M., Mathieu, J. C., Desré, P., and Doucet, Y., *J. Chim. Phys.* **7–8**, 1033 (1973).
13. Redlich, O., Kister, A. T., *Ind. Eng. Chem.*, **40**, 341 (1948).
14. Kohler, F., *Monatsh. Chemie*, **91**, 738 (1960).
15. Jacob, K. T., Fitzner, K., *Thermochim. Acta*, **18**, 197 (1977).
16. Colinet, C., *D.E.S. Fac. des Sci.*, Univ. Grenoble, France (1967).
17. Muggianu, Y.-M., Gambino, M., and Bros, J.-P., *J. Chim. Phys.*, **72**, 83 (1975).
18. Toop, G. W., *Trans. TMS-AIME*, **233**, 850 (1965).
19. Hillert, M., *CALPHAD*, **4**, 1 (1980).
20. Gaune-Escard, M., *CALPHAD*, **3**, 119 (1979).
21. Blander, M., Yosim, S. J., *J. Chem. Phys.*, **39**, 2610, (1963).
22. Hoch, M., Arpshofen, I., *Z. Metallkde*, **75**, 23, (1984).
23. Hoch, M., Arpshofen, I., Predel, B., *Z. Metallkde*, **75**, 30 (1984).
24. Hoch, M., *CALPHAD*, **9**, 59, (1985).
25. Babelot, J. F., Ohse, R. W., Hoch, M., *J. Nucl. Mat.*, **137**, 144 (1986).
26. Hoch, M., *CALPHAD*, **11**, (1987).
27. Hoch, M., *CALPHAD*, **11**, (1987).
28. Hoch, M., *Proc. ASM Meeting*, Orlando, (1966).
29. Hoch, M., *CALPHAD*, **11**, (1987).
30. Huron, E. and Hoch, M., 1987 International Symposium on Metallurgical Slags and fluxes. Gaskell, D. R. and Fine, H. A., editors, *AIME* 305–318 (1984).
31. Holm, J. L., and Kleppa, O. J., *J. Chem. Phys.*, **49**, 2425 (1968).
32. Flood, H., Førland, T. and Grjotheim, K., *Z. Anorg. Allg. Chem.*, **276**, 289 (1954).
33. Førland, T., *Fused Salts*, Sundheim, B. R., ed., McGraw Hill, (1964).
34. Blander, M., and Yosim, S. J., *J. Chem. Phys.*, **39**, 2610, (1963).
35. Longuet-Higgins, H. C., *Proc. Roy. Soc.* (London), **A 205**, 247 (1951).
36. Brown, W., *Proc. Roy. Soc.* (London), **A 240**, 561 (1957).
37. Reiss, H., Katz, J. L., and Kleppa, O. J., *J. Chem. Phys.*, **36**, 144 (1962).
38. Blander, M., Braunstein, J., *Ann. New York Acad. Sci.*, **79**, 838 (1967).
39. Førland, T., *Neorg. Tek. Vitenskapsakad*, **series 2**, N°4, (1957).
40. Flory, J. P., *J. Chem. Phys.*, **49**, 3919 (1971).
41. Hatem, G., *J. Chim. Phys.*, **7–8**, 754 (1977).

42. Hatem, G., *J. Chim. Phys.*, **77**, 925 (1980).
43. Hatem, G. and Gaune-Escard, M., *Thermochimica Acta*, **57**, 351 (1982).
44. Sem, P., Hatem, G., Bros, J. P., Gaune-Escard, M., *J. Chem. Soc.*, Faraday Trans. **1**, 80, 297, (1984).
45. Hatem, G., Gaune-Escard, M., *J. Chem. Thermodynamics*, **19**, 1095 (1987).
46. Gaune-Escard, M., Tabariès, F. and Hatem, G., Molten-Salts Discusson Group, Londres, (1986).
47. Oye, H. A., Molten Salts Discussion Group, Londres, (1986).
48. EUCHEM Conferences (Europe) and Gordon Conferences (USA) from 1966.
49. Dorner, P., Henig, E. Th, Krieg, K., Lukas, H. L., and Petzow, G., *CALPHAD*, **4**, 241 (1980).
50. Gaune-Escard, M., Gambino, M., Hatem, G., Bros, J. P., Fouque, Y., and Juhem, P., *Proc. CALPHAD XVI*, 24 –29 (1987).
51. Juhem, P., Fouque, Y., Chevalier, P. Y., Cheynet, B., Hatem, G., Gaune-Escard, M., *J.E.E.P.*, **83**, (1987).
52. Juhem, P., Hatem, G., Bros, J. P., Fouque, Y., and Gaune-Escard, M., *J.E.E.P.*, **143** (1988).
53. Hatem, G., de Gasquet, B., and Gaune-Escard, M., *J. Chem. Thermodynamics*, **11**, 927 (1979).
54. Hatem, G., Gaune-Escard, M., and Pelton, A. D., *J. Phys. Chem.*, **86**, 3039 (1982).
55. Moore, R. E., Blankenship., F. F., Grimes, W. R., Friedman, H. A., Barton, C. J., and Thomas, R. E., *J. Phys. Chim.*, **62**, 666 (1958).
56. Hatem, G., Tabaries, F., and Gaune-Escard, M., *Thermochim. Acta* (in press)

# A Simulation Model
# to Predict Slag Composition
# in a Coal-Fired Boiler

M. Seapan* and J. Y. Lo

*School of Chemical Engineering, Oklahoma State University,
Stillwater, OK 74078*

## ABSTRACT

A major problem in the operation of coal-fired boilers is the formation of slag and fouling of the boiler tubes. The slagging problem may become detrimental to the operation of the boiler, especially when different coals are blended. The objective of this project was to develop a model to predict the composition of slag or fouling at different locations of a boiler. A simple computer model was developed to predict the rate of deposition and the composition of deposits formed at different locations in the boiler. A multicell, lumped parameter approach was used, assuming that the cells are well mixed and deposition occurs only by condensation of the stable minerals on the boiler walls. An energy balance was used to calculate the coal flame temperature, which was then used to estimate the temperature distribution in the boiler. Chemical equilibrium constants were evaluated to define the stable compounds at different temperatures. The amount of each compound deposited at various locations was determined from the vapor pressure of that compound. Predictions of this model are in general agreement with the power plant data, suggesting that the dominant compounds in the high temperature zone are the aluminosilicates, whereas the major compounds in the low temperature zone are the alkali sulfates.

**Index Entries:** Ash deposition; boiler; coal combustion; fouling; modeling; slag composition.

## INTRODUCTION

Coal has been the main source of energy for generating electricity in the US in the recent years. With the continuous depletion of oil and

---

*Author to whom all correspondence and reprint requests should be addressed.

High Temperature Science, Vol. 26      © 1990 by the Humana Press Inc.

natural gas resources, coal utilization is expected to grow more rapidly in the coming years. All coals contain some amounts of incombustible inorganic material called mineral matter. The mineral matter consists primarily of clays, pyrite, quartz, and calcite with lesser amounts of other minerals (1). Upon combustion, the mineral matter of coal is converted into oxides forming ash.

In coal-fired power plants, the mineral matter must be carried through the combustion process, eventually leaving the boiler either as bottom ash or fly ash. However, inside the furnace, part of the ash can deposit on the boiler tubes, forming slag or fouling deposit. The ash that deposits on the boiler walls in the radiant section of the furnace is usually of liquid form and is called slag. Ash deposition in the convective zone of the furnace is of solid form and is called fouling. The slag and fouling deposits eventually fall down and are removed as the bottom ash.

Deposition of ash on the boiler tubes reduces the heat transfer rate and changes the gas flow patterns around the tubes, often adversely affecting the rate of heat transfer. Furthermore, the deposited ash may contribute to the corrosion and erosion of the boiler tubes. Accumulation of deposits on tube surfaces has been the major cause of outages in coal-fired power plants. Prediction of the slagging propensity of different coals and the conditions under which slagging can occur has been a major problem for utility companies. Formation of slag and its accumulation on the heat transfer surfaces becomes more unpredictable when blends of different coals are burned. An example case occurred at the Public Service Company of Oklahoma's Northeast Power Generation Station in Oologah, OK during 1980 and 1981 (2). While burning blends of two subbituminous Wyoming coals, severe slagging and fouling problems developed, which caused unit derating and shutdown. The burning of the individual coals resulted in no unusual slagging problems. In another case, when two high slagging coals were blended, slagging was unexpectedly reduced (3).

At present there is no reliable method to predict slagging and fouling in coal-fired power plants. Reid (4) and Raask (5) have reviewed the state-of-the-art in predicting the sintering, fusion, and slagging propensities of different coals. Past fouling and slagging problems have been correlated by many investigators with coal ash composition. For example, many past studies have shown coals rich in calcium and sodium cause superheater fouling. Severe slagging has also been observed with coals rich in iron (5). Work on British coal ashes has been correlated with the silica ratio (6). The main problem with these correlations has originated from relating the slagging phenomenon to the ash composition of the parent coal instead of the composition of the deposit formed at different locations in the boiler.

Ash deposits do not have uniform chemical composition throughout a boiler. For example, analyses of superheater deposits are different from those of fly ash, and distinct differences in chemical composition exist between inner and outer layers of a deposit at a given point in a boiler (7). The adverse effects of ash deposits depend on the thickness of deposits and the physical properties of the ash, including thermal conductivity, emissivity, heat capacity, melting point, surface tension, density, and viscosity or mechanical strength of the deposit. These physical properties, in turn, depend on the chemical composition of the ash. Thus, determination of the composition of deposits is essential in predicting the formation and behavior of deposits in coal-fired boilers and any fundamental model to predict the slagging phenomenon should be based on the local composition of the deposit.

In this project, a simple simulation model was developed that can predict the rate of deposition and the distribution of different elements in the deposits formed at different locations in a boiler. This model considers the boiler as a multicomponent equilibrium stage system, where each stage is assumed to be well mixed and ash deposition to occur only by condensation of inorganic vapors from combustion gases.

## FUNDAMENTALS OF ASH DEPOSITION

Virtually all coal-fired electric power generating plants use pulverized coal-fired furnaces. These furnaces are large chambers, where the pulverized coal is blown into the central combustion region. Coal particles burn in the flame zone, radiating heat to the surrounding walls which are made of the boiler tubes. The hot gases rise in the furnace chamber, transferring their heat to the water in the boiler tubes in the waterwall, primary and secondary superheaters, preheaters, and economizers. The gases are further cooled in the air heaters and after passing through the electrostatic precipitator and/or other gas cleaning devices, leave the system through the stack.

In the combustion zone of the pulverized coal-fired furnaces, the temperature may reach as high as 2000 K. The organic constituents of coal essentially completely burn at this temperature, whereas the mineral matter of coal is oxidized, forming inorganic mixed oxide compounds. Part of these oxides vaporize under the high temperature conditions of the flame zone. The unvaporized parts form the residual particles, which may be in liquid or solid form. Laboratory studies have shown that, depending on the temperature and coal particle size, more than one residual particle may form from a single coal particle (8). The residual particles may either fall to the bottom of the furnace or be carried up with the combustion gases. The combustion gases, in rising through the furnace chamber, are cooled, causing the condensation of the inorganic vapors. The condensation of these mixed oxide compounds may occur on the rising particles or on the surfaces of the boiler tubes, which are usually several hundred degrees cooler than the surrounding gas. The particles that rise in the furnace, thus grow by condensation of the vapors. These particles, under the turbulent flow of gases, approach the walls and deposit on the boiler tubes. Depending on the composition of the particles and condensing vapors and the temperature of the wall, the deposit may be in liquid or solid form, thus resulting in slagging or fouling, respectively.

The slag flows down the tube walls to the regions where it solidifies and becomes thick enough to crumble under its own weight or until it is blown off by the soot blowers. The thickness of the flowing slag depends on its viscosity and density, which in turn depend on the chemical composition of the slag. Similarly, the mechanical strength of the solid foulants that determines the ease of removal or falling of these deposits depends on the chemical composition of the foulant. The chemical composition of deposits varies with position in the chamber. Large particles and high boiling point inorganic vapors tend to deposit in the lower sections of the furnace, whereas more volatile compounds condense and deposit at the higher elevations in the boiler. Deposit composition also changes with the thickness of the deposit. Because the deposits usually have low thermal conductivity, the deposit temperature near the wall is lower than away from the wall, causing compounds of different volatility to condense at different distances from the wall. Therefore, in developing a comprehensive model to predict the slagging and fouling phenomena in a boiler, one should be able to predict the composition of the deposits at different locations in the boiler.

The complex nature of ash deposition phenomenon can be classified under the following processes

1. combustion of coal particles and formation of inorganic vapors;
2. formation of residual particles;
3. condensation of vapors on external walls of boiler tubes;
4. formation of new particles by heterogeneous and homogeneous nucleation;
5. growth of particles by condensation of vapors or coagulation of particles;
6. transport of particles in the combustion chamber;
7. capture of particles on the boiler tubes by impaction, interception, Brownian motion, and phoretic forces; and
8. the effect of the previously deposited slag or fouling on the capture of the colliding particles or condensing vapors.

The rates and extents of these processes depend on the physical and chemical characteristics of both particles and vapors at different locations in the boiler and on the temperature of deposition surfaces and combustion gases. In calculating the deposit composition, the problem can be simplified significantly by considering the mechanism that contributes the most to deposition. This approach should be considered as a first approximation to the complex problem of ash deposition in boilers.

## MODEL DEVELOPMENT

The model is developed on the basis of one of the most important deposition mechanisms. In the combustion zone, the coal combustion is assumed to be instantaneous, and the flame temperature is determined from an adiabatic energy balance calculation. At this temperature, the mineral matter of coal is considered to be completely oxidized, forming mixed oxide compounds, and the combustion gases are assumed to be saturated with the vapors of these inorganic compounds. A calculation based on the vapor pressure of the inorganic constituents shows about 50–65% of these oxides are in the vapor phase at the flame temperature. The unvaporized portion of the inorganic compounds will form the residual particles. Although all the vapors will be carried up by the combustion gases, only the small size residual particles will remain entrained in the gas phase. Thus, both the residual particles and vapors contribute to the formation of ash deposits in boilers. However, the contribution of the condensing vapors appears to be more significant than the residual particles, especially at the upper levels of the furnace where most residual particles cannot reach. Therefore, vapor condensation may be considered as the most important mechanism of ash deposition.

The combustion gases that are saturated with inorganic vapors move up in the furnace chamber, losing their heat to the surrounding boiler tubes. Temperature distribution in the furnace is quite complex. Many investigators have developed different models to predict the temperature distribution in boilers. Smoot and Smith (9) present a summary of these models. However, in this project, a simplified one-dimensional temperature profile is considered. The temperature is assumed to be uniform in any horizontal plane in the boiler and is considered to change with the vertical position in the furnace only. The vertical temperature profile of a furnace can be easily measured directly. However, in the absence of a measured temperature profile, the temperature profile suggested by Howard (10) can be modified and used.

As the gases rise in the furnace chamber, their temperature decreases, resulting in the condensation of the inorganic vapors. It is assumed that at each level, all the condensed material deposit on the walls of the furnace at that location. For simulation purposes, the furnace chamber is divided into sections of equal length, called stages or cells, as shown in Fig. 1. Each stage is assumed to be a well-mixed chamber with uniform temperature and composition. The vapors in each cell are in equilibrium with the liquid or solid deposits on the walls. Thus, each cell is considered an equilibrium stage, and by using the material balance and vapor–condensate equilibrium for each compound, the amount of deposit and the amount left in the gas phase are calculated. The deposited amounts of all compounds are then used to calculate the composition of deposit and rate of deposition at each location.

A major problem is development of this model is identification of the deposited compounds and their thermodynamic equilibrium constants. The 10 most important elements in the mineral matter of US coals are Fe, Ca, Mg, Na, K, Si, Al, Ti, S, and P, which are considered in this work. During combustion, these elements are converted to oxides, forming a range of mixed oxide mineral compounds. The extent of formation of these compounds is determined by thermodynamic rather than kinetic constraints. Thermodynamic calculations have been used by Halstead and Rask (*11*), Smith et al. (*12*), Hastie and Bonnell (*13*), and others to predict the formation of ash species during both pulverized coal combustion and deposit formation. In this work, instead of the rigorous thermodynamic calculations, a simplified, though less accurate, approach is used to identify the stable compounds of deposits.

Bryer (*14*) summarizes the transformations that major pure minerals of coal undergo during the heating process. Case (*15*) reported the major crystalline mineral phases found in the pulverized coal combustion products as mullite ($Al_6Si_2O_{13}$), lime (CaO), magnetite ($Fe_3O_4$), hematite ($Fe_2O_3$), quartz ($SiO_2$), and anhydrite ($CaSO_4$). Wibberley and Wall (*16*) showed the order of stability of sodium compounds in the range of 1027–1727°C is NaCl > $Na_2Si_2O_5$ > NaOH. Through thermodynamic calculations, they suggested that sodium sulfate is also a stable compound and is at equilibrium with the other sodium compounds, providing sufficient amount of sulfur is present. Anhydrite is also expected to form in boilers, especially in the temperature range 850–1000°C (*17*). Based on these studies, the major possible compounds in deposits are mullite ($Al_6Si_2O_{13}$), quartz ($SiO_2$), aluminum oxide ($Al_2O_3$), anhydrite ($CaSO_4$), sodium sulfate ($Na_2SO_4$), and hematite ($Fe_2O_3$).

The amounts of these species can be quantified through the evaluation of their equilibrium constants. The mineral species in the combustion gases are assumed to rapidly reach chemical equilibria. Thus, by considering the equilibrium constants at various temperatures, the stable minerals at each temperature can be determined.

For Al and Si elements, the possible compounds in deposits are $Al_2O_3$, $Al_2SiO_5$, $3Al_2O_3 \cdot 2SiO_2$, and $SiO_2$. Stull and Prophet (*18*) show the equilibrium constants of formation of these compounds at various temperatures. The equilibrium constant of $Al_6Si_2O_{13}$ is much higher than that of $Al_2SiO_5$. Thus, for simplicity, we can assume that $Al_2SiO_5$ does not appear in deposits and mullite ($Al_6Si_2O_{13}$) is the only aluminosilicate formed, as long as there is excess of both Si and Al. Once one of these elements is completely consumed, the excess of the other element is assigned to its oxide, i.e., $SiO_2$ or $Al_2O_3$. This agrees with the transformations of Kaolinite, which forms primarily $Al_6Si_2O_{13}$ at temperatures higher than 1400°C (*14*).

The possible sodium compounds are considered to be $Na_2SO_4$, $Na_2SiO_3$, $Na_2CO_3$, and NaCl. A comparison of the equilibrium constants indicates that $Na_2SO_4$, $Na_2SiO_3$, and NaCl are the stable compounds of sodium. However, since Si is removed from gases as mullite and $SiO_2$ at high temperature regions, the stable compounds of Na are assumed to be $Na_2SO_4$ and NaCl. Similarly, the stable compounds of calcium are determined and their relative stability is evaluated as $Ca_3(PO_4)_2$ > $CaSO_4$ > CaO.

A comparison of the equilibrium constants indicates that $Fe_3O_4$ is a more stable compound than $Fe_2O_3$. However, Raask (*5*) states that in an oxidizing atmosphere, iron oxide can exist both as magnetite ($Fe_3O_4$) and hematite ($Fe_2O_3$). Therefore, in this model, the iron is assumed to be distributed in equal moles of $Fe_3O_4$ and $Fe_2O_3$. Similarly, the equilibrium constants show that $Ti_2O_3$ is more stable than $TiO_2$. But in most literature sources, titanium is reported to exist as rutile ($TiO_2$). Therefore, in this model, titanium is assumed to be equally distributed between $TiO_2$ and $Ti_2O_3$.

Magnesium is considered to be predominantly in oxide form, MgO, even though the phase equilibrium model of Hastie and Bonnell (13) shows magnesium–calcium–silicate, $CaMgSi_2O_6$, as a significant compound and some studies have indicated the existence of magnesium aluminate, $MgAl_2O_4$, when MgO is added to coal. Similarly, for potassium element, potassium carbonate is considered as the stable compound.

In summary, both from the review of literature and the study of equilibrium constants, the major compounds in the ash deposits are concluded to be $Al_6Si_2O_{13}$, $Al_2O_3$, $SiO_2$, NaCl, $Na_2SO_4$, $Ca_3(PO_4)_2$, $CaSO_4$, CaO, $Fe_2O_3$, $Fe_3O_4$, $TiO_2$, $Ti_2O_3$, MgO, and $K_2CO_3$.

In order to determine the formation environment of the ash deposits in a pulverized coal-fired boiler, the temperature profile of the boiler must be obtained. The temperature profile can either be obtained by direct measurement or by developing a comprehensive mathematical model to predict the temperature distribution in a boiler. In this work, a simple approximate approach is used to predict the temperature profile in the boiler. An adiabatic flame temperature is calculated for a given coal considering the following input information

1. inlet coal and air temperatures;
2. humidity and percent excess air;
3. the percentage of unburned coal in the bottom ash; and
4. the percentage of carbon monoxide in the combustion gases.

The temperature profile of a typical utility boiler is modified to represent the temperature profile in a given boiler. Howard (10) presents a temperature profile as a function of the residence time in the boiler. By calculating the gas velocity from its flow rate and the cross-sectional area of the furnace, the temperature at any height in the boiler can be calculated. Assuming that most boilers operate with an average gas velocity of 60 ft/sec (= 18.3 m/s), the temperature vs residence time can be converted to a temperature vs height curve. This curve is then normalized by assuming the maximum combustion temperature to be the adiabatic flame temperature. For computer programming purposes, this normalized temperature profile is fitted with two linear equations for radiative and convective zones of the boiler. Thus, the temperature at any height in the boiler can be estimated.

Vaporization of the mineral matter in the combustion zone and subsequent condensation of these vapors in higher elevations are determined by the vapor pressure of the inorganic constituents. For the 14 stable inorganic compounds considered in this model, most vapor pressure correlations could be obtained from the literature. However, the vapor pressure of four of these minerals could not be found; therefore, an

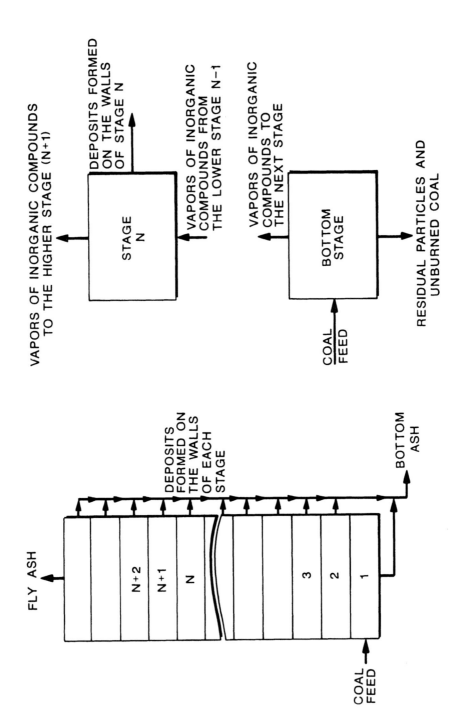

Fig. 1.   Schematic diagram of a multicell model.

approximation technique was developed to estimate their vapor pressure. In most cases, melting point is the only physical property that could be found in the literature. Therefore, the boiling point of these minerals were assumed to be 730°C higher than their melting points. Then by using the Trouton's rule, their enthalpies of vaporization was estimated. The enthalpies of vaporization and the assumed boiling points were then used in the Clausius-Clapeyron equation to obtain an approximate equation for the vapor pressure of these minerals. The validity of this technique was checked against the existing data for $Fe_3O_4$ and was shown that the approximate technique can estimate the vapor pressure of magnetite with 39 and 76% error at 2000 and 1500 K, respectively.

Material balance is used to calculate the deposition of each compound. As Fig. 1 shows, the boiler furnace is divided into a number of cells. Pulverized coal enters into the first stage, where it instantaneously burns, reaching the adiabatic flame temperature. The mineral matter is oxidized generating the fourteen stable compounds, which vaporize and saturate the combustion gases. The amount of each compound required to saturate the combustion gases is calculated from the vapor pressure relation of the compound and is compared with the amount of compound that could be generated from coal combustion. If the existing amount is more than that needed for saturation, the excess is considered to form residual particles, directly forming the bottom ash. The inorganic vapor laden combustion gases rise and enter into the second stage. The temperature of this stage is estimated and used to calculate the vapor pressure of each compound. Again, the amount of each compound required to saturate the gases is calculated and compared with that entering into the cell. The excess amount deposits on the wall, whereas the remaining vapors enter into the next stage. The calculation is thus continued until the exit point of the furnace. The vapors remaining in the exhaust gases constitute the fly ash. The composition of deposit at each stage can be calculated from the amounts of compounds deposited at that stage. The deposits on the walls of all stages eventually fall to the bottom of the furnace forming the bottom ash.

## RESULTS AND DISCUSSION

Based on this model, a computer program was written in FORTRAN and used to simulate the combustion of two subbituminous Wyoming coals in a boiler with a geometry similar to that of the PSO's Oologah power plant. The furnace chamber was considered to be 50 ft (= 15.24 m) wide by 50 ft (= 15.24 m) deep and 200 ft (= 67 m) tall. The analyses of these two coals are given in Table 1. The coal entered at a rate of 300 ton/h and burned with 15% excess air. The inlet coal temperature was assumed to be 30°C and that of entering air 300°C, containing 0.03 kg of water/kg of dry air. The bottom ash was assumed to contain 5% unburned carbon and 2% of the burned carbon was assumed to form carbon monoxide. The height of each cell was considered to be 10 ft (= 3.05 m).

Table 1
Analysis of Coals

| Coal type, as received | Jacobs Ranch | Clovis Point |
|---|---|---|
| **Proximate Analysis Wt.%** | | |
| Moisture | 25.44 | 31.32 |
| Volatile Matter | 36.55 | 33.85 |
| Ash | 6.81 | 6.14 |
| Fixed Carbon | 31.20 | 28.69 |
| Heating Value Btu/lb | 8863.00 | 8014.00 |
| **Ultimate Analysis Wt.%** | | |
| Sulfur | 0.64 | 0.43 |
| Carbon | 48.43 | 44.21 |
| Hydrogen | 3.96 | 3.52 |
| Nitrogen | 0.69 | 0.60 |
| Oxygen | 14.03 | 13.78 |
| Chlorine | 0.00 | 0.00 |
| Moisture | 25.44 | 31.32 |
| Ash | 6.81 | 6.14 |
| **Mineral Analysis Wt.% of Ash** | | |
| Iron as $Fe_2O_3$ | 7.15 | 4.29 |
| Calcium as CaO | 14.22 | 14.64 |
| Magnesium as MgO | 3.15 | 3.12 |
| Sodium as $Na_2O$ | 0.68 | 1.08 |
| Potassium as $K_2O$ | 1.37 | 1.45 |
| Silicon as $SiO_2$ | 46.38 | 50.24 |
| Aluminum as $Al_2O_3$ | 13.81 | 13.55 |
| Titanium as $TiO_2$ | 1.15 | 0.95 |
| Phosphorus as $P_2O_5$ | 1.11 | 1.56 |
| Sulfur as $SO_3$ | 6.05 | 4.53 |

The simulation results are shown in Tables 2 and 3 and Figs. 2–7. These tables show the temperature of the combustion gases at different heights in the boiler, as well as the composition of deposits at these elevations. Furthermore, slagging flux at each cell level and the percent of the total coal ash deposited at that elevation are tabulated in the last two columns. Figures 2–6 show the deposition flux of Si, Al, Fe, Na, and sulfates, and Fig. 7 shows the overall deposition rate at different elevations in the boiler.

Based on these calculations, 37% of the ash of the Jacobs Ranch coal forms residual particles composed of $SiO_2$, $Al_2O_3$, CaO, MgO, and $TiO_2$, whereas the remaining 63% of the ash vaporizes into the gaseous phase. The calculation shows that in the lower 60 ft (= 18.3 m) of the chamber, the wall deposits consist of essentially 99% $SiO_2$. Since the melting point of silica is 1423°C (= 1700 K), the deposits are considered as molten silica that will from the slag at these elevations. As shown in Fig. 2, in the

Table 2
Composition of the Ash Deposits (wt%) from Combustion of Jacobs Ranch Coal

| Height, ft | T, K | Fe as $Fe_2O_3$ | Ca as CaO | Mg as MgO | Na as $Na_2O$ | K as $K_2O$ | Si as $SiO_2$ | Al as $Al_2O_3$ | Ti as $TiO_2$ | P as $P_2O_5$ | S as $SO_3$ | Slagging flux, $kg/h/m^2$ | Contribution as percent of ash in the coal |
|---|---|---|---|---|---|---|---|---|---|---|---|---|---|
| [a] | 1958 | 0.0 | 26.8 | 12.6 | 0.0 | 0.0 | 37.3 | 22.8 | 1.5 | 0.0 | 0.0 | 88.0 | 37.3 |
| 15 | 1928 | 0.0 | 0.0 | 0.1 | 0.0 | 0.0 | 99.4 | 0.3 | 0.2 | 0.0 | 0.0 | 22.2 | 9.4 |
| 25 | 1899 | 0.0 | 0.0 | 0.1 | 0.0 | 0.0 | 99.5 | 0.3 | 0.1 | 0.0 | 0.0 | 14.1 | 6.0 |
| 35 | 1870 | 0.0 | 0.0 | 0.1 | 0.0 | 0.0 | 99.5 | 0.3 | 0.1 | 0.0 | 0.0 | 8.9 | 3.8 |
| 45 | 1840 | 0.0 | 0.0 | 0.1 | 0.0 | 0.0 | 99.5 | 0.3 | 0.1 | 0.0 | 0.0 | 5.5 | 2.3 |
| 55 | 1811 | 0.0 | 0.0 | 0.1 | 0.0 | 0.0 | 99.6 | 0.3 | 0.1 | 0.0 | 0.0 | 3.3 | 1.4 |
| 65 | 1781 | 0.0 | 0.0 | 0.1 | 0.0 | 0.0 | 93.6 | 6.2 | 0.1 | 0.0 | 0.0 | 3.3 | 1.4 |
| 75 | 1752 | 0.0 | 0.0 | 0.0 | 0.0 | 0.0 | 56.7 | 43.2 | 0.0 | 0.0 | 0.0 | 13.7 | 5.8 |
| 85 | 1722 | 0.0 | 0.0 | 0.0 | 0.0 | 0.0 | 53.3 | 46.6 | 0.0 | 0.0 | 0.0 | 9.5 | 4.0 |
| 95 | 1693 | 0.0 | 0.0 | 0.0 | 0.0 | 0.0 | 50.0 | 50.0 | 0.0 | 0.0 | 0.0 | 6.6 | 2.8 |
| 105 | 1663 | 39.6 | 0.0 | 0.0 | 0.0 | 0.0 | 28.3 | 32.1 | 0.0 | 0.0 | 0.0 | 6.5 | 2.7 |
| 115 | 1634 | 60.6 | 0.0 | 0.0 | 0.0 | 0.0 | 17.3 | 22.1 | 0.0 | 0.0 | 0.0 | 6.0 | 2.6 |
| 125 | 1605 | 58.3 | 0.0 | 0.0 | 0.0 | 0.0 | 17.2 | 24.5 | 0.0 | 0.0 | 0.0 | 3.8 | 1.6 |
| 135 | 1575 | 81.9 | 0.0 | 0.0 | 0.0 | 0.0 | 7.1 | 11.1 | 0.0 | 0.0 | 0.0 | 4.2 | 1.8 |
| 145 | 1546 | 84.9 | 0.0 | 0.0 | 0.0 | 0.0 | 5.5 | 9.5 | 0.0 | 0.0 | 0.0 | 3.0 | 1.3 |
| 155 | 1516 | 85.1 | 0.0 | 0.0 | 0.0 | 0.0 | 5.2 | 9.7 | 0.0 | 0.0 | 0.0 | 1.9 | 0.8 |
| 165 | 1487 | 85.2 | 0.0 | 0.0 | 0.0 | 0.0 | 5.0 | 9.9 | 0.0 | 0.0 | 0.0 | 1.7 | 0.5 |
| 175 | 1337 | 5.0 | 37.9 | 0.0 | 0.0 | 0.0 | 0.3 | 0.6 | 0.0 | 5.2 | 51.1 | 22.1 | 9.4 |
| 185 | 1210 | 0.9 | 22.4 | 0.0 | 0.0 | 43.8 | 0.0 | 0.1 | 0.0 | 1.8 | 30.9 | 6.7 | 2.8 |
| 195 | 1083 | 0.1 | 5.4 | 0.0 | 27.3 | 24.2 | 0.0 | 0.0 | 0.0 | 0.3 | 42.7 | 2.8 | 1.1 |
| 205 | 956 | 0.0 | 0.5 | 0.0 | 42.1 | 2.2 | 0.0 | 0.0 | 0.0 | 0.0 | 55.2 | 1.8 | 0.7 |
| 215 | 829 | 0.0 | 0.4 | 0.0 | 42.7 | 1.4 | 0.0 | 0.0 | 0.0 | 0.0 | 55.6 | 0.1 | $3 \times 10^{-2}$ |
| 225 | 701 | 0.0 | 0.2 | 0.0 | 43.1 | 0.8 | 0.0 | 0.0 | 0.0 | 0.0 | 56.0 | 0.0 | $4.2 \times 10^{-4}$ |
| Fly Ash | – | 0.0 | 0.0 | 0.0 | 99.0 | 0.0 | 0.0 | 0.0 | 0.0 | 0.0 | 1.0 | $0.2^b$ | $4.8 \times 10^{-4}$ |

[a] Residual ash that falls to the bottom of the furnace and the slag that is formed in the first stage.
[b] The rate of fly ash leaving the boiler (kg/h).

Table 3
Composition of the Ash Deposits (wt%) from Combustion of Clovis Point Coal

| Height, ft | T, K | Fe as Fe$_2$O$_3$ | Ca as CaO | Mg as MgO | Na as Na$_2$O | K as K$_2$O | Si as SiO$_2$ | Al as Al$_2$O$_3$ | Ti as TiO$_2$ | P as P$_2$O$_5$ | S as SO$_3$ | Slagging flux, kg/h/m$^2$ | Contribution as percent of ash in the coal |
|---|---|---|---|---|---|---|---|---|---|---|---|---|---|
| [a] | 1937 | 0.0 | 22.9 | 9.4 | 0.0 | 0.0 | 50.7 | 16.0 | 1.0 | 0.0 | 0.0 | 102.6 | 48.51 |
| 15 | 1908 | 0.0 | 0.0 | 0.1 | 0.0 | 0.0 | 99.5 | 0.3 | 0.1 | 0.0 | 0.0 | 14.7 | 7.0 |
| 25 | 1879 | 0.0 | 0.0 | 0.1 | 0.0 | 0.0 | 99.5 | 0.3 | 0.1 | 0.0 | 0.0 | 9.3 | 4.4 |
| 35 | 1850 | 0.0 | 0.0 | 0.1 | 0.0 | 0.0 | 99.5 | 0.3 | 0.1 | 0.0 | 0.0 | 5.8 | 2.8 |
| 45 | 1820 | 0.0 | 0.0 | 0.1 | 0.0 | 0.0 | 99.6 | 0.3 | 0.1 | 0.0 | 0.0 | 3.6 | 1.7 |
| 55 | 1791 | 0.0 | 0.0 | 0.1 | 0.0 | 0.0 | 99.6 | 0.2 | 0.1 | 0.0 | 0.0 | 2.2 | 1.0 |
| 65 | 1762 | 0.0 | 0.0 | 0.1 | 0.0 | 0.0 | 67.3 | 32.6 | 0.0 | 0.0 | 0.0 | 8.8 | 4.2 |
| 75 | 1733 | 0.0 | 0.0 | 0.0 | 0.0 | 0.0 | 54.6 | 45.4 | 0.0 | 0.0 | 0.0 | 9.8 | 4.7 |
| 85 | 1704 | 0.0 | 0.0 | 0.0 | 0.0 | 0.0 | 51.2 | 48.7 | 0.0 | 0.0 | 0.0 | 6.8 | 3.2 |
| 95 | 1675 | 0.0 | 0.0 | 0.0 | 0.0 | 0.0 | 48.0 | 51.9 | 0.0 | 0.0 | 0.0 | 4.7 | 2.2 |
| 105 | 1646 | 11.8 | 0.0 | 0.0 | 0.0 | 0.0 | 39.7 | 48.5 | 0.0 | 0.0 | 0.0 | 3.5 | 1.7 |
| 115 | 1617 | 59.3 | 0.0 | 0.0 | 0.0 | 0.0 | 17.2 | 23.5 | 0.0 | 0.0 | 0.0 | 4.1 | 2.0 |
| 125 | 1587 | 56.9 | 0.0 | 0.0 | 0.0 | 0.0 | 17.2 | 26.0 | 0.0 | 0.0 | 0.0 | 2.6 | 1.2 |
| 135 | 1558 | 65.4 | 0.0 | 0.0 | 0.0 | 0.0 | 13.0 | 21.6 | 0.0 | 0.0 | 0.0 | 1.9 | 0.9 |
| 145 | 1529 | 85.0 | 0.0 | 0.0 | 0.0 | 0.0 | 5.4 | 9.6 | 0.0 | 0.0 | 0.0 | 2.1 | 1.0 |
| 155 | 1500 | 75.6 | 3.2 | 0.0 | 0.0 | 0.0 | 4.5 | 8.7 | 0.0 | 8.0 | 0.0 | 1.7 | 0.8 |
| 165 | 1471 | 40.4 | 14.9 | 0.0 | 0.0 | 0.0 | 2.3 | 4.7 | 0.0 | 37.7 | 0.0 | 2.9 | 1.4 |
| 175 | 1323 | 6.4 | 36.8 | 0.0 | 0.0 | 0.0 | 0.3 | 0.8 | 0.0 | 7.6 | 48.2 | 12.9 | 6.1 |
| 185 | 1197 | 0.7 | 19.1 | 0.0 | 0.0 | 50.1 | 0.0 | 0.1 | 0.0 | 1.5 | 26.5 | 5.8 | 2.8 |
| 195 | 1071 | 0.0 | 2.3 | 0.0 | 36.6 | 10.4 | 0.0 | 0.0 | 0.0 | 0.1 | 50.5 | 4.4 | 2.1 |
| 205 | 946 | 0.0 | 0.5 | 0.0 | 42.2 | 2.1 | 0.0 | 0.0 | 0.0 | 0.0 | 55.2 | 1.3 | 0.5 |
| 215 | 820 | 0.0 | 0.4 | 0.0 | 42.7 | 1.3 | 0.0 | 0.0 | 0.0 | 0.0 | 55.7 | 0.1 | $2 \times 10^{-2}$ |
| 225 | 694 | 0.0 | 0.2 | 0.0 | 43.1 | 0.7 | 0.0 | 0.0 | 0.0 | 0.0 | 56.0 | $3 \times 10^{-5}$ | $1.4 \times 10^{-5}$ |
| Fly Ash | — | 0.0 | 0.0 | 0.0 | 99.0 | 0.0 | 0.0 | 0.0 | 0.0 | 0.0 | 1.0 | 0.014[b] | $3.3 \times 10^{-5}$ |

[a]Residual ash that falls to the bottom of the furnace and the slag that is formed in the first stage.
[b]The rate of fly ash leaving the boiler (kg/h).

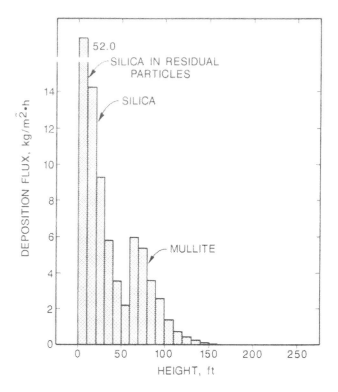

Fig. 2. Deposition flux of Si in kg $SiO_2/m^2/h$ (Jacobs Ranch Coal).

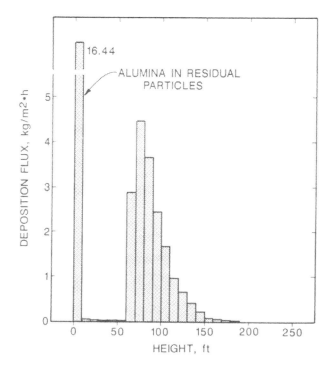

Fig. 3. Deposition flux of Al in kg $Al_2O_3/m^2/h$ (Jacobs Ranch Coal).

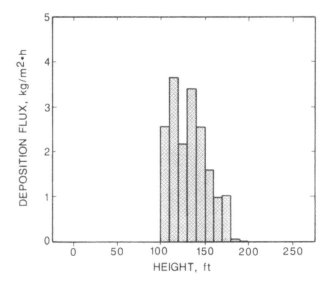

Fig. 4.   Deposition flux of Fe in kg $Fe_2O_3/m^2$/h (Jacobs Ranch Coal).

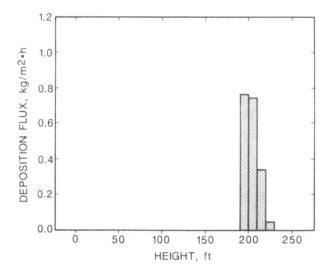

Fig. 5.   Deposition flux of Na in kg $Na_2O/m^2$/h (Jacobs Ranch Coal).

range of 60–100 ft (= 18.3–30.5 m), the predominant compound in deposits will be mullite with a melting point of 1750°C (= 2023 K), forming a solid deposit. Oxides of iron will dominate the deposits of 100–170 ft (= 30.5–51.8 m) height along with some silica and alumina, as shown in Figs. 2–4. Deposition of calcium phosphate and calcium sulfate will occur at 170–190 ft (= 51.8–58 m) elevations, whereas potassium and sodium salts will deposit at 180–220 ft (= 55–61 m) elevations. Figure 5 shows the deposition flux of sodium. Deposition of sodium and calcium sulfates contributes to the total sulfur deposited on the walls, which is shown in Figure 6.

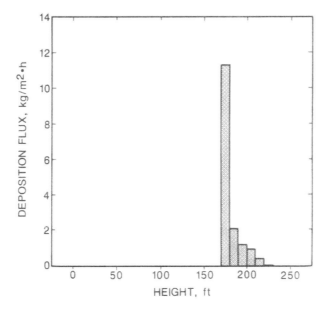

Fig. 6.  Deposition flux of Sulfates in kg $SO_3/m^2/h$ (Jacobs Ranch Coal).

Deposition of sodium and calcium at these elevations agrees with the reported observations in the literature that sodium and calcium deposit on superheater tubes (5). Meanwhile, deposition of liquid silica forming the slag at lower elevations of the boiler agrees with the reports on British coals relating the slagging propensity of coals to their silica contents (6). Although the general deposition of silica, sodium, and calcium agree with the observations reported in the literature, deposition of magnesia does not correspond to the elevations reported in the literature, indicating that the assumed form of the magnesium compound in the deposited ash is probably incorrect. The next phase of this simulation model will incorporate a phase equilibria prediction technique similar to Hastie and Bonnell (13), which is expected to result in a more accurate predictive model.

Figure 7 shows the overall deposition flux at different heights, indicating that the deposition rate is not uniform throughout the boiler chamber. This is an important phenomenon observed by power plant operators, which can be predicted by this multicell simulation model.

Similar conclusions can be drawn from the simulated combustion results of the Clovis Point coal, as shown in Table 3. In this case, about 49% of the coal ash forms residual particles, whereas 51% of the ash vaporizes in the combustion gases. The deposit composition profile appears to be similar to that of Jacobs Ranch coal with some shift in the vertical positions of certain deposits. For example, the slagging height is about 10 ft (= 3 m) shorter and the area of mullite deposits is about 10 ft (= 3 m) wider than with Jacobs Ranch coal.

Calculations for a blend of 50% Clovis Point and 50% Jacobs Ranch coals show a shift in the location of deposits on the walls. When this shift

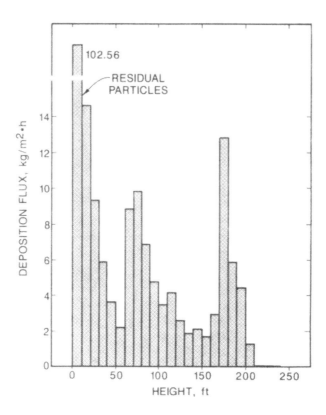

Fig. 7.   Total deposition flux kg/m²/h (Jacobs Ranch Coal).

in deposit composition is combined with the changes in the temperature distribution and dependency of slag viscosity on temperature, it can potentially predict the formation of a high viscosity slag responsible for problems experienced by PSO's power plant.

This simple model, although capable of predicting certain depositon trends in a coal-fired boiler, is too simplified to predict the slagging phenomena accurately. The work is being continued to include other identified minerals and deposition mechanisms with the objective to develop a comprehensive model to predict the slagging phenomenon in a coal-fired boiler.

## ACKNOWLEDGMENT

The support of the University Center for Energy Research of Oklahoma State University in development of this model is acknowledged.

## REFERENCES

1. Gluskoter, H. J., Shimp, N. F., and Ruch, R. R., *Chemistry of Coal Utilization*, second supplementary volume, Elliott, M. A., ed., Wiley, New York, 1981.

2. Lehman, D., *Internal Report*, Public Service Company of Oklahoma, Northeastern Station, Oologah, OK, 1983, and personal communications, 1985.
3. Unsworth, F. J., Cunliffe, F., Graham, C. S., and Morgan, A. P., *Fuel* **66**, 1672 (1987).
4. Reid, W. T., *Chemistry of Coal Utilization*, second supplementary volume, Elliott, M. A., ed., Wiley, New York, 1981.
5. Raask, E., *Mineral Impurities In Coal Combustion*, Hemisphere, New York, 1985.
6. Raask, E., *VGB Kraftwerkstechnik*, **53**, 248 (1973).
7. Babcock and Wilcox Co., *Steam, Its Generation And Use*, 37th edition, 1975.
8. Damle, A. S., Ensor, D. S., and Ranade, M. B., *Aerosol Science and Technology* **1**, 119, 1982.
9. Smooth, L. D. and Smith, P. J. *Coal Combustion and Gasification*, Plenum, New York, 1985.
10. Howard, J. B., PhD Thesis, Pennsylvania State University, Philadelphia, PA, 1965.
11. Halstead, W. D. and Raask, E., *J. of Inst. Fuel* **42**, 83 (1969).
12. Smith, N. Y., Beck, W. H., and Hein K. *Combustion Science Technology* **42**, 115 (1985).
13. Hastie, J. W. and Bonnell, D. W., *High Temp. Sci.* **19**, 275 (1985).
14. Bryers, R. W. and Walchuk, O. R., *International VGB Conference on Slagging, Fouling and Corrosion on Thermal Power Plants*, VGB Technical Association of Large Power Plants Operators, Essen, FRG, 1984.
15. Case, G. D., Farrior, W. L., Kovach, J. J., Lamey, J. J., Lawson, W. F., Lewis, P. S., Mazza, M. H., Oldaker, E. C., Poston, A. M., Shale, C. C., Stewart, G. W., Steinespring, C. D., Waldstein, P. D., and Wallace, W. E., *DOE report No. MERC/SP–78/2*, 1978.
16. Wibberley, L. J. and Wall, T. F., *Fuel* **61**, 87 (1982).
17. Drummond, A. R., Boar, P. L., and Deed, R. G., *Ash Deposits and Corrosion Due to Impurities in Combustion Gases*, Breyer, R. W., ed., Hemisphere, London, 1977.
18. Stull, D. R. and Prophet H., *JANAF Thermochemical Tables*, 2nd edition, NSRDS-NBS 37, Supt. of Documents, 1971.

# A Predictive Thermodynamic Model for Complex High Temperature Solution Phases XI

D. W. BONNELL AND J. W. HASTIE*

*National Institute of Standards and Technology,
Gaithersburg, MD 20899*

## ABSTRACT

A computer-based model has been developed that predicts phase compositions of simple and complex multicomponent, non-ideal, high temperature solutions. Component activities in liquid and solid solutions, and gas phase partial pressures can also be determined from the model. The applicability of the model has been demonstrated for representative test cases with solutions of compounds containing up to eight elements. Examples considered here include various silicate, aluminate, aluminosilicate, and lime aluminosilicates, in addition to soda lime and borosilicate glasses, calcined dolomite and illite minerals, and an alkali-rich coal slag. The model results are compared with mass spectrometrically determined vapor species identities and partial pressures and/or activities. The model has, as its basis, the assignment of complex or associated solution components (e.g., $Na_2SiO_3(\ell)$ and $Na_2Si_2O_5(\ell)$ in $Na_2O$-$SiO_2$ mixtures) that account for the known nonideal interactions. Gibbs energies of formation functions ($\Delta_f G(T)$) for the oxide components, present as simple and complex phases, are explicitly included in an extensive database for use with multicomponent equilibrium codes. Although the components are included explicitly, it is assumed that in most cases, the components model short range order and do not necessarily represent discrete molecular, ionic, or other structural entities.

In this chapter, earlier work performed in our laboratory on the development and application of the model is reviewed. Also, new results are presented for several alkali silicate and borosilicate systems where new experimental data are available.

**Index Entries:** Activity; alkali; Gibbs energy functions; glass; high temperature; mass spectrometry; molten salts; oxides; silicates; slag thermochemistry; thermodynamic modeling; vapor pressure.

*Author to whom all correspondence and reprint requests should be addressed.

## INTRODUCTION

Complex silicate solutions, particularly the generally amorphous slag and glass systems, are important to both established and developing technologies. For instance, in combustion processes, such as magnetohydrodynamic (MHD) combustors, pressurized fluidized bed combustors, coal gasification, and gas turbines, the vapor transport of alkalis and other volatiles from the slag phase is a key factor in hot corrosion and fouling (1). The formation and function of slags involved in metallurgical processes, such as the production of iron, steel, aluminum, copper, and so on, and in glass processing, particularly specialty systems, such as the borosilicate glasses used for the containment of nuclear wastes, represent additional modern technological areas where solution thermodynamics, vapor pressures, and phase distributions are important in process control. In more established technologies, such as soda-lime glass processing, knowledge of the activities of condensed species typically present in the glass-forming process, such as $SiO_2$, $CaCO_3$, $Na_2CO_3$, $K_2CO_3$, $MgCO_3$, and $Na_2SO_4$, and the oxides of these ingredients, is required to describe the transport of volatiles during glass-making (2).

Measurement of high temperature solution and vaporization thermochemistry for such complex, reactive phases is particularly difficult and is not practical for the whole range of compositions and temperatures of technological interest. We have discussed in detail elsewhere, the application of specially developed high temperature mass spectrometric (MS) techniques to this problem (1,3–6). The availability of species partial pressure and activity data for selected systems, obtained using the MS approach, has greatly facilitated the development and testing of a predictive solution model (7). The model is not limited to the conditions of temperature, pressure, and gaseous atmosphere for which limited experimental data exist.

The advantage of mass spectrometrically-based measurements, namely species specificity, often makes this technique the method of choice for high temperature equilibria measurements, particularly where correct identification of the important reaction pathways is critical. Mass spectrometry provides for a number of tests for thermodynamic equilibrium, and can also provide evidence of kinetic limitations. A discussion of modern high temperature mass spectrometry techniques and their application to complex solid/liquid–vapor systems has been given by one of us elsewhere (3). In general, however, it is difficult, or impractical, to determine experimentally all the necessary thermochemistry over the broad composition and temperature ranges typical of practical complex oxide systems, where activities can change by orders of magnitude for composition changes of only a few percent. To guide future experiments and extrapolate the limited existing data to practical conditions, a model is necessary that accounts for these exceptionally strong activity effects.

Polymeric oxide solutions, particularly silicates, aluminosilicates, limesilicates, and borates are known to exhibit major deviations from ideal thermodynamic mixing behavior, e.g., *see* the texts of Richardson

(8) and Turkdogan (9) and articles by Stolyarova and others elsewhere in this volume. High temperature activity data, obtained via Knudsen thermogravimetry, mass spectrometry, electrochemistry, and so on, provide indirect, but compelling evidence that highly complex, strong chemical interactions occur among the component oxides in silicate-based solutions (e.g., *see* Fig. 1 in (7)). For example, the $Na_2O$ activity is much higher in $Na_2CO_3$ and $Na_2SO_4$ before and during glass-forming reaction with $SiO_2$ than it is in the final solution phase where strongly bonded sodium silicates have formed. These complex phases may have a solid, liquid, or glass form. Similar behavior is found with potassium and other alkali silicate solutions.

A variety of modeling techniques exist to treat binary and, in special cases, ternary systems, as may be seen in the chapters by Ansara, Pelton, and others elsewhere in this volume. The majority of these techniques are dependent on the availability, and least squares optimization, of experimental thermochemical and phase equilibria information, using polynomial functions to model the activity relationships of end-member species. Such techniques are necessarily more interpolative than predictive. The extension of these methods to the treatment of ternary systems is usually based on the assumption that, in general, ternary interactions are minimal. Although this assumption is often quite useful in molten salt systems, our database and model strongly indicate that ternary and probably higher-order interactions can be quite important, especially in aluminosilicate systems.

The discussion here treats the development and application of a predictive model that provides generally good agreement with available experimental data with a minimum number of adjustment parameters.

## BASIS OF MODEL

The rationale and theoretical basis for the mixing model have been presented earlier in Parts I (7) and II (10) of this research effort. Modeling high temperature solutions of complex oxides has been a particularly difficult problem using standard nonideal solution model techniques. Such techniques rely on analytical expressions that express the deviation of thermodynamic activity from the physical concentration (nominal molar composition). Redlich et al. (11), Grover (12), and Ansara (13) have discussed and compared various proposed activity model relationships, particularly with $Al_2O_3$ and/or $FeO$ present. Such models, which are often based on Margules power series, or similar polynomial-type expressions for excess properties, are generally satisfactory only for systems where the thermodynamic mixing interactions occur smoothly over the entire composition range, that is, where no discontinuities occur. For acid-base type inorganic systems, particularly those containing alkali oxides, the ratio of formal concentration to thermodynamic activity can vary over the range $\sim 1$–$10^{10}$. Rapid changes in activity can occur over narrow concentration ranges. Consequently sharp slope discontinuities are present in the excess functions that cannot be represented well by analytically tractable expressions. The available models do not appear to be generally applicable to complex oxide systems, with the possible exception of sublattice models (e.g., see Hillert et al. (14) for an example applied to the CaO–SiO$_2$ binary system) and the quasi-chemical approach (e.g., see Pelton elsewhere in this volume).

The model presented here avoids these difficulties by considering that the strong acid-base interactions, typical of many oxide systems, result from the formation of complex liquids and solids with identifiable stoichiometries and Gibbs energies of formation. We refer to these complexes as components since the liquid components are not necessarily independent molecular or ionic species, but serve to represent the local associative order. A listing of these complex components is given in Table 1. In most cases, the liquid complex components are known, neutral, stable, thermodynamically-defined pure liquids appearing in phase diagrams in equilibrium with congruently melting solids. The standard Gibbs energies of formation as a function of temperature ($\Delta_f G(T)$) are either known or can be estimated for these complex liquids and solids. In some cases, $\Delta_f G(T)$ functions for model liquids have been obtained by estimating heats and entropies of melting from known solid phases. In a few cases, notably KCaAlSi$_2$O$_7$($\ell$), Ca$_2$Al$_2$SiO$_7$($\ell$), and some of the potassium- and sodium-aluminosilicates, model liquids that have no known corresponding pure solid phase have been included in the model data base. The presence of such liquids in the model is required to account for the additional associative order present in highly complex solutions. With this basis, the strong solution interactions are contained in the $\Delta_f G(T)$ functions of the complex components, which then mix ideally according to Raoult's law. A database of $\Delta_f G(T)$ functions having the form

$$\Delta_f G(T) = a/T + b + cT + dT^2 + eT^3 + f \cdot T \cdot \ln T \quad J/mol$$

where a–f are fitted coefficients for T in K, has been constructed for use with a multicomponent equilibrium code that has the capability of treating condensed solutions. The SOLGASMIX (15) program is used with this database to calculate the predicted compositions of solids, liquids (nonideal solutions), and the vapor phase. The calculated composition of each individual solution component is then taken as the activity. This is a key assumption in the model. Numerous comparisons between model and experimental activities have confirmed the reliability of this approximation. Thus, in the binary Na$_2$O–SiO$_2$ liquid system, for instance, the presence of complex liquid components, Na$_2$O·2SiO$_2$ (varying from 26 mol% at 95% SiO$_2$ to 0.05 mol% at 50% SiO$_2$) and Na$_2$O·SiO$_2$ (varying from 3.8–99.9 mol%) accounts for the greatly reduced activity of Na$_2$O ($1.4 \times 10^{-9}$ at 95% SiO$_2$ to $6.2 \times 10^{-7}$ at 50% SiO$_2$). The general modeling approach is summarized in Table 2, and is called the Ideal Mixing of Complex Components (IMCC) model. A typical model output listing is

| Name | Formula | S | L | G |
|------|---------|---|---|---|
| Hercynite | $Al_2FeO_4$ | S | L | |
| Alumina | $Al_2O_3$ | S | L | |
| Mullite | $Al_6Si_2O_{13}$ | S | L | |
| Boron oxide | $B_2O_3$ | | L | G |
| Calcium Aluminate | $CaAl_2O_4$ | S | L | |
| Ca-Al Pyroxene | $CaAl_2SiO_6$ | S | | |
| Anorthite | $CaAl_2Si_2O_8$ | S | L | |
| (Calcium leucite) | $CaAl_2Si_4O_{12}$ | | L | |
| | $CaAl_4O_7$ | S | | |
| Calcium ferrite | $CaFe_2O_4$ | S | L | |
| Diopside | $CaMgSi_2O_6$ | | L | |
| Calcium oxide | $CaO$ | S | L | |
| Pseudo-wollastonite | $CaSiO_3$ | S | L | |
| Gehlenite | $Ca_2Al_2SiO_7$ | S | L | |
| Dicalcium ferrite | $Ca_2Fe_2O_5$ | S | L | |
| Akermanite | $Ca_2MgSi_2O_7$ | S | L | |
| Larnite | $Ca_2SiO_4$ | S | L | |
| | $Ca_3Al_2O_6$ | S | | |
| Grossular | $Ca_3Al_2Si_3O_{12}$ | S | | |
| Merwinite | $Ca_3MgSi_2O_8$ | S | | |
| | $Ca_{12}Al_{14}O_{33}$ | S | L | |
| Cesium borate | $CsBO_2$ | S | L | G |
| Cesium oxide | $Cs_2O$ | | L | G |
| Cesium disilicate | $Cs_2Si_2O_5$ | | L | |
| Ferrous oxide | $FeO$ | S | L | |
| Wüstite | $Fe_{0.947}O$ | S | | |
| Hematite | $Fe_2O_3$ | S | | |
| Fayalite | $Fe_2SiO_4$ | S | L | |
| Magnetite | $Fe_3O_4$ | S | L | |
| Potassium aluminate | $KAlO_2$ | S | L | |
| Kaliophilite | $KAlSiO_4$ | S | L | |
| Leucite | $KAlSi_2O_6$ | S | L | |
| Potash feldspar | $KAlSi_3O_8$ | S | L | |
| Potassium borate | $KBO_2$ | S | L | G |
| (Potassium melilite) | $KCaAlSi_2O_7$ | | L | |
| Potassium ferrite | $KFeO_2$ | S | | |
| Potassium $\beta$-alumina | $K_2Al_{18}O_{28}$ | S | L | |
| | $K_2Fe_{12}O_{19}$ | S | | |
| Potassium monoxide | $K_2O$ | | L | |
| Potassium metasilicate | $K_2SiO_3$ | | L | |
| Potassium disilicate | $K_2Si_2O_5$ | | L | |
| Potassium tetrasilicate | $K_2Si_4O_9$ | | L | |
| | $KNaSiO_3$ | | L | |
| | $KNaSi_2O_5$ | | L | |
| Lithium aluminate | $LiAlO_2$ | S | L | |
| | $LiAl_5O_8$ | S | L | |
| Lithium borate | $LiBO_2$ | S | L | |
| Lithium monoxide | $Li_2O$ | S | L | |
| | $Li_5AlO_4$ | S | L | |
| Spinel | $MgAl_2O_4$ | S | | |
| Magnesia | $MgO$ | S | L | |
| Magnesio ferrite | $MgFe_2O_4$ | S | | |
| Clinoenstatite | $MgSiO_3$ | S | L | |
| Forsterite | $Mg_2SiO_4$ | S | | |
| Nepheline | $NaAlSiO_4$ | S | L | |
| Jadeite | $NaAlSi_2O_6$ | S | L | |
| Albite | $NaAlSi_3O_8$ | S | L | |
| Sodium borate | $NaBO_2$ | | L | G |
| | $NaFeSi_2O_6$ | S | | |
| Sodium monoxide | $Na_2O$ | | L | |
| Sodium metasilicate | $Na_2SiO_3$ | | L | |
| Sodium disilicate | $Na_2Si_2O_5$ | S | L | |
| Cristobalite | $SiO_2$ | S | L | |

[a] $\Delta_f G(T)$ coefficients in Eriksson/SOLGASMIX format available in machine-readable form from the authors. Component names in parentheses are proposed.

Table 2
Solution Modeling Approach Using an Ideal Mixing
of Complex Components (IMCC) Approximation

1. Select known solid and liquid components from available phase diagrams and thermodynamic tables, where possible; assign hypothetical components where required by mismatch with experimental activity data.
2. Fit $\Delta_f G(T)$ to known data.
3. Where necessary, estimate $\Delta_f G(T)$ for liquids of known congruently melting solids; assume glass and liquid $\Delta_f G(T)$ to be identical at least to within experimental error.
4. Model as liquid (and solid) solutions of components, with solids as pure precipitable phases.
5. Use free energy minimization calculation (e.g., SOLGASMIX) to determine equilibrium composition and hence activities.
6. Equate activities with mole fractions.

shown in Table 3 for a relatively simple, highly nonideal, ternary system. The column labeled Nominal is the input overall atomic composition. The detailed equilibrium compositions are given by the Amount column, and since the model is based on ideal mixing, the mole fractions and activities are equal.

As the model has a thermodynamic basis, it does not rely on knowing which specific complex components are important in particular mixture phases. Also unnecessary is a knowledge of experimental information, such as phase boundaries, eutectic, peritectic, and similar invariant compositions, or vapor pressures, although for detailed work where solid solutions are important, some information regarding the composition range of solid solutions can be useful.

A general presumed weakness in this model is its apparent nonapplicability to systems exhibiting positive deviations from Raoult's law and systems having liquid (or solid solution) miscibility gaps. We have not pursued this aspect in detail since the systems studied by us have generally shown large negative departures from ideality. However, it appears likely that for an appropriate choice of component species, using self-associate model components {e.g., $(SiO_2)_n$}, antimixing systems could also be accommodated by the model. Modeling immiscibility effects would require prior knowledge of the existence of multiple solution phases and some rough estimates of the composition boundaries of the solutions. It is clear that the reliability of the model, given that the mixing assumption is reasonably valid, depends intimately on the inclusion of all appropriate complex components and on the accuracies of the $\Delta_f G(T)$ functions. As a tool for modeling the major features of systems, particularly high-order mixtures where available data are generally extremely sparse, this model has a simplicity and robustness that can make it particularly effective as the thermochemical "engine" in many large-scale process models.

Table 3
Typical Model Composition Calculation for a Melt Na$_2$O (0.116 mol%)$^a$–K$_2$O (0.223)–SiO$_2$ (0.661)

| T = 1430.00 K | Input composition # 1 | | P$_{total}$ = 1.000 atm | | | |
| | Nominal$^b$ mol | Amount mol | Pressure, atm | Activity | Δ$_f$G(T), kcal/mol | Log, K$_p$ |
| Species | | | | | | |
| --- | --- | --- | --- | --- | --- | --- |
| **Gas Mixture** | | | | | | |
| Ar(g) | 5.00000E + 00 | 5.00000E + 00 | 9.99996E − 01$^c$ | 9.99996E − 01 | 0.000 | 0.000 |
| K(g) | 4.46000E − 01$^c$ | 1.34077E − 05 | 2.68153E − 06 | 2.68153E − 06 | 0.000 | 0.000 |
| Na(g) | 2.32000E − 01 | 4.20078E − 06 | 8.40152E − 07 | 8.40152E − 07 | 0.000 | 0.000 |
| O$_2$(g) | 8.30500E − 01 | 4.40212E − 06 | 8.80420E − 07 | 8.80420E − 07 | 0.000 | 0.000 |
| Si(g) | 0.00000E + 00 | 2.13753E − 27 | 4.27504E − 28 | 4.27504E − 28 | 56.933 | −8.701 |
| SiO(g) | 0.00000E + 00 | 1.38673E − 13 | 2.77345E − 14 | 2.77345E − 14 | −53.251 | 8.139 |
| SiO$_2$(g) | 0.00000E + 00 | 1.88651E − 13 | 3.77299E − 14 | 3.77299E − 14 | −73.936 | 11.300 |
| Si$_2$(g) | 0.00000E + 00 | 3.77649E − 49 | 7.55295E − 50 | 7.55295E − 50 | 77.119 | −11.786 |
| Si$_3$(g) | 0.00000E + 00 | 1.36904E − 68 | 2.73807E − 69 | 2.73807E − 69 | 82.175 | −12.559 |
| | | | | Mole Fraction | | |
| **Liquid Mixture** | | | | | | |
| K$_2$O(l) | 0.00000E + 00 | 1.87683E − 10 | 4.63581E − 10 | 4.63581E − 10 | −31.649 | 4.837 |
| K$_2$SiO$_3$(l) | 0.00000E + 00 | 5.44861E − 02 | 1.34582E − 01 | 1.34582E − 01 | −248.790 | 38.023 |
| K$_2$Si$_2$O$_5$(l) | 0.00000E + 00 | 1.56750E − 01 | 3.87176E − 01 | 3.87176E − 01 | −413.559 | 63.205 |
| K$_2$Si$_4$O$_9$(l) | 0.00000E + 00 | 1.17573E − 02 | 2.90409E − 02 | 2.90409E − 02 | −729.733 | 111.527 |
| Na$_2$O(l) | 0.00000E + 00 | 1.64884E − 09 | 4.07267E − 09 | 4.07267E − 09 | −44.420 | 6.789 |
| Na$_2$SiO$_3$(l) | 0.00000E + 00 | 5.18742E − 02 | 1.28130E − 01 | 1.28130E − 01 | −255.246 | 39.010 |
| Na$_2$Si$_2$O$_5$(l) | 0.00000E + 00 | 6.41237E − 02 | 1.58387E − 01 | 1.58387E − 01 | −417.615 | 63.825 |
| SiO$_2$(l) | 0.00000E + 00 | 6.58632E − 02 | 1.62684E − 01 | 1.62684E − 01 | −156.607 | 23.934 |
| **Pure Phases** | | | | | | |
| Na$_2$Si$_2$O$_5$(s) | 0.00000E + 00 | 0.00000E + 00 | | 8.47202E − 02$^d$ | −415.837 | 63.553 |
| SiO$_2$(s) | 0.00000E + 00 | 0.00000E + 00 | | 1.88786E − 01$^d$ | −157.029 | 23.999 |
| Si(s or ℓ) | 6.61000E − 01 | 0.00000E + 00 | | 2.14860E − 19$^d$ | 0.000 | 0.000 |

$^a$Nominal or gross composition, i.e., without complex phases considered.
$^b$Composition in terms of the reference elements.
$^c$Computer notation for numeric powers of 10, i.e., E−01 = 10$^{-1}$.
$^d$Activities less than one here represent relative stabilities of pure phases, but the phase is *not* present until unit activity is reached.

## TESTING OF MODEL

### Experimental Approach

There is relatively little emphasis in the literature on the thermo-dynamics of amorphous systems, such as slags and glasses, undoubtedly because these systems are not in complete thermodynamic equilibrium (although the departures are usually relatively small). Also, such systems are extremely complicated compared to the pure chemical compounds for which most thermodynamic data have been amassed. Detailed exper-imental studies are difficult because of the composition changes resulting from vapor transport losses during measurements. In addition, oxide solutions are particularly difficult to contain because of their reactivity and propensity for creeping.

To obtain activities in silicate and other oxide liquids, one of the more common and reliable techniques is the measurement of vapor pressures. Mass spectrometric methods exploit the activity definition,

$$a_i(T) = P_i(T)/P_i^o(T)$$

where $a_i$ is the activity of species $i$ at temperature $T$ in the solution, $P_i^o$, the saturation partial pressure for $i$ over the solution, and $P_i^o$, the saturation vapor pressure over pure $i$ (at unit activity). The primary experimental methods used in these studies are the Knudsen Effusion Mass Spec-trometric (KMS) and Transpiration Mass Spectrometric (TMS) methods, described elsewhere ((4) KMS; (5) TMS). These methods involve mass spectrometric (MS) analysis of modulated molecular beams that allows for phase-sensitive detection. Both techniques have been shown to yield accurate measurements of both gaseous and condensable species partial pressures for complex systems (1,4,5).

Basically, the KMS method consists of measuring mass analyzed ion currents from electron impact ionization of a collimated molecular beam from a Knudsen cell. At pressures where the molecular beam effuses by molecular flow and where suitable precautions are taken to insure satura-tion (equilibrium vaporization) within the cell, the partial pressures can be obtained via

$$P_i = k_i I_i^+ T$$

where $I_i^+$ is the observed ion current and k is a mass spectrometer calibration "constant" containing geometric factors and transmission, cross-section, and detector response functions for species $i$. These factors are obtained by a combination of direct measurement, literature data, and calibration. The KMS method, requiring effusive conditions, is gen-erally limited to maximum total cell pressures of ca. $10^{-4}$ atm (1 atm = 101,325 Pa).

The TMS technique substitutes a transpiration cell for the Knudsen source, extracting a molecular beam from much higher source pressures with a supersonic nozzle. This approach provides for sampling from regions of one atm or more and allows for internal calibration by direct comparison of the observed ion current and the known carrier gas pressure, as well as the gravimetric techniques appropriate for the KMS method.

It is not always possible, in practice, to measure the pressures of all the solution constituents since the equilibrium pressure of one or more of the constituents may not have sufficient vapor pressure to allow measurements to be made in the temperature region of interest. Generally, however, activity data can be obtained by measurement of other equilibrium reactions by marker additions (*see* Rudnyi, et al. (16)), or by using data analysis techniques such as Gibbs-Duhem integration to obtain activities.

## Validation

In the development of the model database, a systematic effort has been made to validate the assignment of complex components and their $\Delta_f G(T)$ functions. This has been done through comparisons of model predictions with experimental activity and phase equilibria information, beginning with binary and ternary and then progressing to higher order systems. Representative examples of such comparisons are given here.

### System $Na_2O-SiO_2$

For relatively simple binary systems, a significant number of experimental studies have been carried out by various workers (*see* Fig. 1). However, the results typically show considerable scatter among laboratories even for precise measurements. Figure 1 shows a wide selection of literature experimental data obtained by a variety of reliable methods (*see* (10) for references to experimental data). The model clearly represents the data as well as any statistical fit and correctly predicts the known location of the phase boundary (17). It should be noted that the model calculation contains no parameters derived from the displayed data, except insofar as those data may have influenced compilations such as JANAF (18) in their evaluation of $\Delta_f G(T)$ for $Na_2O$ (s and *l*), $Na_2SiO_3$ (s and *l*) and $Na_2Si_2O_5(l)$. A similar comparison for the $K_2O-SiO_2$ system has been made elsewhere (7).

Recently, Rudyni et al. (16) have obtained activity data for binary $Na_2O-SiO_2$ mixtures by a new mass spectrometric technique that utilizes pressure ratios obtained from negative ions observed with chromate additives present in equilibrium with gaseous oxygen over the melts. Figure 2 shows their reported data and the model prediction. It should be emphasized that the model curve is in no way a "fit" to the experimental observations. In comparing Figs. 1 and 2, it should be noted that the error differences are more noticeable in Fig. 2, owing to the scaling factor relation between activity and pressure, i.e.

$$a(Na_2O) \propto P_{Na}^{2.5}$$

### System $Na_2O-K_2O-SiO_2$

Although sufficient experimental activity data have been available for model validation tests for the systems $Na_2O-SiO_2$ and $K_2O-SiO_2$,

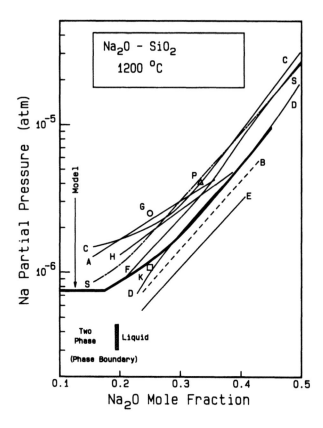

Fig. 1.   Comparison of model and literature data for Na partial pressure vs total $Na_2O$ content in $Na_2O$–$SiO_2$ mixtures at 1473 K. See ref. *10* for citations of experimental data.

until recently, few data have been available for the ternary system formed from these binaries. Some relative data comparing the binary and ternary systems were presented by Choudary et al. (*19*), and, more recently, by Chastel et al. (*28*), showing the presence of a relatively small ternary interaction. Rudnyi et al. (*16*) recently obtained data for this ternary system using the same approach as for the $Na_2O$–$SiO_2$ case discussed above. Figure 3 shows a comparison of the model predictions with their data. The agreement is very good for $K_2O$ and satisfactory for $Na_2O$, where the experimental data are systematically higher than the model results. This trend may also be present in the binary case, at least at a similar $Na_2O$ concentration (*see* Fig. 2). From Table 3, it can be seen that the predominant component at 1430 K is $K_2Si_2O_5(l)$ (39 mol %), with the other alkali silicate liquids all being between 13 and 16 mol%, whereas the simple oxide mole fractions (activities) are within an order of magnitude of being $10^{-10}$.

One of the experimental difficulties with measurements of these alkali silicates is the presence of residual alkali carbonate used in the preparation of silicate mixtures. As we have shown elsewhere (*1*), even where the residual carbonate concentration is below the one mol percent level, this is a much higher activity form of alkali that leads to higher

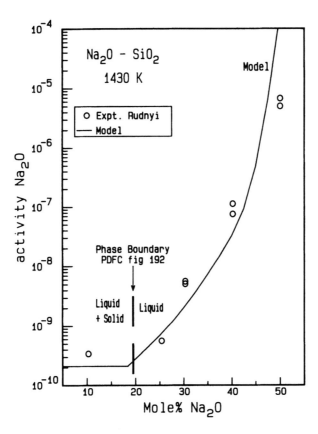

Fig. 2. Comparison of model calculations in the $Na_2O$–$SiO_2$ system with recent experimental data obtained using a novel negative ion-forming additive technique.

alkali pressures. Alternatively, $\Delta_fG(T)$ for the added $NaCrO_4^-$, which was used (*16*) to convert experimental ion ratios to Na pressures may be slightly in error. Likewise, a slight error in the literature stability of $Na_2Si_2O_5(l)$ in the model database could account for the observed difference. There are still insufficient activity and $\Delta_fG(T)$ data to resolve the issue of apparent systematic differences in model and experimental $Na_2O$ activities.

Based on the good agreement for $K_2O$ and satisfactory agreement for the $Na_2O$ model and experimental activities in this system, there appears to be no basis, at present, for inclusion of ternary mixed alkali species, such as $KNaSiO_3(\ell)$, in the database. Such species may still be important, however, for compositions where the $K_2O$ and $Na_2O$ activities are more nearly equal.

*System FeO–Al₂O₃*

A primary advantage of the model is as a direct compositional mapping tool. Although rather tedious using general free energy minimization codes (e.g., SOLGAMIX (*15*)), it is possible to calculate phase diagrams in some detail, as shown in Fig. 4, where a model calculation is

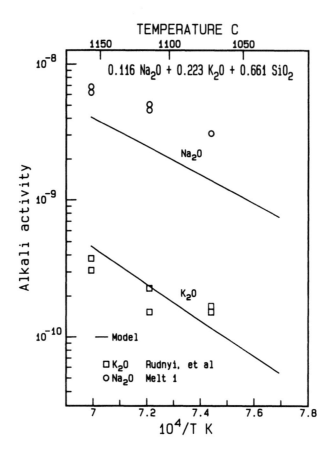

Fig. 3.   Comparison of model and experimental activities with recent experimental data for a $K_2O$ (0.223 mol%)–$Na_2O$ (0.116)–$SiO_2$ melt.

shown together with a literature (20) diagram. In the liquidus, the main complex component is $Al_2FeO_4(l)$, which varies between 20 and 40 mol% through much of the composition range. The stability of this component plays a major role in determining the activity of $FeO(l)$.

Even though the phase boundaries predicted by the model are in good agreement with the experimental diagram, the model does not rely on any data from the literature diagram, except where it may have influenced the literature selection of thermodynamic functions. The most sensitive comparison for this model is the composition location of the eutectics. Here the uncertainties in the known $\Delta_f G(T)$ are much larger than the free energy differences among the various phases near the invariant compositions, and the nearly parallel nature of the phase free-energy surfaces magnifies the model sensitivity. It should be noted that the other boundary differences are comparable with those found among the various experimental determinations.

In the process of applying the model to this system, it was noted that FeO(s) appeared unexpectedly. Since FeO(s) is found experimentally to be unstable under all conditions, apparently the $\Delta_f G(T)$ function fitted to

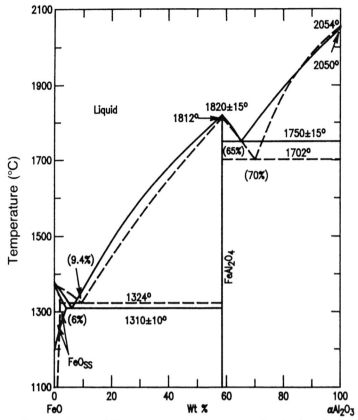

Fig. 4.   Comparison of literature (solid curve) and calculated (dash curve) phase diagram of the "FeO"–$Al_2O_3$ system. Ordinate is degrees Celsius.

the JANAF (*18*) data is too stable. This argument is borne out by considering the two following simple equilibria

$$FeO = Fe_xO + (1 - x)\, Fe \quad (x \simeq 0.947,\ wüstite)$$

$$4\, FeO = Fe_3O_4 + Fe$$

For FeO(s) to be less stable than $Fe_xO(s)$, over the range 800–1800 K, at least 7.1 to 9.2 kJ/mol must be added to $\Delta_fG(T, FeO(s))$. Relative to $Fe_3O_4(s)$, from 5.9 to 19.2 kJ/mol must be added to $\Delta_fG(T, FeO(s))$ over this temperature range. This correction is similar to that anticipated by the editors of JANAF (*21*) based on their reevaluation of the current literature data. To make the correction, the adjustment

$$\Delta(\Delta_fG(T)) = 8400 + 1.185 \cdot T \cdot \ln T \ \text{J/mol}$$

was added to the FeO(s) coefficient set b and c values (*22*).

### System $K_2O$–$Al_2O_3$–$CaO$–$SiO_2$

This system is one of a number of quaternary systems tested. In general, experimental test data are sparse for quaternaries, but this system has received systematic experimental study by us because of its importance as an MHD slag subsystem.

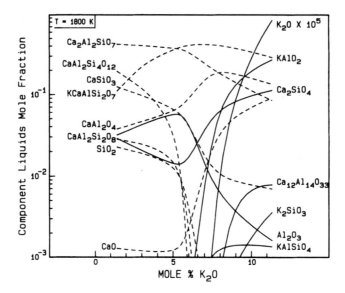

Fig. 5. Example of model calculation of liquid component mole fractions (activities) for $K_2O$–CaO (37.3 mol%)–$Al_2O_3$ (20.5 mol%)–$SiO_2$ mixtures at 1800 K. Solid curves labeled on RHS.

In (23), we presented comparisons of the model with experimental data for a selection of slag compositions in the $K_2O$–$Al_2O_3$–CaO–$SiO_2$ system. The model agreement with the various cases was well within the experimental error, indicating that the model has general applicability in higher order systems. Figure 5 shows a compositional map for a liquidus in this system along the $K_2O$ composition line. Of note is the complexity of the melt, including significant activities of several calcium aluminosilicate ternary components and the quaternary component, $KCaAlSi_2O_7(l)$ (*see* (10) and discussion of dolomite below). Note also the dramatic change in which components are important over the relatively narrow composition range from five to 10 mol% $K_2O$.

## MODEL APPLICATIONS

### Mixed-Alkali Glass

Figure 6 shows a comparison of KMS experimental data (points) and calculated model curves for NIST (NBS) Standard Reference Material (SRM) 621, a carefully homogenized and characterized soda-lime-silica glass. This comparison was originally presented elsewhere (10), but Fig. 6 shows the results of a more recent calculation using a current database containing Fe components. The excellent agreement is strong evidence of the practicality of this modeling procedure and its usefulness in supplementing and optimizing experimental measurements in very complicated glass mixtures.

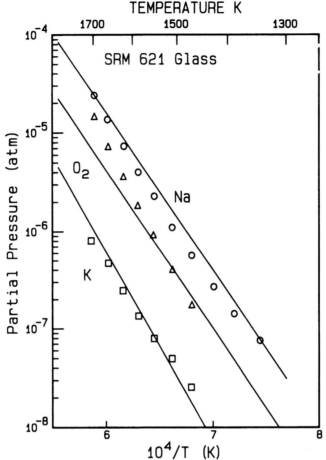

Fig. 6.   Comparison of model (solid curves) and experimental partial pressure vs temperature data for the seven component SRM-621 glass with the nominal composition: $Na_2O$ (12.54 mol%//12.75 wt%), $K_2O$ (1.3//2.02), $CaO$ (11.68//10.75), $MgO$ (0.41//0.27), $Al_2O_3$ (1.66//2.78), $SiO_2$ (72.4//71.39), $Fe_2O_3$ (0.02//0.04). $\square$ is K, $\triangle$ is $O_2$, and $\circ$ is Na.

## Dolomitic Limestones

Further advantages of the model in treating higher order systems are illustrated in Figs. 7a and b. Figure 7a shows both model calculations and experimental measurements for a dolomitic limestone, NIST (NBS) SRM 88a. Compared with other systems we have studied, dolomites have very high ratios of CaO to $K_2O$. Therefore, this composition emphasizes the model sensitivity to predicted higher order interactions of K and Ca. In particular, in ref. (*10*), we presented arguments supporting the existence of a liquid association, $KCaAlSi_2O_7(l)$, paralleling a reported (*24*) potassium melilite phase that, when included in the database, compensates nearly perfectly for a systematic variation between experiment and the model K-pressure predictions over almost three decades of $CaO/K_2O$

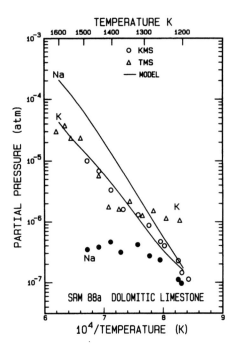

Fig. 7a. Calculated alkali partial pressure data for NIST SRM 88a dolomitic limestone as a function of temperature (solid curves). Points are experimental data. Composition of SRM 88a (wt% certified) is $Al_2O_3$ (0.19 wt%//0.17 mol%), CaO (30.10//49.23), $Fe_2O_3$ (0.28//0.16), $K_2O$ (0.12//0.12), MgO (21.30//48.47), $Na_2O$ (0.01//0.015), and $SiO_2$ (1.20//1.83).

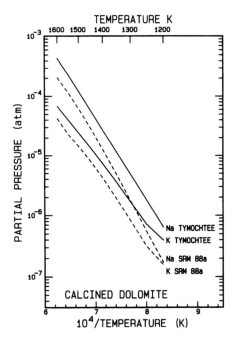

Fig. 7b. Comparison of calculated alkali partial pressures for SRM 88a and Tymochtee dolomites, showing effect of composition variations. Composition of the Tymochtee dolomite is $Al_2O_3$ (4.57 wt%//8.91 mol%), CaO (43.58//46.73), $Fe_2O_3$ (0//0), $K_2O$ (0.16//0.0.29), MgO (40.5//31.2), $Na_2O$ (0.004//0.005), and $SiO_2$ (11.15//12.81). SRM 88a composition is given in Fig. 7a.

composition ratios. The major complex liquid components for SRM-88a are $Ca_2SiO_4(l)$ at 30 mol% and $KCaAlSi_2O_7(l)$ at 16 mol%, with MgO and CaO being 13 and 11 mol%, respectively, at 1800 K. The complex components, $KCaAlSi_2O_7(l)$ and $KAlO_2(l)$, 3 mol%, largely control the $K_2O(l)$ activity in this system.

It should be noted in Fig. 7a that the experimental Na pressure data are quite uncertain, with the higher temperature tailing resulting from sample depletion by volatilization during the experimental run; the run chronology is from lower to higher temperatures. In many cases, this kind of preferential sample loss precludes quantitative measurement of activities. The model Na curve parallels the initial data, and provides a good indication of the sodium volatility at high temperatures for process conditions where the steady-state Na concentration is that of the initial dolomite composition.

Figure 7b compares calculations for SRM 88a with a Tymochtee dolomite selected for use in fluidized bed coal combustors. The sensitivity of the K- and Na-pressures to dolomite composition can be readily seen in this comparison.

The addition of oxides with multiple metal oxidation states (e.g., Fe–oxides) provides a particularly stringent test of the model. Figures 6 and 7a show well-characterized systems where small amounts of (nominally) $Fe_2O_3$ are present. These calculations now include new Fe-containing species (*see* Table 1) and demonstrate that in well-behaved systems, redox processes are readily modeled by the IMCC approach.

## Illite Mineral

In the case of the common clay mineral illite, the model does not appear to agree with experiment, even using the experimental $O_2$ pressures (*see* Fig. 8). Examination of the overall composition of the system indicates the likely problem in comparison of model and experimental species pressures. For this mineral, the model K-pressure is critically dependent on the $FeO_x$ stoichiometry. As with other Fe-containing systems, the presence of competing equilibria controls the K-pressure for fixed $K_2O$ activity, i.e.

$$K_2O(\ell) = 2K + \tfrac{1}{2}O_2$$

$$Fe_3O_4 = 3\,FeO_x + (2 - 1.5x)O_2$$

The K-pressure values can, therefore, be particularly sensitive to the value of $x$, as may be seen in Fig. 8. Careful examination of the sample shows inhomogeneities and likely variations in $x$. In fact, the variations shown in the figure are much smaller than the inhomogeneities observed in the sample. The model now provides readily testable expectations that can be used in further experimental investigations of this complex system. Additional experimental and model details for this system are given elsewhere (25).

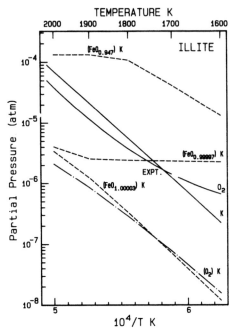

Fig. 8. Experimental and model partial pressure data for K and $O_2$ over Illite. Nominal mineral composition is $Al_2O_3$ (26.0 wt%), $Fe_2O_3$ (4.4), $K_2O$ (7.4), $Na_2O$ (0.2), MgO (2.1), $SiO_2$ (60.2), and S (not modeled, 0.1). Solid curves are experiment; dot-dash curve is model K-pressure with $O_2$-pressure set equal to the experimental $O_2$ value; dash curves are K-pressure data calculated based on slight variations in condensed phase iron–oxide O/Fe atom ratios as shown in parentheses.

## Potassium-Enriched Coal Slag (MHD)

Typically, in systems controlled by redox processes, particularly where both solid and liquid solutions coexist, there often can be considerable uncertainty about the actual composition responsible for observed experimental vapor pressures. In the MHD slag of Fig. 9 (*see (10)* for more details), if the model assumes an approximately $Fe_3O_4$ overall stoichiometry, the model prediction is low. However, the model agrees within experimental error when the experimentally observed $O_2$ pressure is used to control the model system oxygen content, indicating that the analysis of oxygen content in the slag is critical to understanding its behavior. In this system, the main model complex component that controls the $K_2O$ activity is $KAlSi_2O_6(l)$.

## Borosilicate Glass (Nuclear Waste Host)

Borosilicate glass systems are currently of high interest because of their planned use in encapsulation and storage of nuclear wastes. A preliminary borate database has been developed. The model prediction of alkali borate volatility is compared with experimental TMS and KMS data (*26*) in Fig. 10. The agreement for $LiBO_2$ is very good, and the agreement is satisfactory for $NaBO_2$. It is perhaps noteworthy that the

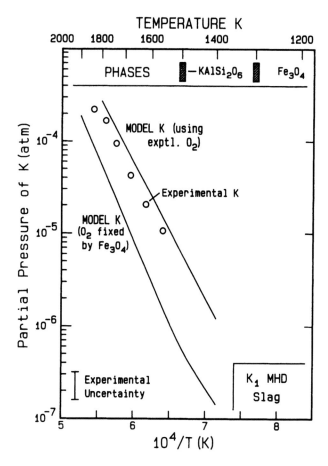

Fig. 9. Experimental data (this lab) for an MHD slag of composition $Al_2O_3$ (12.1 wt%), CaO (3.8), $Fe_2O_3$ (14.3), $K_2O$ (19.5), MgO (1.0), $Na_2O$ (0.5), and $SiO_2$ (46.8); model calculations show the $Fe_xO_y$ composition uncertainty effects.

model prediction is less than the experimental result for the Na case, as was also found in the $Na_2O$–$SiO_2$ and $Na_2O$–$K_2O$–$SiO_2$ systems discussed above. Further experimental and modeling work is needed on this and related $Na_2O$-containing systems.

## CONCLUSIONS

A predictive silicate solution model and an extensive thermodynamic database have been developed, validated through comparison with experiment, and applied to a variety of oxide systems containing up to seven oxides. These oxides (containing Al, B, Ca, Fe, K, Li, Mg, Na, and Si) span many of the technologically important glass, slag, ceramic, and mineral systems.

Over 100 complex liquid and solid components are included as $\Delta_f G(T)$ functions. Although even modest uncertainties in $\Delta_f G(T)$ (*see* Kaufman (27) and Fig. 4) can lead to inconsistencies in phase diagram

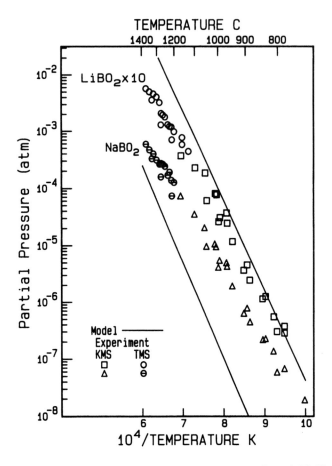

Fig. 10. Comparison of model prediction with TMS and KMS experimental data for a simulated nuclear waste borosilicate glass with nominal composition of: $SiO_2$ (59.4 mol%), $Na_2O$ (9.20), $Li_2O$ (8.80), $B_2O_3$ (6.19), $Al_2O_3$ (3.09), $MnO_2$ (2.72), $ZrO_2$ (2.04), NiO (1.60), CaO (1.19), MgO (0.88), $RuO_2$ (0.05), SrO (0.03), $Cs_2O$ (0.02), and $Re_2O_7$ (0.01).

calculations, for high order systems not readily amenable to experimental study, phase composition and activity predictions using the present database and model should be reasonably reliable, and particularly valuable in the absence of experimental data.

Oxide slags, glasses, and mineral substances clearly can be very successfully modeled by the Ideal Mixing of Complex Components (IMCC) model. The model has the advantages that it is:

1. Predictive—in general, additional activity data for specific systems are not necessary;
2. Conceptually simple, based on known thermochemistry;
3. Effective as a means for interpretation and extrapolation of sparse experimental data;
4. Equally facile with high order, as well as low order systems— it has been applied successfully to oxide systems with from two to seven constituents;

5. Generally applicable—suitable for inclusion as the "composition engine" in process codes since there are minimal adjustable parameters; and
6. Able to handle the gas phase naturally, giving detailed phase compositions.

The model has some possible weaknesses, in that it is:

1. Difficult to get "traditional outputs," e.g., phase diagrams directly;
2. An overdetermined system, which is not easily optimized;
3. Dependent on sparse thermochemical data, some with relatively high uncertainties;
4. Sometimes necessary to make rough estimates of solid solution boundaries and the existence of immiscibility;
5. Not typically as accurate as optimized models, where good experimental data exist; and
6. Computationally intensive—requires a large database of critically evaluated data.

Recent advances in coupled thermodynamic-phase diagram optimization of relatively simple oxide systems (*see* Pelton, and Jonsson, Sundeman, and Hillert elsewhere in this volume) should lead to significant reductions in uncertainties in $\Delta_f G(T)$ for key compounds in the IMCC database. It is also possible, though increasingly less likely as new test cases arise, that the present database is not unique. By unique, we mean that an alternative set of complex components or significant changes in the $\Delta_f G(T)$ for the existing complex components would not equally well represent the experimental data used for validation tests.

Future work is concentrating on expanding and refining the oxide thermodynamic database and providing experimental data for validation. This effort includes developing multi-range coefficient sets for $\Delta_f G(T)$ functions, investigating means for handling antimixing phenomena, and experimental investigations of a wide variety of technologically important slag, glass, ceramic, and mineral systems. The database is currently available in machine-readable form directly from the authors on request.

## ACKNOWLEDGMENTS

Collaborations with W. S. Horton and E. R. Plante in the early phases of this work are gratefully acknowledged.

## REFERENCES

1. Hastie, J. W., Plante, E. R., and Bonnell, D. W., *Metal Bonding and Interactions in High Temperature Systems with Emphasis on Alkali Metals*, Gole, J. L. and Stwalley W. C., eds., ACS Symposium Series 179 American Chemical Society, Washington DC 1982, pp. 543–600.

2. Kirkbride, B. J., *Glass Tech.* **20,** 174 (1979).
3. Hastie, J. W., *Pure and Applied Chem.* **56,** 1583 (1984).
4. Plante, E. R., *Characterization of High Temperature Vapors and Gases*, Hastie, J. W., ed., NBS SP561/1, US Gov. Printing Office, Washington, DC, 1979, p. 265.
5. Bonnell, D. W. and Hastie, J. W., *ibid.* ref. 4, p. 357.
6. Hastie, J. W., Plante, E. R., and Bonnell, D. W., *Alkali Vapor Transport in Coal Conversion and Combustion Systems*, NBSIR 81-2279, NIST, Gaithersburg, MD, 1981.
7. Hastie, J. W., Horton, W. S., Plante, E. R., and Bonnell, D. W., *High Temp.-High Press.* **14,** 669–679 (1982).
8. Richardson, F. D., *Physical Chemistry of Melts in Metallurgy*, Academic Press, New York, NY, 1974.
9. Turkdogan, E. T., *Physical Chemistry of High Temperature Technology*, Academic Press, New York, NY, 1980.
10. Hastie, J. W. and Bonnell, D. W., *High Temp. Sci.* **19,** 275 (1985).
11. Redlich, O., Kister, A. T., and Turnquist, C. E., *Phase Equilibria* **2,** 49 (1952).
12. Grover, J., *Thermodynamics in Geology*, Fraser, D. G., ed., Reidel, Dordrecht, Holland, 1977, p. 67.
13. Ansara, I., *Int. Met. Rev.* **1,** 20 (1979).
14. Hillert, M., Sundeman, B., and Wang, X., *TRITA-MAC-0355*, Materials Research Center, Royal Institute of Technology, Stockholm, 1987.
15. Eriksson, G., *Chemica Scripta* **8,** 100 (1975); SOLGASMIX, version 3, with a NIST-modified input module is currently in use.
16. Rudnyi, E. B., Vovk, O. M., Siderov, L. N., Stolyarova, V. L., Shakhmatkin, B. A., and Rakhimov, V. I., *Fiz. Khim. Stekla* **14,** 218 (1988). (See also elsewhere in this volume)
17. Levin, E. M., Robbins, C. R., and McMurdie, H. R., *Phase Diagrams for Ceramists* **1,** diagram 192 (1964).
18. *JANAF Thermochemical Tables*, 3rd Edition, Chase, M. W., Jr., Davies, C. A., Downey, J. R., Jr., Frurip, D. J., McDonald, R. A., & Syverud, A. N. eds., *J. Phys. Chem. Ref. Data* **14** (1985).
19. Choudary, U. V., Gaskell, D. R., and Belton, G. R., *Met. Trans. B* **8b,** 67 (1977).
20. Levin, E. M., Robbins, C. R., and McMurdie, H. R., *Phase Diagrams for Ceramists*, diagram 2070 (1969).
21. Chase, M. W., Jr., private communication.
22. Horton, W. S., *Interim Contract Report WSH018* **1** (1987).
23. Hastie, J. W., Bonnell, D. W., and Plante, E. R., *Proc. Symp. on High Temperature Materials Chemistry II*, Munir, Z. A. and Cubicciotti, D., eds., The Electrochemical Society, Pennington NJ, 1983, pp. 349–359.
24. Levin, E. M., Robbins, C. R., and McMurdie, H. R., *Phase Diagrams for Ceramists* **1,** diagram 800 (1964).
25. Hastie, J. W., Bonnell, D. W., and Plante, E. R., *Proc. of Engineering Foundation Conf. on Mineral Matter and Ash Deposition from Coal*, Vorres, K., ed., Santa Barbara, CA, 1989.
26. Hastie, J. W., Plante E. R., and Bonnell, D. W., *Vaporization of Simulated Nuclear Waste Glass*, NBSIR 83-2731, NIST, Gaithersburg, MD, 1983.
27. Kaufman, L., *CALPHAD* **3,** 27 (1979).
28. Chastel, R., Bergman, C., Rogez J., and Mathieu, J.-C., *Chem. Geol.* **62,** 19 (1987).

# Structural and Thermodynamic Evidence for the Survival of Zintl Ions Through the Melting Process

## The Alkali-Lead Alloys

MARIE-LOUISE SABOUNGI,*,[1] G. K. JOHNSON,[1] D. L. PRICE,[1] AND H. T. J. REIJERS[2]

[1]Argonne National Laboratory, 9700 South Cass Ave., Argonne, IL 60439; and [2]Laboratory of Solid State Physics, University of Groningen, Melkweg 1, 9718 EP Groningen, The Netherlands

## ABSTRACT

Recent heat capacity and neutron diffraction measurements carried out on liquid equiatomic alkali-lead (APb) alloys give unambiguous evidence for the survival of Zintl ions in the liquid. The presence of polyvalent anionic species, $Pb_4^{4-}$, is responsible for the anomalous temperature dependence of the heat capacity, whose magnitude decreases from ~80 J mol$^{-1}$K$^{-1}$ in the vicinity of the melting point to ~40 J mol$^{-1}$K$^{-1}$ at ~300 K above the melting point. A first sharp diffraction peak (FSDP) is obtained at wavevectors of Q ~1 Å$^{-1}$ in the total structure factor, which is a signature of intermediate-range ordering. Models based on the presence of $A_4Pb_4$ structural units in the liquid are used to interpret the changes in the width and position of the FSDP when the alkali metal, A, is varied from Na to Cs.

**Index Entries:** Structure; heat capacity; neutron scattering; diffraction; alkali metals.

## INTRODUCTION

Transport and thermodynamic properties of some liquid alloys exhibit unusual behavior at the equiatomic composition corresponding to important structural changes. This is the case for the alkali-lead alloys (1–

---

*Author to whom all correspondence and reprint orders should be addressed.

*8*) with the exception of Li-Pb (*9–10*). It has been shown that the electrical resistivity reaches a maximum at equiatomic APb, where A is K, Rb, or Cs (*1–3*); depending on the alkali metal, this maximum takes either typical metallic values (e.g., 920 $\mu\Omega$ cm for KPb) with a negligible temperature coefficient or typical nonmetallic values (e.g., 7000 $\mu\Omega$ cm for CsPb) with a large negative temperature coefficient (e.g., $-115$ $\mu\Omega$ cm $K^{-1}$). The electrical resistivity for NaPb reaches a maximum at $Na_4Pb$ and a shoulder at NaPb, with a value of 445 and 290 $\mu\Omega$ cm, respectively. According to Mott and Davis' classification (*11*), NaPb is metallic, KPb and RbPb belong to regime II, and CsPb can be regarded as a liquid semiconductor.

The thermodynamic properties have remarkable features that are in consonance with those of the electrical transport properties (*4–7*). Extrema are reported in the entropy, the Darken excess stability, and the heat capacity at the equiatomic composition. More striking is the anomalous behavior of the excess heat capacity as the temperature increases from the melting point (*7*). The magnitudes far exceed values reported for any liquid alloy in the literature.

Total structure factors [S(Q)] deduced from neutron diffraction measurements carried out recently (*12*) at the Intense Pulsed Neutron Source (IPNS) at Argonne National Laboratory show the presence of a first sharp diffraction peak (FSDP) at low wavevectors, $Q \sim 1$ Å$^{-1}$, which is a signature of intermediate-range ordering in a liquid (*13*).

In this review, we report on an attempt to analyze the thermodynamic and structural results in terms of a progressive dissociation of chemical units similar to those present in the corresponding crystalline compounds. The dissociation is enhanced by increasing the temperature; a dilution of these chemical units is obtained by adding alkali metal or Pb to the equiatomic composition. In the solid, it was suggested by Zintl (*14*) over 50 years ago that the intermetallic compound contains molecular-like entities made up of $Pb_4$ tetrahedra. We are proposing that these entities persist in the liquid. The questions that we address include: Can the Zintl ions be experimentally detected in the liquid? How is their stability affected by temperature and size of the constituent atoms? Can we reconcile structure and thermodynamic measurements and derive a representation of the liquid?

## EXPERIMENTAL

In this section, we will describe briefly the experimental methods used to derive the results discussed below. For more specific information, the reader is referred to previous publications (*5,7,8*).

The thermodynamic results were deduced from two different approaches. For Na-Pb, K-Pb, and Rb-Pb, electromotive force (emf) measurements of the activity of the alkali metal were made using a coulometric titration technique with $\beta''$-alumina as the solid electrolyte. The composition of the alloys can be accurately controlled to the ppm level;

thus, this method is ideally suited to precisely measure the compositional dependence of the Gibbs free energy of mixing, the entropy, and the excess heat capacity, $\Delta C_p$. As mentioned in the Introduction, since $\Delta C_p$ exhibits unusually large values at the equiatomic composition, the temperature dependence of $\Delta C_p$ was investigated using high-temperature drop calorimetry. The variation of the enthalpy function, defined as $H_T - H_{298}$, with temperature was measured below and above the melting point. The heat capacity was deduced as the first derivative of $H_T - H_{298}$. This method has proven to yield accurate and precise heat capacities.

The structure of the equiatomic alloys was investigated using the Special Environment Powder Diffractometer (SEPD) at IPNS. Time-of-flight diffraction measurements were carried out on the alloys in the solid and liquid phase. The alloys were contained in thin-walled vanadium containers, and corrections for these were included. For practical reasons and experimental self-consistency, the samples used in the calorimetric experiments were pulverized before loading in the containers. Inelastic neutron diffraction experiments were also conducted to probe the internal vibrations of the polyanionic species in the solid and liquid using the Low Resolution Medium-Energy Chopper Spectrometer (LRMECS) at IPNS (*12*).

## RESULTS AND DISCUSSION

In this section, we review the most striking features in the thermodynamic and structural properties of these alloys. Analysis of the data is conducted using a thermodynamic model (*7*) based on a dissociation scheme, and two structural models based on the presence of structural units in the liquid (*15,16*). Following the crystallographic structures obtained for the solids, these units were taken to be $A_4Pb_4$ and consist of $Pb_4$ tetrahedra surrounded by $A_4$ tetrahedra oriented in the opposite direction.

Perhaps the most remarkable illustration of the unusual behavior of the alkali-lead alloys is achieved by plotting the entropy, the Darken excess stability

$$ES = RT/(1 - X_A) \, (\partial \ln \gamma_A / \partial X_A)$$

and the excess heat capacity as a function of alloy composition. Figures 1–3 show the variations with composition of these properties for K-Pb. The entropy goes through a sharp, well-defined minimum, which is a signature for ordered systems. Note that Rb-Pb mimics K-Pb to a large extent. Figure 2 gives the excess stability for all the alkali-lead alloys (with the exception of Cs-Pb where no thermodynamic data on the activity are available). The curves in this figure exhibit major peaks at $Li_4Pb$, $Na_4Pb$, NaPb, KPb, and RbPb. At the 4:1 composition, the maximum excess stability is related to a charge transfer between Pb and the alkali metal, giving the alloy a quasi-ionic character. This charge transfer can be easily explained on the basis of simple chemical valence rules. At the equi-

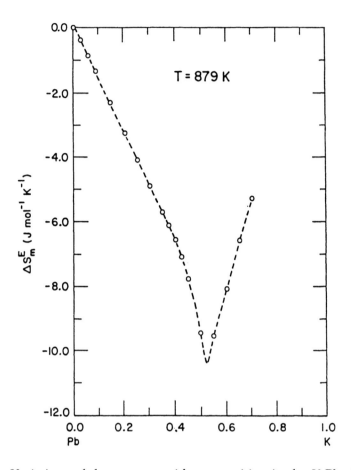

Fig. 1.   Variations of the entropy with composition in the K-Pb alloys.

atomic composition, the maximum is related to charge transfer leading to the formation of Zintl ions in the liquid. It is interesting to point out that, even though solid LiPb exists, its crystal structure differs from APb (A = Na, K, Rb, and Cs). There are many intermetallic compounds in each of these systems, but the chemistry is such that only the "octet" compounds ($A_4Pb$) and the equiatomic compounds play a role in the behavior of the liquid. Interestingly, $A_4Pb$ exists only for A = Li and Na; its absence is conspicuous for K, Rb, and Cs and can be explained by size arguments. According to Geertsma et al. (*17*), the alkali metal plays a stabilizing role in the persistence of $Pb_4$ tetrahedra: a small alkali ion leads to destabilization of the $Pb_4$ clusters, which become stable entities as the size of the alkali ion increases. This scenario successfully interprets the electrical resistivity results, which remarkably parallel those of the excess stability. Finally, the zero-wavevector limit of the concentration-concentration correlation structure factor, $S_{CC}(0)$, introduced by Bhatia and Thornton (*18*), is related to the inverse of the excess stability; in the case of Li-Pb, we have shown (*9*) that $S_{CC}(0)$ is directly obtainable from thermodynamic measurements and is in excellent agreement with direct experimental determinations.

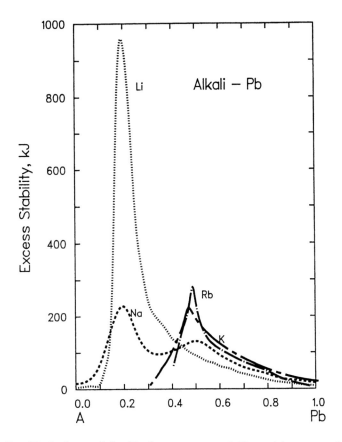

Fig. 2.   Variations of the Darken excess stability with composition for the A-Pb alloys, where A = Li, Na, K, and Rb.

The excess heat capacity of K-Pb, shown in Fig. 3, has two unusual features: first, it is significant and not negligible as for most alloys and, second, it has a pronounced maximum at the 1:1 composition. The Rb-Pb alloys have a remarkably similar behavior to K-Pb. The temperature dependence of the maximum in the heat capacity has been investigated by calorimetry for all the equiatomic alloys. The dramatic increase in $\Delta C_p$ for the cases of K-Pb, Rb-Pb, and Cs-Pb as the temperature decreases is contrasted by an almost constant and small $\Delta C_p$ in the case of Na-Pb (Fig. 4). No data are available for Li-Pb, but from emf measurements it is expected that $\Delta C_p$ will be even smaller than for Na-Pb. The heat capacity of NaPb can be explained if the concentration of $Na_4Pb_4$ structural units does not vary much with temperature. This would occur if the species are already largely dissociated. For KPb, RbPb, and CsPb, a chemical equilibrium scheme between $A_4Pb_4$ and "free" A and Pb atoms can be invoked to derive the relative proportions of each entity as a function of temperature.

The total structure factors, $S(Q)$, derived from diffraction measurements on molten NaPb, KPb, RbPb, and CsPb alloys at temperatures close to their melting points exhibit well-defined FSDP's around $Q \sim 1$

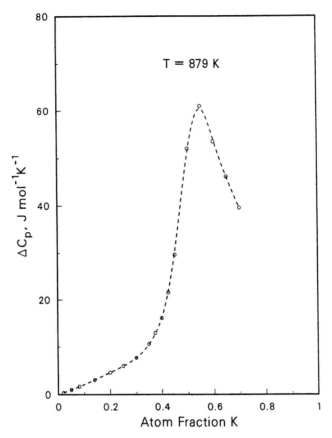

Fig. 3. Variations of the excess heat capacity with composition for the K-Pb alloys at an average temperature of 879 K.

$\text{Å}^{-1}$ (12). Such peaks have never been observed as clearly in other metallic alloys. The most striking example is that of CsPb, shown in Fig. 5, where for temperatures slightly above the melting point the magnitude of the FSDP exceeds that of the main peak. As expected, the magnitude of the FSDP decreases with increasing temperature. In the case of NaPb, it is almost a small shoulder, which is consistent with experimental observations of a metallic behavior and the absence of a large number of species $Na_4Pb_4$. The strong FSDPs can be interpreted in terms of the presence of structural units in the liquid, namely, the polyvalent anions, which have been shown to be responsible for the conduction mechanism in the alkali-lead alloys (8). The structure of the corresponding crystalline compounds, established by neutron and X-ray scattering experiments at room temperature, can be pictured in terms of lead tetrahedra surrounded by alkali tetrahedra oriented in the opposite direction.

Two models were used to analyze the APb structure, both based on the presence of $A_4Pb_4$ interlocking tetrahedra with the positions of the atoms taken from the corresponding crystal structure. One of the models was developed by Moss and Price and consists of a random packing of structural units (RPSU) with random uncorrelated orientations (15). The

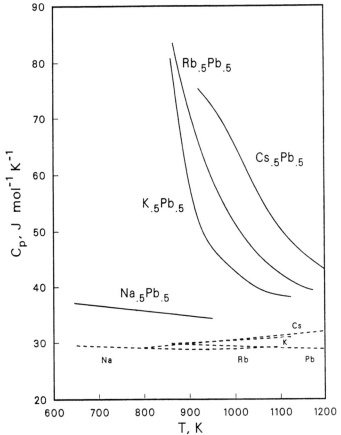

Fig. 4. Anomalous heat capacity of liquid alkali-lead alloys at the equiatomic composition as a function of temperature.

other one, the reference interaction site model (RISM), was developed by Andersen, Weeks, and Chandler (*16*) and is similar to the first, but hard-sphere interactions between pairs of individual atoms are taken into account and the orientations are therefore correlated. Once the structural units are specified, the structure factor depends only on two parameters in the case of RPSU: the packing fraction and the hard sphere diameter $\sigma$ of the unit; and on three parameters in the case of RISM, the density and the hard sphere diameters of the two atoms. Both models lead to calculated structure factors that are in consonance with measurements in reciprocal and real spaces. Figure 5 shows the good agreement between measurements and RPSU and RISM calculations for S(Q) of CsPb, and Fig. 6 shows the good agreement between measurements and RPSU calculations for the radial distribution function of KPb.

Results from inelastic neutron scattering experiments conducted on LRMECS were used to calculate the structure factor S(Q) from the scattering function S(Q,E) using the relation

$$S_\Delta(Q) = \int_{-\Delta}^{\Delta} S(Q,E)dE$$

For $\Delta \to \infty$, $S_\Delta(Q)$ reduces to S(Q) as measured by diffraction. Figure 7 plots the S(Q) results for liquid KPb measured with SEPD and $S_\Delta(Q)$

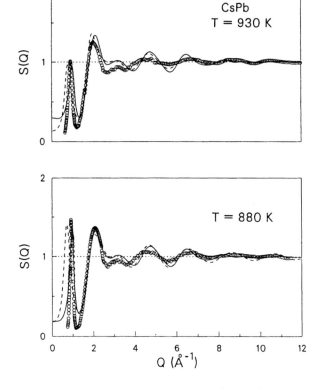

Fig. 5.  Total structure factor, S(Q), of liquid CsPb. The circles refer to measurements, solid line to RPSU model (15), and dashed line to RISM (16).

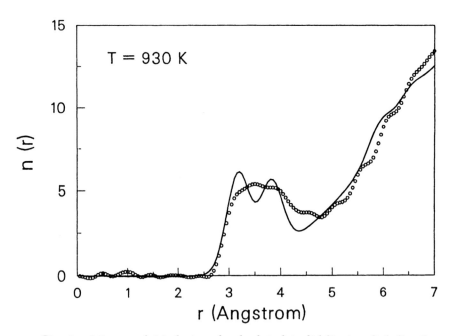

Fig. 6.  Measured (circles) and calculated (solid line) radial distribution function, n(r), of liquid KPb. Calculations from RPSU model (15).

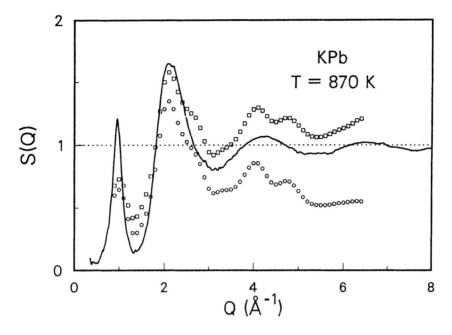

Fig. 7. Structure factors, S(Q), for liquid KPb. Solid line: S(Q) from diffraction measurements on SEPD; points: $S_\Delta(Q) = \int_{-\Delta}^{\Delta} S(Q,E)dE$ from inelastic scattering measurements on LRMECS: ($\square$) $\Delta$ = 40 meV ($\bigcirc$) $\Delta$ = 5 meV.

results obtained from the LRMECS measurements for $\Delta$ = 40 and 5 meV, respectively. The first case represents the closest approach to S(Q) possible in this measurement, i.e., the instantaneous (t = 0) density correlations in the structure, whereas the second gives a measure of the contributions from low energy transfers, representing the long-time average of the structural correlations. The agreement between the inelastic results for $\Delta$ = 40 meV and the diffraction result is reasonable, considering the lower Q resolution inherent in the inelastic measurements and the very different experimental procedures. The FSDP at $Q \cong 1.0$ Å$^{-1}$ shows up clearly in the inelastic measurements at both values of $\Delta$, indicating that the structural correlations responsible for this peak persist for times as long as $t = \hbar/\Delta \cong 10^{-13}$ s and possibly longer.

## CONCLUSION

This work represents a striking example of the way in which neutron diffraction, supported by inelastic scattering measurements, can provide structural information to explain a particular set of physical properties. In the case of the alkali-lead alloys, the observation of remarkably strong first sharp diffraction peaks in the neutron diffraction patterns establishes the existence of complex ions that are responsible for the dramatic changes in transport and thermodynamic properties near the equiatomic composition.

## ACKNOWLEDGMENTS

This work was performed under the auspices of the US Department of Energy, Division of Materials Sciences, Office of Basic Energy Sciences, under Contract W-31-109-ENG-38. Accordingly, the U. S. Government retains a nonexclusive, royalty-free license to publish or reproduce the published form of this contribution, or allow others to do so, for U. S. Government purposes. Discussions with W. van der Lugt have been illuminating. The assistance of R. Kleb, K. J. Volin, and the Operations Staff at IPNS is gratefully acknowledged.

## REFERENCES

1. Meijer, J. A., Geertsma, W., and van der Lugt, W., *J. Phys. F: Metal Phys.* **15**, 899 (1985).
2. van der Lugt, W. and Geertsma, W., *Can. J. Phys.* **65**, 326 (1987).
3. van der Lugt, W. and Meijer, J., *Amorphous and Liquid Materials*, Lüscher, E., Fritzsch, G., and Jaccuci, G., eds., Martinus Nijhoff (1987), Dordrecht.
4. Saboungi, M. -L., Herron, S. J., and Kumar, R., *Ber. Bunsenges. Phys. Chim.* **89**, 375 (1985).
5. Saboungi, M. -L., Leonard, S. R., and Ellefson, J., *J. Chem. Phys.* **85**, 6072 (1986).
6. Tumidajski, P. T., Petric, A., Takenake, T., Saboungi, M. -L., and Pelton, A. D., in press.
7. Saboungi, M. -L., Reijers, H. T. J., Blander, M., and Johnson, G. K., *J. Chem. Phys.* **89**, 5869 (1988); Johnson, G. K. and Saboungi, M. -L., *J. Chem. Phys.* **86**, 6376 (1987).
8. Saboungi, M. -L., Blomquist, R., Volin, K., and Price, D. L., *J. Chem. Phys.* **87**, 2278 (1987); Reijers, H. T. J., van der Lugt, W., and van Dijk, C., *Physica B* **144**, 404 (1987).
9. Saboungi, M. -L., Marr, J. J., and Blander, M., *J. Chem. Phys.* **68**, 1375 (1978).
10. Nguyen, V. T. and Enderby, J. E., *Philos. Mag.* **35**, 1013 (1977).
11. Mott, N. F. and Davis, E. A., *Electronic Processes in Noncrystalline Materials*, Oxford, Clarendon Press (1979).
12. Reijers, H. T. J., Saboungi, M. -L., Richardson, J. W., Volin, K., Price, D. L., and van der Lugt, W., in press.
13. Price, D. L., Moss, S. C., Reijers, R., Saboungi, M. -L., and Susman, S., *J. Phys. C: Solid State Phys.* **21**, L1069 (1988).
14. Zintl, E., Goubeau, J., and Dullenkopf, W., *Z. Physik. Chem. Abt A* **154**, 1 (1931).
15. Moss, S. C. and Price, D. L., *Physics of Disordered Materials*, Adler, D., Fritzsche, H., and Ovshinsky, S. R., eds., Plenum, New York (1985), p. 77.
16. Andersen, H. C., Weeks, J. D., and Chandler, D., *Phys. Rev. A* **4**, 1597 (1971).
17. Geertsma, W., Dijkstra, J., and van der Lugt, W., *J. Phys. F: Metal Phys.* **14**, 1833 (1984).
18. Bhatia, A. B. and Thornton, D. E., *Phys. Rev.* **B2**, 3004 (1970).

# Thermodynamic Properties of NiAs-Type $Co_{1 \pm x}Sb$ and $Ni_{1 \pm x}Sb$

HERBERT IPSER, REGINA KRACHLER, GERALD HANNINGER, AND KURT L. KOMAREK*

*Institute of Inorganic Chemistry, University of Vienna, Währingerstr. 42, A-1090 Wien, Austria*

## ABSTRACT

A statistical model is developed that describes the composition dependence of the thermodynamic activities in phases with the NiAs-(B8-) structure. The model is based on vacancies in the transition metal sublattice (octahedral sites) and additional transition metal atoms in interstitial positions (trigonal-bipyramidal sites) as intrinsic disorder. It can be applied to both types of defect mechanisms, i.e., NiAs-type (where vacancies are assumed to be distributed statistically over all transition metal sites) and $CdI_2$-type (where the vacancies are restricted to alternate layers of the transition metal sublattice); any kind of intermediate behavior with different vacancy concentrations on alternate transition metal layers can also be described. The model is applied to the two phases $Co_{1 \pm x}Sb$ and $Ni_{1 \pm x}Sb$ for which antimony activities were determined by an isopiestic vapor pressure method.

**Index Entries:** NiAs-phases, statistical model for; NiAs-phases, defect concentrations in; thermodynamics of $Co_{1 \pm x}Sb$; thermodynamics of $Ni_{1 \pm x}Sb$; $Co_{1 \pm x}Sb$, thermodynamic properties of; $Ni_{1 \pm x}Sb$, thermodynamic properties of; Co-Sb; Ni-Sb; vapor pressure measurements in the Co-Sb system; vapor pressure measurements in the Ni-Sb system.

*Author to whom all correspondence and reprint orders should be addressed.

## INTRODUCTION

The two phases $Co_{1\pm x}Sb$ and $Ni_{1\pm x}Sb$ are the only known examples of the NiAs-type (B8$_1$, hP4) with substantial ranges of homogeneity on both sides of stoichiometry. Deviation from stoichiometry is achieved by vacancies in the regular tansition metal sites (octahedral holes) on the antimony-rich side and by interstitial transition metal atoms (in the trigonal-bipyramidal holes) on the other side. This defect mechanism has been confirmed in the literature, for example by the composition dependence of the lattice parameters (1–6), of the pycnometric density (1,3,4,6), and of the electrical conductivity (1,2,4,7).

Different theoretical models have been derived in the past, based on the pioneering work of Wagner and Schottky (8), to describe the composition dependence of the thermodynamic properties in NiAs-type phases (9–11) assuming a statistical distribution of the defects. However, there are two possible ways of arranging the vacancies in the transition metal sublattice: They may either be distributed over all sublattice sites or they may be restricted to alternate layers resulting in a partially filled up CdI$_2$-type structure (12). This, of course, requires different theoretical equations, and the shapes of the resulting curves differ considerably with increasing vacancy concentration as has been demonstrated by Ipser and Komarek for the corresponding phase in the Fe-Te system (13).

It was the intention of the present work to derive a single model which can be applied to both defect mechanisms, i.e., vacancies in all or in alternate layers of the transition metal sublattice. The theoretical equations will be tested using experimental activity data for the NiAs-type phases in the two systems Co-Sb and Ni-Sb.

## EXPERIMENTAL METHOD

The experimental details of the isopiestic method used for the vapor pressure measurements in the Ni-Sb system have been described previously by Leubolt et al. (14). A very similar method was employed in the Co-Sb system using crucibles made of quartz instead of graphite. A detailed description of this apparatus can be found in (13). Starting materials were cobalt wire (0.1 mm ø; 99.99%; Johnson-Matthey, Vienna, Austria) and antimony (lumps; 99.99 + %; ASARCO, New York).

The evaluation of the results for both systems was based on the assumption that in the temperature range of our investigation antimony vapor contains only Sb$_2$ and Sb$_4$ in comparable amounts, whereas the contribution of Sb and Sb$_3$ to the total vapor pressure can be completely neglected (15). A full account of this evaluation can be found in Ref. (14).

## THE THEORETICAL MODEL

The ideal NiAs-lattice can be visualized as a hexagonally close packed array of atoms of a main group element (metalloid atoms) where all octahedral holes are filled with transition metal atoms. As discussed above, deviation from stoichiometry is accomplished by additional transition metal atoms in the trigonal-bipyramidal holes or by vacant octahedral sites. The so-called intrinsic disorder, i.e., disorder present at nonzero temperatures at the stoichiometric composition, is thought to be caused by a certain number of transition metal atoms that leave the octahedral positions and enter the trigonal-bipyramidal holes, frequently designated as interstitial positions.

For the derivation of the model it is assumed that the metalloid sublattice (β-sublattice) remains undisturbed. The transition metal sublattice (α-sublattice) is divided into two parts (α1 and α2) consisting of alternate layers. The interstitial sites form a third sublattice, the i-sublattice. It is assumed that all defects are strictly statistically arranged.

Following the concept of Wagner and Schottky (8), there are two equilibria for the exchange of transition metal atoms (A-atoms) between octahedral sites and interstitial sites.

$$A_{\alpha 1} \rightleftharpoons A_i \tag{1}$$

$$A_{\alpha 2} \rightleftharpoons A_i \tag{2}$$

where $A_{\alpha 1}$, and $A_{\alpha 2}$, and $A_i$ designate an A-atom on an α1-, α2-, or i-sublattice site. The total Gibbs energy of the crystal is given by

$$G = (1/N_L) (N^t \mu^* + N_A^i \mu_A^i + N_\square^{\alpha 1} \mu_\square^{\alpha 1} + N_\square^{\alpha 2} \mu_\square^{\alpha 2}) - TS_{conf} \tag{3}$$

where $N_L$ = Avogadro's number, $N^t$ = total number of real lattice sites (without i-sublattice, i.e., $N^t = N^\alpha + N^\beta = N^{\alpha 1} + N^{\alpha 2} + N^\beta$), $N_A^i$ = number of A-atoms on interstitial sites, $N_\square^{\alpha 1}$, $N_\square^{\alpha 2}$ = number of vacancies on α1- or α2-sites, resp.; $\mu^*$ = Gibbs energy of an ideally ordered crystal of 1 mole (A + B) with stoichiometric composition $\mu_A^i$ = change of Gibbs energy, referred to 1 mole, for the insertion of A-atoms into interstitial sites, $\mu_\square^{\alpha 1}$, $\mu_\square^{\alpha 2}$ = change of Gibbs energy for the removal of 1 mole of A-atoms from the α1- or α2-sublattice, resp. All these Gibbs energy terms are independent of composition.

Keeping in mind that the β-lattice remains undisturbed, the configurational entropy can be obtained in the following way:

$$S_{conf} = k \ln$$

$$\left( \frac{N^{\alpha 1}!}{N_\square^{\alpha 1}!(N^{\alpha 1} - N_\square^{\alpha 1})!} \quad \frac{N^{\alpha 2}!}{N_\square^{\alpha 2}!(N^{\alpha 2} - N_\square^{\alpha 2})!} \quad \frac{N^i!}{N_A^i!(N^i - N_A^i)!} \right) \tag{4}$$

The equilibrium conditions for the equilibria (1) and (2) are:

$$(\partial G/\partial N_A^i) \, dN_A^i + (\partial G/\partial N_\square^{\alpha 1}) \, dN_\square^{\alpha 1} = 0;$$
$$dN_A^i = dN_\square^{\alpha 1} \tag{5}$$

$$(\partial G/\partial N_A^i) \, dN_A^i + (\partial G/\partial N_\square^{\alpha 2}) \, dN_\square^{\alpha 2} = 0;$$
$$dN_A^i = dN_\square^{\alpha 2} \tag{6}$$

Evaluating the partial derivatives with the aid of Eqs. (3) and (4) one obtains:

$$\frac{1}{N_L}(\mu_A^i + \mu_\square^{\alpha 1}) + \frac{RT}{N_L} \ln \frac{N_A^i N_\square^{\alpha 1}}{(N^i - N_A^i)(N^{\alpha 1} - N_\square^{\alpha 1})} = 0 \qquad (7)$$

$$\frac{1}{N_L}(\mu_A^i + \mu_\square^{\alpha 2}) + \frac{RT}{N_L} \ln \frac{N_A^i N_\square^{\alpha 2}}{(N^i - N_A^i)(N^{\alpha 2} - N_\square^{\alpha 2})} = 0 \qquad (8)$$

It is obvious that the second terms in both equations must be constants for a given temperature since $\mu_A^i$, $\mu_\square^{\alpha 1}$, and $\mu_\square^{\alpha 2}$ were assumed to be composition independent.

The vacancy concentrations (referred to the total number of lattice sites $N^t$) at the stoichiometric composition are defined as the so-called disorder parameters $\alpha$ and $\beta$:

$$\alpha = (N_\square^{\alpha 1}/N^t)_{stoich} \text{ and } \beta = (N_\square^{\alpha 2}/N^t)_{stoich} \qquad (9)$$

With this the number of defects at stoichiometry can be expressed as:

$$N_\square^{\alpha 1} = \alpha N^t; \ N_\square^{\alpha 2} = \beta N^t; \ N_A^i = (\alpha + \beta)N^t$$

If Eqs. (7) and (8) are now calculated for the stoichiometric composition itself, and if it is kept in mind that the Gibbs energy terms in these equations are constants one obtains the following relationships:

$$\frac{N_A^i N_\square^{\alpha 1}}{[(N^t/2) - N_A^i][(N^t/4) - N_\square^{\alpha 1}]} = \frac{\alpha(\alpha + \beta)}{[(\frac{1}{2}) - \alpha - \beta][(\frac{1}{4}) - \alpha]} \qquad (10)$$

$$\frac{N_A^i N_\square^{\alpha 2}}{[(N^t/2) - N_A^i][(N^t/4) - N_\square^{\alpha 2}]} = \frac{\beta(\alpha + \beta)}{[(\frac{1}{2}) - \alpha - \beta][(\frac{1}{4}) - \beta]} \qquad (11)$$

where the numbers of the different sublattice sites have been expressed in terms of $N^t$, i.e., $N^i = N^t/2$ and $N^{\alpha 1} = N^{\alpha 2} = N^t/4$

For one mole of alloy the total number of atoms as well as the number of A-atoms (transition metal atoms) are given by

$$N_L = N^t - N_\square^{\alpha 1} - N_\square^{\alpha 2} + N_A^i \qquad (12)$$

$$N_A = (1 - x_B)N_L = (N^t/2) - N_\square^{\alpha 1} - N_\square^{\alpha 2} + N_A^i \qquad (13)$$

This yields now a system of four independent equations (Eqs. (10)–(13)), which can be solved for the four unknowns $N^t$, $N_A^i$, $N_\square^{\alpha 1}$, and $N_\square^{\alpha 2}$.

It is clear that for $\alpha = \beta$, the vacancies in the $\alpha$-sublattice must be distributed over all sites, whereas increasing difference between $\alpha$ and $\beta$ means that the number of vacancies in alternate layers starts to differ approaching a partially filled $CdI_2$-type lattice for which one of the disorder parameters must become zero.

With the well known thermodynamic relationship

$$\mu_A = RT \ln a_A = N_L(\partial G/\partial N_A)_{N_{B, \ p, \ T}} \qquad (14)$$

(where $\mu_A$ = chemical potential, $a_A$ = thermodynamic activity of component A) and an equivalent equation for component B one can derive (as shown in the APPENDIX) the following expressions for the thermodynamic activities of A and B relative to their value at the stoichiometric composition ($a_{A,0}$ and $a_{B,0}$):

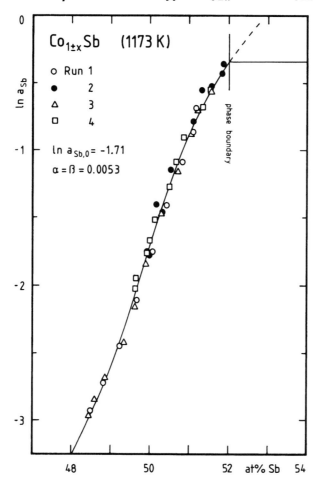

Fig. 1.   Activity of antimony at 1173 K in the NiAs-phase of the Co-Sb system; standard state: Sb(l). The theoretical curve was calculated with $\alpha = \beta = 0.0053$ and $\ln a_{Sb,0} = -1.71$.

$$\ln \frac{a_A}{a_{A,0}} = \ln \frac{n_A^i}{x_B - n_A^i} - \ln \frac{2(\alpha + \beta)}{1 - 2(\alpha + \beta)} \tag{15}$$

$$\ln \frac{a_B}{a_{B,0}} = \ln \frac{(x_B - N_A^i)^2}{x_B n_A^i} + \frac{1}{2} \ln \frac{(x_B - 2n_\square^{\alpha 1})(x_B - 2n_\square^{\alpha 2})}{x_B^2}$$

$$- \ln \frac{[1 - 2(\alpha + \beta)]^2}{2(\alpha + \beta)} - \frac{1}{2} \ln [(1 - 4\alpha)(1 - 4\beta)] \tag{16}$$

where the numbers of defects are now expressed in the form of defect concentrations, i.e., $n_\square^{\alpha 1} = (N_\square^{\alpha 1}/N_L)$, $n_\square^{\alpha 2} = (N_\square^{\alpha 2}/N_L)$, and $n_A^i = (N_A^i/N_L)$.

## APPLICATION OF THE MODEL

The results of the vapor pressure measurements in the Co-Sb system are shown in Fig. 1, where $\ln a_{Sb}$ at 1173 K is plotted vs composition in the range of the NiAs-phase. The data points can be described with a

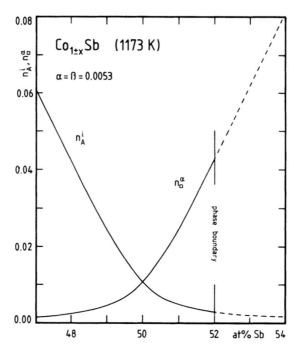

Fig. 2.   Variation of the defect concentrations at 1173 K in the NiAs-phase of the Co-Sb system with $\alpha = \beta = 0.0053$.

theoretical curve according to Eq. (16) using the parameters $\alpha = \beta = 0.0053$ and $\ln a_{Sb,0} = -1.71$. The identical value for $\alpha$ and $\beta$ indicates—as discussed above—equal vacancy concentrations on the two partial $\alpha$-sublattices, i.e., the vacancies are distributed with equal probability over *all* cobalt-sublattice sites. Fig. 2 illustrates the variation of the defect concentrations $n_\square^\alpha = n_\square^{\alpha 1} + n_\square^{\alpha 2}$ and $n_A^i$ with composition.

The composition dependence of $\ln a_{Sb}$ at 1173 K in the corresponding NiAs-type phase in the Ni-Sb system is shown in Fig. 3. Here the inflection point, which occurs at the "concentration of order" (8) and which would thus be expected exactly at the stoichiometric composition (as in the Co-Sb system), is shifted to about 50.6 at% Sb. A possible explanation for this phenomenon was offered by Leubolt et al. (6,14): It was speculated that the change from the defected to the partially "filled up" NiAs-structure (where nickel-atoms start to enter the interstitial positions) takes place before all regular nickel sites are occupied, i.e., a certain amount of octahedral holes remains vacant. The data in Fig. 3 could be perfectly described by a theoretical curve with $\alpha = \beta = 0.0055$ and $\ln a_{Sb,0} = -1.49$, if the stoichiometric composition is shifted by 0.6 at% to the antimony-rich side.

An alternate explanation for this shift of the inflection point is offered in Fig. 3: A possible second-order transition around 50 at% from a NiAs-type vacancy distribution (over all nickel-sublattice sites) with $\alpha = \beta = 0.008$ to a CdI$_2$-type distribution (vacancies restricted to alternate nickel-sublattice layers) with $\alpha = 0.008$ and $\beta = 0$ and with a hypothetical $\ln a_{Sb,0} = -2.08$. This would mean that one of the partial sublattices ($\alpha 1$ or $\alpha 2$) loses the ability to incorporate vacancies around this composi-

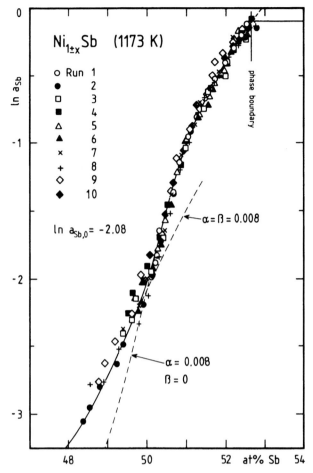

Fig. 3.   Activity of antimony at 1173 K in the NiAs-phase of the Ni-Sb system; standard state: Sb(l). The theoretical curves were calculated with $\alpha = \beta = 0.008$ and with $\alpha = 0.008$ and $\beta = 0$, resp., and with a hypothetical ln $a_{sb,0} = -2.08$ (*see text*).

tion. Although such second-order transitions are well known, for example in the Ni-Te system (*16,17*), one would expect them to occur at a higher content of the main group element, i.e., at a higher vacancy concentration on the transition metal-sublattice.

Both Figs. 1 and 3 show that the experimental data can be described very well by theoretical curves without the necessity of assuming an interaction energy between like defects. This is in accord with the results of Leubolt et al. (*14*), who reported very small interaction energies (of the order of $-2$ kJ g-atom$^{-1}$) for the NiAs-phase in the Ni-Sb system based on an evaluation of the experimental data using Libowitz' model (*9*).

## ACKNOWLEDGMENT

Partial financial support of this investigation by the "Hochschul-jubiläumsstiftung der Stadt Wien" is gratefully acknowledged.

## REFERENCES

1. Makarov, E. S., *Izv. Sekt. Fiz-Khim. Anal., Akad. Nauk SSSR* **16**, 149 (1943).
2. Ageev, N. V. and Makarov, E. S., *Izv. Akad. Nauk SSSR, Otd. Khim. Nauk,* No. 2, 87 (1943).
3. Kjekshus, A. and Walseth, K. P., *Acta Chem. Scand.* **23**, 2621 (1969).
4. Gal'perina, T. N., Zelenin, L. P., Bashkatov, A. N., Gel'd, P. V., and Oblasov, A. K., *Izv. Akad. Nauk SSSR, Neorg. Mater.* **12**, 392 (1976); *Inorg. Mater.* **12**, 339 (1976).
5. Chen, T., Mikkelsen Jr., J. C., and Charlan, G. B., *J. Crystal Growth* **43**, 5 (1978).
6. Leubolt, R., Ipser, H., Terzieff, P., and Komarek, K. L., *Z. Anorg. Allg. Chem.* **533**, 205 (1986).
7. Chen, T., Rogowski, D., and White, R. M., *J. Appl. Phys.* **49**, 1425 (1978).
8. Wagner, C. and Schottky, W., *Z. Phys. Chem.* **B11**, 163 (1931).
9. Lightstone, J. B. and Libowitz, G. G., *J. Phys. Chem. Solids* **30**, 1025 (1969).
10. Geffken, R. M., Komarek, K. L., and Miller, E., *J. Solid State Chem.* **4**, 153 (1972).
11. Grønvold, F., *J. Chem. Thermodyn.* **8**, 757 (1976).
12. Kjekshus, A. and Pearson, W. B., *Progress in Solid State Chemistry*, H. Reiss, ed., Pergamon Press, Oxford, vol. 1, 83 (1964).
13. Ipser, H. and Komarek, K. L., *Monatsh. Chem.* **105**, 1344 (1974).
14. Leubolt, R., Ipser, H., and Komarek, K. L., *Z. Metallk.* **77**, 284 (1986).
15. Hultgren, R., Desai, P. D., Hawkins, D. T., Gleiser, M., Kelley, K. K., and Wagman, D. D., *Selected Values of the Thermodynamic Properties of the Elements*, American Society for Metals, Metals Park, OH (1973).
16. Coffin, P., Jacobson, A. J., and Fender, B. E. F., *J. Phys. C: Solid State Phys.* **7**, 2781 (1974).
17. Carbonara, R. S. and Hoch, M., *Monatsh. Chem.* **103**, 695 (1972).

## APPENDIX

In the following the derivation of the expression for $\ln (a_A/a_{A,0})$ given in Eq. (15) is demonstrated starting from the thermodynamic relationship in Eq. (14). A similar procedure yields the thermodynamic activity of component B.

Considering that G is a function of the four variables $N^t$, $N_A^i$, $N_\square^{\alpha 1}$, $N_\square^{\alpha 2}$, one can write the total differential dG, and from this one obtains (for constant p and T)

$$(\partial G/\partial N_A)_{N_B} = (\partial G/\partial N^t)(\partial N^t/\partial N_A)_{N_B} +$$
$$(\partial G/\partial N_A^i)(\partial N_A^i/\partial N_A)_{N_B} + (\partial G/\partial N_\square^{\alpha 1})(\partial N_\square^{\alpha 1}/\partial N_A)_{N_B}$$
$$+ (\partial G/\partial N_\square^{\alpha 2})(\partial N_\square^{\alpha 2}/\partial N_A)_{N_B} \qquad (A1)$$

Since in the NiAs-lattice the total number of lattice sites is determined by the number of B-atoms and is consequently independent of the number of A-atoms (transition metal atoms):

$$(\partial N^t/\partial N_A)_{N_B} = 0 \qquad (A2)$$

the first term on the right-hand side of Eq. (A1) becomes zero. Differentiation of Eq. (13) with respect to $N_A$ yields the following expression:

$$(\partial N_A^i/\partial N_A)_{N_B} = 1 + (\partial N_\square^{\alpha 1}/\partial N_A)_{N_B} + (\partial N_\square^{\alpha 2}/\partial N_A)_{N_B} \tag{A3}$$

and inserting Eqs. (A2) and (A3) into (A1) results in

$$
\begin{aligned}
(\partial G/\partial N_A)_{N_B} = {}& (\partial G/\partial N_A^i) + (\partial N_\square^{\alpha 1}/\partial N_A)_{N_B}[(\partial G/\partial N_A^i) + (\partial G/\partial N_\square^{\alpha 1})] \\
& + (\partial N_\square^{\alpha 2}/\partial N_A)_{N_B}[(\partial G/\partial N_A^i) + (\partial G/\partial N_\square^{\alpha 2})]
\end{aligned} \tag{A4}
$$

However, according to the equilibrium conditions defined in Eqs. (5) and (6) the terms in the square brackets must be zero in thermodynamic equilibrium, and Eq. (A4) becomes simply

$$(\partial G/\partial N_A)_{N_B} = (\partial G/\partial N_A^i) \tag{A5}$$

This can be easily computed from Eqs. (3) and (4) to give an expression for $\mu_A$. If $\mu_{A,0}$, i.e., its value at the stoichiometric composition, is subtracted one obtains $\ln(a_A/a_{A,0})$ as given in Eq. (15).

# The High Temperature Vaporization and Thermodynamics of the Titanium Oxides (XVIII)

## Temperature Coefficients and Oxygen Potentials over $Ti_{10}O_{19}$

GLEN F. KESSINGER AND PAUL W. GILLES*

*Department of Chemistry, University of Kansas, Lawrence, KS 66045*

## ABSTRACT

The substance $Ti_{10}O_{19}$ has been synthesized and studied mass spectrometrically in the temperature range 1753–1946 K. The melting point and the partial molal enthalpies of vaporization for TiO(g), and $TiO_2$(g) are reported, as are the partial pressures of O(g) and $O_2$(g) at 1800 K.

**Index Entries:** High temperature; vaporization; thermodynamics; titanium oxides; $Ti_{10}O_{19}$ oxygen potentials; $Ti_{10}O_{19}$; $Ti_{10}O_{19}$ melting point.

## INTRODUCTION

The titanium–oxygen system has been renowned for the richness of its chemistry. Five gaseous species, Ti, TiO, $TiO_2$, O, and $O_2$ and many solid phases are well known. Among the latter are $TiO_{1 \pm x}$, $Ti_2O_3$, $Ti_3O_5$, the Magneli phases $Ti_nO_{2n-1}$ with $n$ running from 4 to 9, $Ti_{10}O_{19}$, and $TiO_2$. The present work is part of a continuing study of the high temperature vaporization and thermodynamic properties over the entire composition range.

The phase diagram for the Ti-O system has been evaluated by Murray and Wriedt (1). The structures of the solid phases have been established by Andersson, Magneli, and coworkers (2–4) and LePage

---

*Author to whom all correspondence and reprint orders should be addressed.

High Temperature Science, Vol. 26    © 1990 by the Humana Press Inc.

and Strobel (5,6). Zador and Alcock (7) included $Ti_{10}O_{19}$ as the high oxygen end-member in their high temperature emf studies in which their highest temperature is near our lowest. Merritt, Hyde, Bursill, and Philp (8) worked at much lower temperatures.

The principal difficulties in the study of this system arise because of the low vapor pressures, the relatively low melting points, and the extreme range of oxygen potentials. Thus, no one crucible material is satisfactory. In the metal region, tungsten appears to be reasonably satisfactory, although some solubility occurs. Molybdenum and tungsten are satisfactory in the monoxide region, and the latter is satisfactory to $Ti_3O_5$. In the higher oxygen potential regions, tungsten is oxidized and molybdenum even more so. Platinum reacts. Iridium was used in the present work and appears satisfactory.

Mass spectrometrically measured intensities of $TiO^+$ and $TiO_2^+$ arising from $Ti_{10}O_{19}$ were measured as a function of temperature. Pressures of $O(g)$ and $O_2(g)$ at 1800 K and partial molal enthalpies of vaporization of the gases have been deduced. The melting point is reported.

## EXPERIMENTAL

### Sample Synthesis

All measurements were made with portions of the same sample of $Ti_{10}O_{19}$, laboratory designation GFK-19. It was synthesized from an approximately 3 to 1 molar mixture of anatase titanium dioxide $TiO_2$, J. T. Baker Chemical Co. reagent grade, lot #3585, previously described by Lin (9), and trititanium pentoxide $Ti_3O_5$, by heating under high vacuum, in an inductively heated platinum-lined tungsten Knudsen cell, laboratory designation #138.

The trititanium pentoxide was synthesized by arc melting an approximately 1 to 5 molar mixture of iodide titanium, laboratory designation Ti-6, previously described by Hampson and Gilles (10), and titanium dioxide from the lot described above. Approximately 3 g of the mixture was pressed into a pellet that was melted under about 0.5 atmosphere $Ar(g)$, turned over, and melted again. The product gave an X-ray powder diffraction pattern that matched that of $Ti_3O_5$ (2). Part of this material was used for the preparation of $Ti_{10}O_{19}$. No combustion analysis was performed on this $Ti_3O_5$ sample.

The mixture of $TiO_2$ and $Ti_3O_5$ was heated twice. The first was for 6.5 h at 1750 ± 30 K. After this first heating the product was ground well with an alumina mortar and pestle, and then heated again for 4.5 h at 1400 ± 30 K. The background pressure was less than $1 \times 10^{-5}$ Torr during each heating.

The resulting product was characterized by Guinier X-ray powder diffraction and by oxygen combustion analysis. The X-ray pattern matched that of the Magneli phase, $Ti_{10}O_{19}$ (2), without additional lines. Three portions of the sample were analyzed by combustion in air at

925 ± 25°C until a constant mass was reached. The samples were contained in platinum boats that had previously been heated in air to constant mass at about 1100°C. The sample masses were in the range 50–100 mg. Larger portions were not used because of the small amount of sample available. The product of each of the analyses was identified by the Debye-Sherrer powder diffraction patterns of the product. In all three instances the product was finely divided, cream colored, rutile $TiO_2$. The analyses showed the composition to be $TiO_{1.913 \pm 0.013}$.

## Crucibles

The two tungsten Knudsen cells used for sample containment during this study were those with laboratory designations #4 and #138. Cell #4 was used for the melting temperature experiments; cell #138 was used both for synthesis experiments and mass spectrometer experiments.

Cell #4 is the same cell that was previously used by Wahlbeck and Gilles (11) and Sheldon and Gilles (12) during studies of less oxygen-rich titanium oxides. The present work was with highly oxygen-rich samples. In order to minimize sample-crucible interactions the cell was equipped with an iridium liner similar to the ones previously used by Conard, Bennett, and Gilles during their Run T (13). The iridium liner was kept centered in the Knudsen cell by a tungsten aligning assembly. The height of the liner was such that the bottom of the lid of the Knudsen cell rested on the top edge of the liner when the lid was in place. The tungsten lid was the same one used by Sheldon and Gilles (12), who reported that the cylindrical orifice had an area 0.02164 ± 0.00003 $cm^2$, length to radius ratio 0.3929, and a Clausing factor 0.8367 ± 0.0015. A hole in the bottom of the crucible served as a black body cavity for measuring the temperature of the crucible.

Cell #138 was machined from a tungsten rod. Its outer dimensions were approximately 19 mm in diameter by 32 mm tall. The inside chamber was about 13 mm in diameter by 21 mm high. The orifice was conical, widening in the direction of flow, with semiapex angle $\theta = 0.43°$, length $1 = 2.33 \times 10^{-1}$ cm, radius r(small end) = $5.89 \times 10^{-2}$ cm. The Clausing factor, W = 0.371, was interpolated from the calculated values of Freeman and Edwards (14). During the synthesis experiments the cell was equipped with a platinum liner, roughly of the same volume as the inside of the cell, that was fabricated from a sheet of 0.05 mm thick platinum. During the mass spectrometer experiments the cell was equipped with an iridium liner that had previously been used by Conard et al. in cell B of the crucible with laboratory designation #132 in their Run T (13). The iridium liner was kept centered in the Knudsen cell by a tungsten aligning assembly that also elevated the liner so that the lid of the Knudsen cell rested snugly on the top edge of the liner when the lid was in place. A hole drilled in the bottom of the cell served as a black body cavity for temperature measurement.

## Temperature Measurement

The temperatures were measured with a Leeds and Northrup single color optical pyrometer, laboratory designation #3. During the mass spectrometer experiments the optical path contained a shuttered window in the bottom of the Knudsen furnace and a prism. The vacuum line was similarly equipped.

The pyrometer was calibrated by comparing the readings taken with it to the readings taken with a similar pyrometer, laboratory designation #1, that had been previously calibrated at the US National Bureau of Standards. A correction factor for the attenuation of radiation by the window and prism in the optical path was calculated from the expression

$$C' = T^{-1} \text{ (window \& prism)} - T^{-1} \text{ (without window and prism)} \quad (1)$$

Temperature readings of a tungsten strip lamp were taken with the window and prism in and out of the optical path of the pyrometer. All reported temperatures have been corrected to IPTS-68.

## Melting Temperature Experiments

Five experiments were performed to establish the melting temperature of the $Ti_{10}O_{19}$ phase. The iridium-lined cell #4 was used to contain the samples. These constant temperature experiments were each about 1 h in length. The sample masses were in the range 3–10 mg. Visual examination of the residues subsequent to the experiment revealed whether melting had occurred.

## Mass Spectrometry Experiments

The mass spectrometry experiment designated Run BL was performed in a Nuclide Analysis Associates first-order, single-focusing magnetic mass spectrometer. The experimental data for $TiO^+$ and $TiO_2^+$ were collected by counting pulses from an electron multiplier. Ionization was by electron impact with 70 eV electrons. This electron energy has previously been shown by Conard et al. to produce negligible $TiO^+$ by fragmentation of $TiO_2^+$ (13). A more complete description of the instrument and the data acquisition system is given elsewhere (12). The sample was contained in the iridium-lined Knudsen cell #138. The cell was heated to temperature up to about 1100°C by radiation from tungsten filaments and to temperatures above 1100°C by radiation and electron bombardment.

The pulse counting efficiencies for $TiO^+$ and $TiO_2^+$ ions were assumed to be the same, and for the purposes of the present work silver calibration and multiplier gains that were about $3 \times 10^5$ were not required.

Ionization efficiency curves were collected for mass 80 and mass 62 with the shutter alternately open and closed; the net digital signal, open shutter-closed shutter, was evaluated to determine the appearance potentials of $TiO_2^+$ and $TiO^+$. The electron beam energy setting was calibrated by collecting ionization efficiency curves of the mass 202 Hg$^+$ background signal in the mass spectrometer and comparing the appearance potential to a published value of the first ionization energy of mercury and applying the appropriate correction to the measured values.

The TiO$^+$ measurements were made by monitoring the mass 62 shutterable signal because the high background at mass 64, arising from sulfur contamination of the instrument, precluded reliable measurement of the shutterable signal at mass 64.

## RESULTS

### Melting Temperature Experiments

The melting point measurements consisted of the five constant temperature experiments at 1627 ± 10, 1681 ± 10, 1693 ± 10, 1707 ± 10, and 1734 ± 10°C. Mass losses amounted to 2–3%. Visual inspection of the residues revealed that the samples heated to temperatures greater than or equal to 1707°C melted and the samples heated to temperatures less than or equal to 1693°C did not melt.

The residue from the highest temperature experiment, $T = 1734$°C, was not recoverable. The residues from the experiments at the two lowest temperatures gave Guinier powder diffraction patterns that were the same as those of the starting material. The recoverable residues from the experiments at temperatures greater than or equal to 1693°C gave Guinier powder patterns that were poor in quality, and showed lines that could be attributed to Ti$_8$O$_{15}$ and Ti$_9$O$_{17}$ as well as Ti$_{10}$O$_{19}$. The melting temperature of Ti$_{10}$O$_{19}$ is taken to be 1707 ± 10°C.

### Mass Spectrometer Experiments

The mass spectrometry measurements were performed with another portion of the sample. The residue after the measurements showed only lines that could be attributed to the phase Ti$_{10}$O$_{19}$. The measured mass loss of the crucible and contents that could be attributed to the vaporization of the titanium oxide sample was 0.19 mg, or 0.36% of the initial sample mass of 52.48 mg.

The appearance potential of TiO$_2^+$ was 10.0 ± 0.5 eV, the value of TiO$^+$ was 8.8 ± 0.5 eV; however, the uncertainty for TiO$^+$ may be larger than indicated because the signal was so low.

The ionization cross section ratio $\sigma(TiO)/\sigma(TiO_2)$ was taken to be 2.2. This value was experimentally obtained by Hampson and Gilles (10) and Sheldon and Gilles (15) from experiments involving the congruent vaporization of Ti$_3$O$_5$.

The data for TiO$^+$ and TiO$_2^+$ are plotted on Fig. 1 as the logarithm of the intensity-temperature product, the product being proportional to the pressure, vs the reciprocal temperature. The points were taken in an order determined by lot. The upper set of 15 points are the TiO$_2^+$ data;

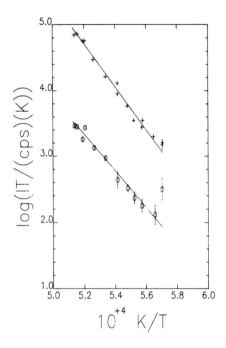

Fig. 1.    Plot of log (IT) vs $T^{-1}$ for the vaporization of $Ti_{10}O_{19}$. The upper set of 15 points is for $TiO_2^+$ (mass 80), +. The lower set of 12 points is for $TiO^+$ (mass 62), 0. The lowest temperature point for $TiO^+$ and the less intense of the two lowest temperature points for $TiO_2^+$ were not included in the least squares treatment.

the lower set of 12 points are the $TiO^+$ data. The data were subjected to a weighted least-squares treatment. The weight of the each log(IT) value was taken to be the square of the reciprocal of the standard deviation in the logarithm of the net signal. The parameters of the equation

$$\log(IT) = A/T + B \qquad\qquad (2)$$

where $I$ is the ion intensity in counts per second and $T$ is the temperature in Kelvin, were calculated separately for $TiO^+$ and $TiO_2^+$. For $TiO_2^+$, $A = -32879 \pm 904$ and $B = 21.813 \pm 0.473$; for $TiO^+$, $A = -27871 \pm 1846$ and $B = 17.828 \pm 0.962$. The uncertainties are probable errors. These results are represented by the solid lines on the two plots. The least squares calculation for $TiO^+$ did not include the lowest temperature point on that plot; the calculation for $TiO_2^+$ did not include the less intense of the two lowest temperature points on that plot.

### Thermodynamic Quantities

The partial molar enthalpies of vaporization at 1800 K of the gaseous titanium species, along with the appropriate chemical equations, are listed below. The values for TiO(g) and $TiO_2$(g) were calculated by the second law method from the experimental data shown in Fig. 1 for $TiO^+$ and $TiO_2^+$.

$$\text{TiO(in Ti}_{10}\text{O}_{19}) = \text{TiO(g)} \quad \Delta\bar{H} = 127.5 \pm 8.4 \text{ kcal} \tag{3}$$

$$\text{TiO}_2\text{(in Ti}_{10}\text{O}_{19}) = \text{TiO}_2\text{(g)} \quad \Delta\bar{H} = 150.4 \pm 4.1 \text{ kcal} \tag{4}$$

The ion intensity ratio, $\text{TiO}_2{}^+/\text{TiO}^+$, along with ionization cross sections ($\sigma$), isotopic fractions ($f$), and ion source emission current readings ($e_c$) associated with the two ions, may be used to calculate the pressures of O(g) and $\text{O}_2$(g) in the temperature range of interest if the standard free energies of formation of the vapor species are known. The pressure ratio is given by the equation.

$$P(\text{TiO}_2)/P(\text{TiO}) = (I/[\sigma f e_c])(\text{TiO}_2{}^+)/(I/[\sigma f e_c])(\text{TiO}^+) \tag{5}$$

Consider the two equations,

$$\text{TiO(g)} + \tfrac{1}{2} \text{O}_2\text{(g)} = \text{TiO}_2\text{(g)}, \quad P(\text{O}_2) = (P(\text{TiO}_2)/[K_pP(\text{TiO})])^2 \tag{6}$$

$$\text{TiO(g)} + \text{O(g)} = \text{TiO}_2\text{(g)}, \quad P(\text{O}) = P(\text{TiO}_2)/[K_pP(\text{TiO})] \tag{7}$$

For each, an expression for $\log K_p$ was obtained from $\Delta H°$ and $\Delta S°$ values at 1800 K for the individual substances given in the JANAF Tables (*16*). This expression was coupled to the equation for the logarithm of the pressure ratio in Eq. (5) obtained from the parameters of Eq. (2) and values of the ratios of $\sigma$, $f$, and $e_c$ to obtain equations for the logarithms of the pressures of $\text{O}_2$(g) and O(g) near 1800 K,

$$\log P(\text{O}_2) = -(47582 \pm 4402)/T + 15.784 \pm 2.318 \tag{8}$$

$$\log P(\text{O}) = -(37104 \pm 2201)/T + 11.371 \pm 1.159 \tag{9}$$

The actual $\Delta H°$ and $\Delta S°$ values used at 1800 K were (in calories) $-77691 \pm 3000$ and $84.822 \pm 1$ for $\text{TiO}_2$(g); $+8254 \pm 2000$ and $71.402 \pm 0.06$ for TiO(g); zero and $63.262 \pm 0.008$ for $\text{O}_2$(g); and $+60916 \pm 24$ and $47.548 \pm 0.005$ for O(g). The $\text{TiO}/\text{TiO}_2$ ratios of the values of $\sigma$, $f$, and $e_c$ were 2.2 ($1 \pm 0.15$), $8.0/73.8$, and $3.456 \pm 0.030$.

These equations give at 1800 K the pressures of $\text{O}_2$(g) and O(g) as $2.236 \times 10^{-11}$ and $5.728 \times 10^{-10}$ atm, to within about 15%.

The values in Eq. (3) and (4) and the enthalpies of formation for $\text{TiO}_2$(g) and TiO(g) given above yield for the formation of $\text{Ti}_{10}\text{O}_{19}$ $\Delta H°_{1800} = -217.3 \pm 4.5$ kcal/g-atom Ti.

## DISCUSSION

### Phase Behavior of *Ti₁₀O₁₉(s)*

The number of solid phases existing at high temperature is unknown. That the Guinier X-ray powder diffraction patterns of the recoverable residues from the melting temperature experiments above 1693°C have lines that can be attributed to the phases $\text{Ti}_8\text{O}_{15}$, $\text{Ti}_9\text{O}_{17}$, and $\text{Ti}_{10}\text{O}_{19}$ suggests that $\text{Ti}_{10}\text{O}_{19}$ preferentially lost oxygen, as is consistent with its known incongruent vaporization. The masses of the melting point residues were so small, less than 10 mg in each instance, that combustion analysis of the residues was not possible.

## Pressures of O(g) and O₂(g)

The value of $P(O_2)$ calculated for 1773 K is $8.85 \times 10^{-12}$ atm, which is about 2.8 times smaller than the value $2.43 \times 10^{-11}$ atm read from the highest temperature plot reported by Zador and Alcock (7). Because the values come from such drastically different techniques, we regard this agreement as excellent.

## Thermodynamics of Ti₁₀O₁₉

The mass of the mass spectrometric sample was chosen large enough so that inappreciable decomposition occurred. Even though the number of phases present at the high temperature is not known, the partial molal enthalpies of vaporization of TiO and $TiO_2$ relate to removing the gases from the condensed material, regardless of what it is. Or, we are treating the condensed material as a solid solution whose unchanging composition is $Ti_{10}O_{19}$. The calculated value of the partial molar enthalpy of vaporization of TiO(g) is indistinguishable from the value of Sheldon and Gilles (12) derived from vapor pressure measurements over stoichiometric TiO(s) in the temperature interval 1694 to 1919 K, the values of Heideman, Reed and Gilles (17) derived from vapor pressure measurements over various compositions of the monoxide solution region in the temperature interval 1679 to 1948K, and the values of Hampson and Gilles (10) and Wahlbeck and Gilles (11) derived from measurements over congruently vaporizing $Ti_3O_5$ over roughly the same temperature range as the present work.

Within our limits of error, our value of $-217.3 \pm 4.5$ kcal/g-atom Ti for the formation of $Ti_{10}O_{19}$ is in agreement with $-213.0$ given by Zador and Alcock (7), who do not list an uncertainty.

## Sources of Error

The major sources of error in this work arise from the choices of ionization cross sections for TiO(g) and $TiO_2$(g), and the problems encountered measuring the $TiO^+$ intensities.

The choice of ionization cross section ratio also introduces uncertainty into the calculated result of $P(O_2)$. The value chosen was that from Hampson and Gilles (10) who studied the congruent vaporization of $Ti_3O_5$ in the same instrument used in the present work. The congruency condition and their measured intensities set a constraint on the value of the cross section ratio. At the congruently vaporizing composition in the $Ti_3O_5$ phase the escaping mole ratio $n(TiO_2)/n(TiO)$ is equal to two. They deduced a value of the ionization cross section ratio that, when combined with their measured intensities and estimated multiplier gains, produced a mole ratio equal to two. The maximum uncertainty of this cross section ratio is probably about 50% (15).

It was not possible to monitor the $TiO^+$ mass 64 signal owing to sulfur contamination of the mass spectrometer; therefore, the second most abundant $TiO^+$ signal, at mass 62, was monitored. The shutterable $TiO^+$ signal was reduced by nearly an order of magnitude. The problems arising from the weak signal were a lengthening of the time necessary to make measurements, a much greater relative uncertainty in the measurements as compared to $TiO_2^+$, and a less certain value of the appearance potential.

## ACKNOWLEDGMENTS

We thank the Wesley Medical Center of Wichita, KS for the gift of computing equipment and Robert I. Sheldon for discussions and the loan of some computer programs. We also thank the University of Kansas Graduate Research fund. We acknowledge helpful conversations with J. G. Edwards and H. F. Franzen.

## REFERENCES

1. Murray, J. L. and Wriedt, H. A., *The O-Ti (Oxygen-Titanium) System, Bulletin of Alloy Phase Diagrams*, **8**, 148 (1987).
2. Andersson, S., Collen, B., Kruuse, G., Kuylenstierna, U., Magneli, A., Pestmalis, H., and Asbrink, S., *Acta Chem. Scand.* **11**, 1653 (1957).
3. Andersson, S., Collen, B., Kuylentierna, U., and Magneli, A., *Acta Chem. Scand.* **11**, 1641 (1957).
4. Andersson, S. and Jahnberg, L., *Arkiv Kemi* **21**, 413 (1963).
5. LePage, Y. and Strobel, P., *J. Solid State Chem.* **43**, 314 (1982).
6. LePage, Y. and Strobel, P., *J. Solid State Chem.* **44**, 273 (1982).
7. Zador, S. and Alcock, C. B., *High Temp. Sci.* **16**, 187 (1983).
8. Merritt, R. R., Hyde, B. G., Bursill, L. A., and Philp, D. K., *Philos, Mag.* **274**, 628 (1973).
9. Lin, S., *Titanium-Vanadium-Oxygen System*, Thesis, University of Kansas, p. 22 (1966).
10. Hampson, P. J. and Gilles, P. W., *J. Chem. Phys.* **55**, 3712 (1971).
11. Wahlbeck, P. G. and Gilles, P. W., *J. Chem. Phys.* **46**, 2465 (1967).
12. Sheldon, R. I. and Gilles, P. W., *J. Chem. Phys.* **66**, 3705 (1977).
13. Conard, B. R., Bennett, J. E., and Gilles, P. W., *J. Chem. Phys.* **63**, 5502 (1975).
14. Freeman, R. D. and Edwards, J. G., *The Characterization of High Temperature Vapors*, Margrave, J. L., ed., Wiley, New York, 1967, p. 508.
15. Sheldon, R. I. and Gilles, P. W., *NBS Special Publication 561* **231**, (1979).
16. Chase, M. W., Curnutt, J. L., Prophet, H., McDonald, R. A., and Syverud, A. N. "JANAF Thermochemical Tables, 1975 Supplement," *J. Phys. Chem. Ref. Data* **4**, 1–175 (1975).
17. Heideman, S. A., Reed, T. B., and Gilles, P. W., *High Temp. Sci.* **13**, 79 (1980).

# Thermodynamics, Kinetics, and Interface Morphology of Reactions Between Metals and GaAs

JEN-CHWEN LIN AND Y. AUSTIN CHANG*

University of Wisconsin-Madison, Department of Materials Science and Engineering, 1509 University Avenue, Madison, WI 53706

## ABSTRACT

The chemical stability of interfaces between metals and GaAs was discussed in terms of reaction sequence and diffusion path concepts. The factors that determine interface morphology were also given. These general ideas can be applied to any interfacial reactions between two dissimilar materials.

**Index Entries:** Thermodynamics; kinetics; interface.

## INTRODUCTION

Essential to the designing of microelectronic devices is the need for connections between active elements and, therefore, the need for active contacts. The need for reliable, low-resistance, reproducible, and stable ohmic contacts and Schottky barriers is particularly great for III-V compound semiconductors such as GaAs and for solid solutions of III-V compound semiconductors. In addition to exhibiting the appropriate electrical properties, a successful contact needs to be chemically stable. However, the chemical stability and, hence, the electrical properties of a contact are governed by the thermodynamics, kinetics, and interface morphology of phase formation between the metallizing elements and the compound semiconductors.

During the past few years, a large number of experimental results have been reported in the literature concerning the metallization of III-V compound semiconductors, primarily GaAs. These studies have includ-

*Author to whom all correspondence and reprint orders should be addressed.

High Temperature Science, Vol. 26    © 1990 by the Humana Press Inc.

365

ed the identification of the phases formed and, to some extent, the resulting morphologies when M/GaAs contacts were exposed to specific environments. Frequently, owing to the paucity of relevant phase equilibria, thermodynamic, and kinetic data for ternary Ga-M-As systems, a rationalization of these results (some of which appear to be contradictory) has been difficult, if not impossible. The recent studies by us (1–15), Williams and coworkers (16–20), Beyers, Kim, and Sinclair (21), and Sands (22,23) have demonstrated the importance of phase diagrams in the understanding of interfacial reactions between thin-film metals and GaAs. In addition, the concept of diffusion path in ternaries has been introduced by us (1,5–7,9,10,13,15) to rationalize the phase formation in M/GaAs contacts. The factors that determine phase formation sequences and phase morphologies in terms of thermodynamic and kinetic viewpoints have also been discussed by us (5). These ideas have been applied to M/GaAs diffusion couples in both the bulk and thin-film forms. We have found that the reaction sequences and diffusion paths for all M/GaAs couples investigated to date are similar for the bulk and thin-film cases.

## INTERFACIAL STABILITY

### Thermodynamic Considerations

It is well known by now that most metals are not chemically compatible with GaAs (1,3,10,21,22,24). Figure 1 shows three typical phase diagrams of Ga-M-As when M is not in thermodynamic equilibrium with GaAs and when the only stable binary phases are $Ga_3M$, $GaM$, $GaM_3$, $MAs$, $M_2As_3$, and GaAs. The diagram in Fig. 1(a) depicts the case when the binary phases exhibits limited solubilities of the third component elements. The diagram in Fig. 1(b) depicts the case when the binary phase MAs dissolves a considerable amount of the counter phase "GaM." The symbol "GaM" denotes the unstable phase GaM that exhibits the MAs structure. The diagram in Fig. 1(c) depicts a case where, in addition to extensive solubilities of MAs, GaM, $GaM_3$, and (M), there exists a ternary phase T with a composition lying along the GaAs-M join. The appearances of the three types of phase diagrams depend on the relative stabilities of the competitive phases, i.e., $Ga_3M$, $GaM$, $GaM_3$, $MAs$, $M_2As_3$, GaAs, and T, the lattice stabilities of the component elements and the intermediate phases, and the thermodynamic solution behaviors of the solution phases. In order to be able to accurately calculate the phase diagrams of Ga-As-M, it is necessary to know the Gibbs free energies of the phases and the solution behaviors of the phases with extensive solubilities. Since most of the data are not available, accurate phase diagram determination is necessary. Several ternary Ga-M-As systems in the composition range of relevance to the metallization of GaAs have been determined by us. It is evident from Fig. 1 that in all three cases, the M/GaAs contacts will undergo chemical reactions when subjected to sufficiently high temperatures. In some cases, a sufficiently high temperature may be only 100°C or even lower. However, in view of

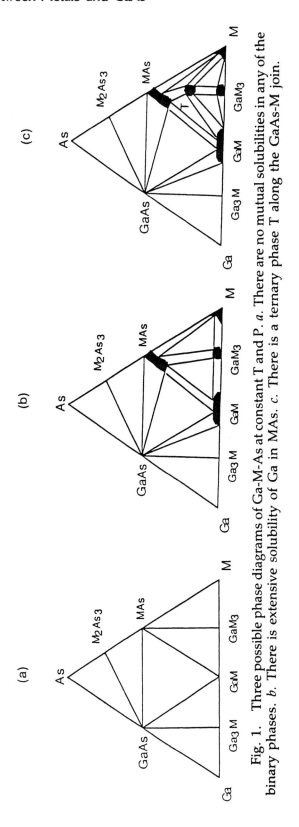

Fig. 1. Three possible phase diagrams of Ga-M-As at constant T and P. *a*. There are no mutual solubilities in any of the binary phases. *b*. There is extensive solubility of Ga in MAs. *c*. There is a ternary phase T along the GaAs-M join.

the many types of phase equilibria exhibited by Ga-M-As, the kinetics of the reactions for the M/GaAs couples are quite different, as will be discussed in the next section.

## Kinetic Considerations

### Diffusion Paths

Thermodynamics tells us what will happen when equilibrium conditions are achieved but does not tell us what combinations of phases may form under actual conditions. For instance, we cannot tell what phases will form when M is in contact with GaAs from the phase diagram given in Fig. 1(a). Only one path is possible in a binary couple, but this is not true for a ternary couple. When M is in contact with Ga for a binary Ga-M shown in Fig. 1, the three intermediate phases $Ga_3M$, $GaM$, and $GaM_3$ will form, the only possible diffusion path. On the other hand, when M is in contact with GaAs, there is more than one possible path. Figure 1(a) is reproduced as Fig. 2(a) with two possible paths given in Fig. 2(b,c). When M is in contact with GaAs, several phases may form, such as GaM, $GaM_3$, and MAs. According to Kirkaldy and Brown (37), given a specific ternary diffusion couple there is only one diffusion path. Recent experimental studies by van Loo and coworkers (43–46) on oxide/metal systems, Leute (47) on pseudoternary compound semiconductor/compound semiconductor systems, and by us (1,5–8,10,15) on GaAs/M systems yield results in accordance with this statement. Although it is possible in principle to calculate the diffusion path, it is practically impossible given the current state of our understanding of ternaries such as Ga-M-As. We mst depend on experiments to determine the diffusion paths of GaAs/M couples. If the diffusion path is that given as path I (Fig. 2(b)), MAs would be in contact with GaAs; on the other hand, if path II (Fig. 2(c)) is the diffusion path, GaM would be in contact with GaAs. From a device point of view, it is important to know which of these phases is in contact with GaAs under equilibrium conditions.

Because microelectronic devices are made in the form of thin films, it is equally important to study the kinetics of thin-film phase formation on GaAs. Let us now refer to Fig. 1(a) or 2(b) for discussion. When a thin-film metal M is deposited on GaAs, the final equilibrium mixture will be GaAs, GaM, and MAs. If the diffusion path is that of path I, we will have the final configuration of GaAs|MAs|GaM. On the other hand, if path II is the diffusion path, the final configuration of thin-metal film deposited on GaAs would be GaAs|GaM|MAs. Although there are indications that the results obtained from some bulk and thin-film M/Si binary couples may not be the same (25–30), the results of our studies on GaAs/M couples are similar for both the bulk and thin-film cases (1,5–8,10,15). If discrepancies between the bulk and thin-film studies do occur, additional investigation will be carried out to ascertain whether they are intrinsic in nature.

### Interface Morphology

Up to this point, we have not discussed the interface morphology, which is determined by the growth kinetics of the phases in a couple.

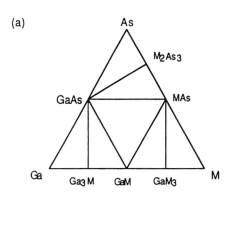

(a)

(b) Path I

| GaAs | MAs | GaM | GaM₃ | M |

(c) Path II

| GaAs | GaM | MAs | M |

Fig. 2. *a*. A hypothetical Ga-M-As ternary isothermal section. *b*. A possible diffusion path for a couple of GaAs/M. *c*. Another possible diffusion path for a couple of GaAs/M.

Wagner (*31*) has considered the morphological and kinetic aspect of displacement reactions in the solid state. In a later paper, Wagner (*32*) established the criteria for the stability of a flat growth interface. Rapp and coworkers (*33–35* have utilized this concept to study reactions in M/oxide and M/sulfide couples. Let us now apply these criteria to a GaAs/M diffusion couple under two specific conditions as discussed by Lin et al. (*5*). In Fig. 3(a), assuming that the initially predominant moving species is Ga and that As diffuses the slowest, the growth of $GaM_3$ and MAs would necessarily occur at the $GaM_3$/M and GaAs/MAs interfaces, respectively. A moving interface is referred to as a growth front. The growth of $GaM_3$ and MAs is controlled by the diffusion of Ga and M, respectively. The flux of Ga arriving at position I exceeds that at position II, resulting in the formation of a planar $GaM_3$/M interface. Similarly, the flux of M arriving at position III exceeds that at position IV, again resulting in the formation of a planar GaAs/MAs interface. On the other hand, in the second case, as is shown in Fig. 3(b), the species M is the predominant moving element for the growth of $GaM_3$ and MAs. In this case, the growth fronts of $GaM_3$ and MAs are at the MAs/$GaM_3$ and

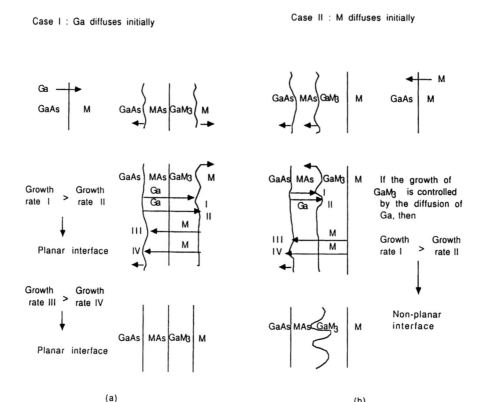

Fig. 3.  Possible interface morphologies for a GaAs/M diffusion couple.

GaAs/MAs interfaces, respectively. If the rate-controlling step for the growth of GaM₃ is the diffusion of Ga, then the growth rate at position I is higher than that at position II. Under these circumstances, a planar MAs/GaM₃ interface would be unstable. The situation for the GaAs/MAs interface is the same as that in Fig. 3(a) and, therefore, the interface remains planar. From the above discussion, it may readily be seen that a knowledge of the predominant moving element and the rate controlling steps to phase growth are the key points to understanding and predicting interface morphology. In fact, we have successfully applied the above concept to the preliminary results obtained for GaAs/Pt (*10,11*), GaAs/Ir (*10,15*), GaAs/Co (*6*), GaAs/Ni (*8*), and GaAs/Nb (*10*).

*Phase Formation Sequence*

The diffusion path of a ternary diffusion couple such as GaAs/M represents the stable phase arrangement between the end phases, i.e., GaAs and M for the GaAs/M couples. However, depending on the types of phase equilibria and the relative mobilities of the component elements, the phases formed initially and even the phases formed subsequently may not correspond to the stable phase arrangements. Although it is difficult to know in advance precisely the phase formation sequence, we may follow the ideas suggested by Lin et al. (*5*) to forecast possible phase formation sequences. The three phase diagrams given in Fig. 1 are

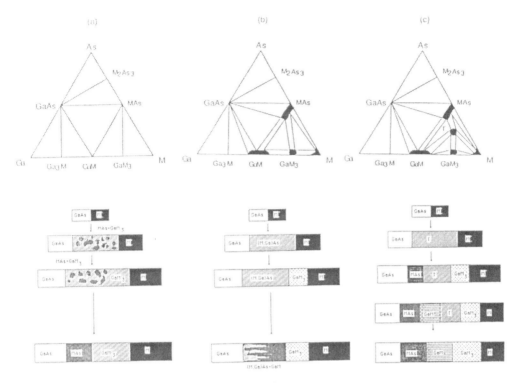

Fig. 4.  Possible phase formation sequences for three types of Ga-M-As phase diagrams.

reproduced in Fig. 4. Let us assume that the diffusion path corresponds to GaAs|MAs|GaM₃|M for the case shown in Fig. 4(a) and to GaAs |MAs|GaM|GaM₃|M for the cases shown in Fig. 4(b,c). Three possible reaction sequences are given in Fig. 4. Let us now discuss the rationale for forecasting the reaction sequences for the phase diagram in Fig. 4(c).

When a GaAs/M couple is exposed to a high enough temperature, interdiffusion will occur at the interface. Initially, the overall composition near the interface should be close to the GaAs and M connection line. Once the amount of foreign elements reaches the sustainable limit of the crystalline solid, new phases nucleate at the interface. According to the phase diagram in Fig. 4(c), the T phase would most likely form first. This behavior is a result of kinetics. The composition of the T phase is close to the initial composition at the interface, and, therefore, less time would be necessary to redistribute the elements in order to nucleate this phase. We would then have an initial configuration of GaAs|T|M, but none of the interfaces would be at thermodynamic equilibrium. Consequently, more reaction evolution would take place. Although several phases may nucleate and form at the GaAs|T and T|M interfaces, we believe MAs would be the first one to nucleate and grow at the GaAs|T interface. Again, it would take less time to redistribute the elements from T to nucleate and form MAs than would be required to form GaM. At the same time, the phase GaM₃ would probably form at the T|M interface since it is in equilibrium with T and M. The configuration would then be GaAs |MAs|T|GaM₃|M with all of the interfaces at thermodynamic equilibrium.

Although the phases between GaAs and M are now thermo-dynamically stable, some of these phases may be kinetically unstable. The kinetic stabilities of the phases depend on the relative fluxes at the various interfaces. Let us suppose that T is kinetically unstable. This is a reasonable assumption since the growth of a ternary phase or a solution phase in the ternary region requires specific elemental flux ratios. Then MAs and GaM$_3$ would grow at the expense of T with subsequent formation of GaM. The most likely evolution of the reaction sequence would be GaAs|MAs|GaM|T|GaM$_3$|M with the final stable configuration of GaAs |MAs|GaM|GaM$_3$|M, the diffusion path.

The preliminary results we have obtained concerning the reaction sequences for GaAs/Co (2,4,6,9), GaAs/Ni (8), GaAs/Pd (1,10) and GaAs/Pt (10,11) are consistent with the arguments presented above. We are currently conducting more extensive experimental studies on GaAs/Co, GaAs/Ni, and GaAs/Pt to firmly establish the reaction already found in light of the above discussion.

### Quantitative Layer-Growth of Phases

Up to this point, we have discussed only qualitative arguments for the formation of phases in a ternary diffusion couple such as GaAs/M. We now present the basic quantitative formulation for the growth of phases formed in a bulk diffusion couple in terms of ternary diffusion theory. Onsager (48–50) has shown that the diffusional flux, J$_i$, in a multicomponent system may be expressed as a linear function of the chemical potential gradients. However, because chemical potential gradients are not convenient for experimental analysis, these equations are transformed, making use of the concentration gradients as independent variables. This leads to the generalized form of Fick's first law for a ternary system

$$\tilde{J}_1 = - \tilde{D}_{11} \, \partial C_1/\partial x \, - \, \tilde{D}_{12} \, \partial C_2/\partial x \qquad (1.A)$$

$$\tilde{J}_2 = - \tilde{D}_{21} \, \partial C_1/\partial x \, - \, \tilde{D}_{22} \, \partial C_2/\partial x \qquad (1.B)$$

where $\tilde{D}_{11}$ and $\tilde{D}_{22}$ reflect the effect of the concentration gradient of a gradient of a given component of its own flux, and $\tilde{D}_{12}$ and $\tilde{D}_{21}$ represent the crosseffects, or the ternary diffusional interactions. The subscript 1 refers to component A (or 1) and 2 refers to component B (or 2). $C_1$ and $C_2$ are concentrations and $x$ is a distance coordinate. Combining the above equations with the equations of continuity yields

$$\partial C_1/\partial t \, + \, \partial \tilde{J}_1/\partial x = O \qquad (2.A)$$

$$\partial C_2/\partial t \, + \, \partial \tilde{J}_2/\partial x = O \qquad (2.B)$$

where $t$ denotes time. Substituting Eqs. (1) into Eqs. (2) and neglecting the compositional dependencies of the $\tilde{D}_{ij}$s, we have Fick's second law for a ternary system

$$\partial C_1/\partial t \;=\; \bar{D}_{11}\, \partial^2 C_1/\partial x^2 \;+\; \bar{D}_{12}\, \partial^2 C_2/\partial x^2 \tag{3.A}$$

$$\partial C_2/\partial t \;=\; \bar{D}_{21}\, \partial^2 C_1/\partial x^2 \;+\; \bar{D}_{22}\, \partial^2 C_2/\partial x^2 \tag{3.B}$$

Parametric solutions to these equations are of the form

$$C_1 \;=\; C_1(\lambda) \tag{4.A}$$

$$C_2 \;=\; C_2(\lambda) \tag{4.B}$$

where $\lambda = x/\sqrt{t}$: provided that the boundary conditions of a semi-infinite couple are met. Two important conclusions can be made from these equations. First, a unique solution exists for a given set of boundary conditions, i.e., only one diffusion path exists for a given temperature and pressure. Second, the growth of a phase varies with the square root of time.

Kirkaldy (36–41) has solved the basic ternary diffusion equations and has applied the solution to the growth of a ternary two-phase system. He and his coworkers have not applied the solution to the growth of multiphase ternary systems. In addition to the works of Kirkaldy and coworkers, other investigators have tackled ternary diffusion in a number of single-phase and simple two-phase systems.

In the case of GaAs/M, the growth of the intermediate phases is more complicated since several phases exist in addition to GaAs and M. Moreover, diffusion coefficients for the component elements in most of the phases are unknown. Our approach to this problem is to obtain diffusion coefficients from the analysis of the layer growth and the concentration gradients within the single-phase regions, if measurable ranges of homogeneity exist for these phases. Jan (42) has applied the solution to ternary diffusion equations to multiphase ternary systems. With certain simplifications, he has been able to obtain satisfactory results using the limited experimental results obtained by Schulz (10) for GaAs/Ni and GaAs/Nb.

## Bulk vs. Thin-Film Couples

In bulk diffusion couples, an infinite supply of the two end phases is realized. This is not the case for a thin-metal film deposited on GaAs. The metal thin film may be consumed during the growth of the first phase. This situation was discussed briefly earlier. Except for the limited supply of M in the thin-film couple, similar phase growth sequences and arrangements would occur in bulk and thin-film couples. This correlation has been observed by us for a number of GaAs/M couples, with M being Pd, Pt, Co, Ni, Rh, or Ir (1,5–8,10,11,13,15). This implies that chemical stability is the primary factor governing phase formation in these systems. The initial and transient phase configurations may be influenced by strain energy (lattice mismatch) and interfacial energies especially in the case of thin films. The final, stable phase configuration, however, will be determined primarily by chemical stability, i.e., equilibrium thermodynamics.

## SUMMARY

Interfacial stability is a key issue and a challenge for the development of new contact materials for GaAs. In general, a kinetically stable interface may be maintained at room temperature. However, in GaAs technology, processing temperatures can be as high as 800°C. At such high temperatures, one cannot rely on kinetic factors to stabilize the interfaces. In this paper, a general approach has been provided for the understanding of interfacial stability. The discussions show that thermodynamic and kinetic factors actually determine the reaction sequences, interface morphologies and final phase arrangements, i.e., diffusion paths. Further study is needed to correlate the chemistry and morphology with electrical properties of interfaces. Such information is very important in the design of electronic devices.

The general concepts of reaction sequence and diffusion path are not only suitable for GaAs/metal interface study but are also applicable to other materials with heterogeneous interfaces. For instance, the interactions between surface film and bulk usually determine the life of coatings for wear or corrosion resistance; the interface between fiber and matrix are highly related to the mechanical properties of structural composites. A basic understanding of interfacial stability is urgently needed for the development of new materials and the improvement of processing parameters.

## ACKNOWLEDGMENTS

The authors wish to thank K. J. Schulz, X.-Y. Zheng, F.-Y. Shiau, C.-H. Jan, and D. J. Swenson for help and discussion concerning this work. They also wish to thank the Department of Energy for financial support through Grant No. DE-FG02-86ER452754.

## REFERENCES

1.  Lin, J.-C., Hsieh, K.-C., Schultz, K. J., and Chang, Y. A., *J. Mater. Res.* **3**, 148 (1988).
2.  Shiau, F. Y., Chang, Y. A., and Chen, L. J., *J. Electron Mat.* **17**, 433 (1988).
3.  Zheng, X.-Y., Schulz, K. J., Lin, J.-C., Chang, Y. A., *J. Less Common Metals* **146**, 233 (1989).
4.  Shiau, F.-Y., Zuo, Y., Lin, J.-C., Zheng, X.-Y., and Chang, Y. A., *Z. Metallk.* 1989, accepted for publication.
5.  Lin, J.-C., Schulz, K. J., Hsieh, K.-C, Chang, Y. A., *High Temperature Materials Chemistry IV*, Munir, Z. A., Cubicciotti, D., and Tagawa, H., eds., the Electrochem. Soc., Princeton, NJ, 477 (1988).
6.  Shiau, F.-Y., Zuo, Y., Zheng, X.-Y., Lin, J.-C., Chang, Y. A., *Adhesion in Solids*, Mattox, D. M., Baglin, J. E. E., Gottschall, R. J., and Batich, C. D., eds., MRS Symposium Proc. **119**, 171 (1988).
7.  Schulz, K. J., Zheng, X.-Y., and Chang, Y. A., *Electronic Packaging Materials Science III*, Jaccodine, R., Jackson, K. A., Sundahl, R. C., eds., MRS Symposium Proc., **108**, 455 (1988).

8. Lin, J.-C., Zheng, X.-Y., Hsieh, K.-C., and Chang, Y. A., *Epitaxy of Semiconductor Layered Structure*, Tung, R. T., Dawson, L. R., and Gunshor, R. L., eds., MRS Symposium Proc., **102**, 233 (1988).

9. Shiau, F. Y., Chang, Y. A., and Chen, L. J., *Microstructural Science for Thin Film Metallizations in Electron in Applications*, Sanchez, J., Smith, D. A., and Delanerolle, N., eds., The Minerals, Metals, and Materials Soc., Warrendale, PA, **15086**, 57 (1988).

10. Schulz, K. J., PhD Thesis, University of Wisconsin-Madison, Madison, WI (1988).

11. Schulz, K. J., Zheng, X.-Y., Lin, J.-C., and Chang, Y. A., *Acta Met.* 1989, under review.

12. Zheng, X.-Y., Schulz, K. J., and Chang, Y. A., *Bull. Alloy Phase Diagram* 1989, under review.

13. Schulz, K. J., Zheng, X.-Y., and Chang, Y. A., *Mat. Sci. and Engin.* 1989, under review.

14. Zheng, X.-Y., Lin, J.-C., Swenson, D. J., Hsieh, K.-C., and Chang, Y. A., *Mat. Sci. and Eng.*, **B**, 1989, accepted for publication.

15. Schulz, K. J., and Chang, Y. A., *Advances in Materials, Processing and Devices in III-V Compound Semiconductors*, Sadana, D. K., Dupois, R., and Eastman, L., eds., MRS Symposium Proc., 1989, **144**, in press.

16. Tsai, C. T., and Williams, R. S., *J. Mater. Res.* **1**, 820 (1986).

17. Tsai, C. T., and Williams, R. S., *J. Mater. Res.* **1**, 352 (1986).

18. Lince, J. R., Tsai, C. T., and Williams, R. S., *J. Mater. Res.* **1**, 537 (1986).

19. Pugh, J. H. and Williams, R. S., *J. Mater. Res.* **1**, 343 (1986).

20. Lince, J. R., and Williams, R. S., *J. Vac. Sci. Technol.* **B3**, 1217 (1985).

21. Beyers, R., Kim, K. B., and Sinclair, R., *J. Appl. Phys.* **61**, 2195 (1987).

22. Sands, T., *Mat. Sci. Eng. B: Solid-State Materials for Advanced Technology*, 1988, in print.

23. Sands, T., *J. Metals* **38**, 31 (1986).

24. Schmid-Fetzer, R., *J. Electron. Mater.* **17**, 193 (1988).

25. Tu, K. N., Ottaviani, G., Goselle, U., and Foll, H., *J. Appl. Phys.* **54**, 756 (1983).

26. Ottaviani, G., *Thin Films and Interfaces II*, Baglin, J. E. E., Campbell, D., and Cho, W. K., eds., North-Holland, NY, **21**, (1984); *J. Vac. Sci. Technol.* **16**, 1112 (1979).

27. Gosele, U. and Tu, K. N., *J. Appl. Phys.* **53**, 3252 (1982).

28. Tu, K. N., Research Report, IBM T. J. Watson Res. Center, Yorktown Hts., NY, 1984.

29. Colgan, E. G., and Mayer, J. W., Poster Epl. 21, presented at the 1985 MRS Meeting in Boston, MA, Dec. 2–7, 1985.

30. Majni, G., Costato, M., and Panini, F., *Thin Solid Films* **125**, 71 (1985).

31. Wagner, C., *Z. Anorg. Allgem. Chem.* **236**, 320 (1938).

32. Wagner, C., *J. Electrochem. Soc.* **103**, 571 (1956).

33. Rapp, R. A., Ezis, A., and Yurek, G. J., *Metall. Trans.* **4**, 1283 (1973).

34. Yurek, G. J., Rapp, R. A., and Hirth, J. P., *Metall. Trans.* **4**, 1293 (1973).

35. Shatynski, S. R., Hirth, J. P., and Rapp, R. A., *Metall. Trans.* **10A**, 591 (1979).

36. Kirkaldy, J. S., and Young, D. J., *Diffusion in the Condensed State*, Institute of Metals, London (1985).

37. Kirkaldy, J. S. and Brown, L. C., *Can. Met. Quart.* **2**, 89 (1963).

38. Kirkaldy, J. S. *Can. J. Phys.* **36**, 899 (1958).

39. Kirkaldy, J. S., *Can. J. Phys.* **36**, 907 (1958).

40. Kirkaldy, J. S., *Can. J. Phys.* **36**, 917 (1958).

41. Kirkaldy, J. S. and Fedak, D. G., *Trans. TMS-AIME* **224**, 490 (1962).

42. Jan, C. H., a graduate student working under the direction of Y. A. Chang, unpublished research, University of Wisconsin-Madison, 1988.

43. van Loo, F. J. J., van Beck, J. A., and Bastin, G. F., *Solid State Ionics* **16,** 131 (1985).
44. van Beck, J. A., Kok, Pimit, and van Loo, F. J. J., *Oxid Met.* **22,** 147 (1984).
45. Vosters, P., Laheij, M., van Loo, F. J. J., and Metselaar, R., *Oxid. Met.* **20,** 147 (1983).
46. Laheij, M., van Loo, F. J. J., and Metselaar, R., *Oxidation Met.* **14,** 207 (1980).
47. Leute, V., *Solid State Ionics* **17,** 185 (1985).
48. Onsager, L., *Phys. Rev.* **38,** 2265 (1931).
49. Onsager, L., *Phys. Rev.* **37,** 4305 (1937).
50. Onsager, L., *Ann. NY Acad. Sci.* **46,** 241 (1941).

# Recent Studies on Thermochemistry and Phase Equilibria in Alkali Metal Systems

C. K. MATHEWS

*Radiochemistry Programme, Indira Gandhi Centre for Atomic Research, Kalpakkam, Tamil Nadu 603 102, India*

**Index Entries:** Thermochemistry of; alkali metal; ternary compounds; phase equilibria; electrochemical meters.

## INTRODUCTION

Liquid alkali metals are being used increasingly as heat transfer fluids in energy technology (1). Liquid sodium is the coolant of choice in fast breeder nuclear reactors. Potassium is being considered as the working fluid in the topping cycle of multirankine cycle energy conversion systems. Sodium, potassium, and lithium have potential for use in high temperature heat pipes.

The structural materials with which liquid alkali metals come into contact under the operating conditions of the above-mentioned systems include stainless steels and ferritic steels. Corrosion processes taking place in the alkali metal-steel systems are largely influenced by the presence of dissolved oxygen. In order to understand the corrosion mechanism, one must have detailed information on the A-M-O ternary system, where A represents an alkali metal and M an alloying element in the steel. Detailed investigations carried out in our laboratory on the phase relationships and the thermochemistry of the relevant ternary oxides in these ternary systems are summarized in this paper. Electrochemical online meters developed in our laboratory for measuring the activities of oxygen, carbon, and hydrogen in liquid sodium are also briefly described. Based on the carbon meter, a new technique has been developed for measuring the carbon potential of the structural materials as well as metal carbides. Some of these results are also briefly discussed in view of their importance in the transport of carbon in sodium-steel systems. Similarly, the hydrogen meter was used to investigate the

Na-O-H system, which becomes important in the event of any water ingress into the sodium loops of fast reactors.

## STUDIES ON A-M-O SYSTEMS
### (A = Na, K; M = Fe, Cr, Mo, and Ni)

### Na-Cr-O System

$NaCrO_2(s)$ is the ternary oxygen compound that is generally observed under normal operating conditions in sodium-steel circuits. Formation of this compound has a bearing on the corrosion rate of austenitic stainless steels and on the tribological behavior of moving components in sodium systems. Precise thermochemical data on $NaCrO_2(s)$ is needed to understand the conditions of its formation in these systems. This was obtained by carrying out the following experiments in 2 different phase fields, viz, $Na(l)$-$Cr(s)$-$NaCrO_2(s)$ and $NaCrO_2(s)$-$Cr(s)$-$Cr_2O_3(s)$, in the ternary Na-Cr-O system (2). (a) Equilibrium oxygen potentials were measured in liquid sodium equilibrated with chromium and $NaCrO_2(s)$, using YDT-based solid electrolyte galvanic cells. The measured oxygen potentials can be represented as

$$\Delta \bar{G}_{O_2} = -800847 + 147.85T \text{ J/mol} (\pm 1350) (657\text{–}825 \text{ K}) \quad (1)$$

(b) Equilibrium partial pressures of sodium gas over $NaCrO_2(s)$-$Cr(s)$-$Cr_2O_3(s)$ phase field were measured by Knudsen cell mass spectrometry. The equilibrium studied can be represented as

$$Cr(s) + 3 NaCrO_2(s) \leftrightarrows 2 Cr_2O_3(s) + 3 Na(g) \quad (2)$$

The measured sodium pressures in conjunction with the Gibbs energy data on $Na(g)$ and $Cr_2O_3(s)$, yielded the Gibbs energy of formation of $NaCrO_2(s)$.

$$\Delta G^o_{f,T} <NaCrO_2>$$
$$= -870,773 + 193.171 T \text{ J/mol} (\pm 1300) (825\text{–}1025 \text{ K}) \quad (3)$$

The threshold oxygen concentration for the formation of $NaCrO_2(s)$ in sodium-austenitic stainless steel circuits were computed using the above data and by considering the following equilibria

$$2 Na(l) + \tfrac{1}{2} O_2(g) \rightleftharpoons [Na_2O]_{Na} \quad (4)$$

$$2 [Na_2O]_{Na} + Cr(s) \rightleftharpoons NaCrO_2(s) + 3 Na(l) \quad (5)$$

The values obtained from in-sodium emf measurements using YDT cells were higher by an order of magnitude than those calculated with data from Knudsen cell mass spectrometric data. Consideration of other data reported in literature on $NaCrO_2(s)$, shown in Table 1, also indicated identical trends. This discrepancy was experimentally traced to the interaction of carbon impurity, present in sodium, with chromium to form thin layers of chromium carbides. The carbides participate in the in-sodium equilibrium reactions in sodium according to the equation

Table 1
Equilibrium Oxygen Potential Data in the Phase Field
Na(1)-Cr(s)-NaCrO$_2$ (s)

| Technique | Temp. range, K | $\Delta\bar{G}_{O_2}$ J/mol O$_2$ = A + BT(K) | | Precision kJ | Ref. |
|---|---|---|---|---|---|
| | | A | B | | |
| Direct in-sodium measurements | | | | | |
| YDT emf on Na-Cr-NaCrO$_2$ | 573–773 | $-1131797$ | 510.5 | — | (3) |
| -do- (5 data points) | 876–1072 | — | — | — | (4) |
| -do- | 800–980 | $-841985$ | 186.881 | $\pm$ 1.26 | (5) |
| -do- | 657–825 | $-800847$ | 147.85 | $\pm$ 1.35 | (2) |
| Other measurements | | | | | |
| CaF$_2$ emf on NaF-NaCrO$_2$-Cr-Cr$_2$O$_3$ | 800–1040 | $-863911$ | 203.743 | $\pm$ 5.64 | (6) |
| K-cell on NaCrO$_2$-Cr$_2$O$_3$-Cr | 700–1100 | $-865591$ | 188.231 | $\pm$11.11 | (7) |
| YSZ[a] emf on NaCrO$_2$-Cr$_2$O$_3$-Na$_2$CrO$_4$ | 784–1012 | $-869980$ | 185.75 | $\pm$ 1.86 | (8) |
| -do- | 820–1006 | $-852308$ | 167.206 | $\pm$ 1.1 | (9) |
| CaF$_2$ emf on NaF-NaCrO$_2$-Cr-Cr$_2$O$_3$ | 782–1040 | $-930750$ | 254.85 | $\pm$2.59 | (10)[b] |
| K-cell on NaCrO$_2$-Cr$_2$O$_3$-Cr | 825–1025 | $-870773$ | 193.171 | $\pm$ 1.3 | (2) |

[a] = Yttria stabilized zirconia.
[b] Averaged data.

$$CrC_x + 2\,[Na_2O]_{Na} \rightleftharpoons NaCrO_2(s) + Na(l) + x[C]_{Na} \qquad (6)$$

Figure 1 shows the phase stability diagram in the Na-Cr-O-C system constructed to highlight the influence of carbon potential on the equilibrium oxygen potentials. The threshold oxygen potential, and, thus, the threshold oxygen concentration, for the formation of NaCrO$_2$(s) increases with increasing carbon potential, the latter dictating the nature of the carbide taking part in the reaction (2). These results could be used to explain the high oxygen levels that were needed to observe NaCrO$_2$(s) in operating sodium circuits and also in in-sodium oxygen potential measurements.

## Na-Mo-O System

Molybdenum metal and its alloys are considered for use as structural materials in high temperature heat pipes that use liquid sodium as the working fluid. It is also added as a minor alloying element in austenitic stainless steels to improve their high temperature creep strength. Data available in the literature on Na-Mo-O system, which would give an insight on corrosion of Mo in sodium systems, are not consistent with

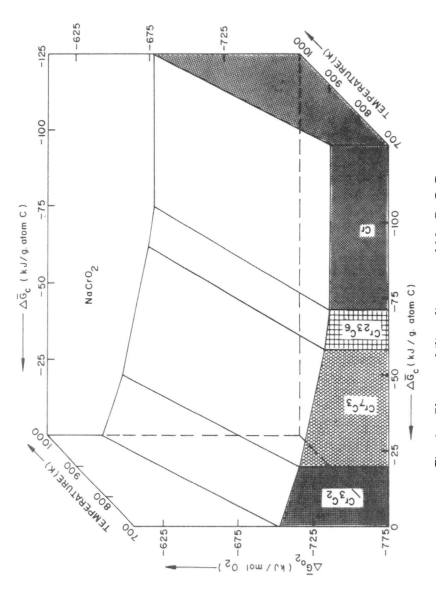

Fig. 1.   Phase stability diagram of Na-Cr-O-C system.

each other. Observations indicating the stability or instability of various ternary oxygen compounds have been reported (*11,12*). In order to investigate the system in a systematic manner, experiments involving various techniques were conducted (*13*). Long term in-sodium equilibrations and pseudoisopiestic equilibrations of oxides of molybdenum with sodium vapor were conducted. Solid state reactions between $Na_2O(s)$ and $Mo(s)$ under high vacuum and experiments for the confirmation of the coexistence of condensed phases were carried out. YDT based solid oxide electrolyte galvanic cells were used to measure equilibrium oxygen potentials in liquid sodium containing oxygen and molybdenum. The results indicated that at $T < 681.1$ K, Na(l) and Mo(s) coexists with $Na_2O(s)$. At $T > 681.1$ K, the ternary compound $Na_4MoO_5(s)$, appears as the coexisting phase with these two metals. From the results of oxygen potential measurements, Gibbs energy of formation of $Na_4MoO_5(s)$ was derived as a function of temperature.

$$\Delta G^o_{f,T} <Na_4MoO_5> =$$
$$-1907223 + 436.13\ T\ \text{J/mol}\ (\pm 5300)\quad (681–773\ \text{K})\qquad (7)$$

In order to establish the other phase fields in the Na-Mo-O system, studies involving the reduction of polymolybdates under a flow of hydrogen gas, followed by confirmation of the coexistence of various ternary phase fields, were undertaken (*14*). The information obtained from the above experiments could be used to construct the isothermal crossections of Na-Mo-O system, shown in Fig. 2, in the temperature range of 673–923 K.

### Na-Ni-O System

Among the alloying constituents of structural steels used in sodium circuits, nickel does not form a ternary oxygen compound stable in liquid sodium. However, ternary oxygen compounds such as $Na_3Ni_2O_5(s)$, $Na_2NiO_2(s)$, and $NaNiO_2(s)$ have been prepared and characterized. The only reported thermochemical data on $NaNiO_2(s)$ by Shaiu et al. (*6*), from $CaF_2$ solid electrolyte galvanic cells, have been held unreliable by the authors themselves. Further, the phase diagram of Na-Ni-O system has not been established unequivocally. Various experimental approaches were adopted for establishing the phase relationships in this system.

Varying molar ratios of $Na_2O(s)$ and Ni(s) were heated under high vacuum [$<1.33 \times 10^{-3}$Pa] at 773 K for prolonged periods. The characterization of reaction products indicated the presence of a ternary phase field with $Na_2NiO_2(s)$ and Ni(s) as two of the three coexisting condensed phases. Solid state equilibrations involving (a) $Na_2O_2(s)$ and Ni(s) were carried out at 773 K. The reaction products indicated the coexistence of $Na_2NiO_2(s)$-NiO(s), on the basis of X-ray diffraction measurements. Since NiO(s)/Ni(s) is a stable metal-metal oxide couple, the presence of a phase field with $Na_2NiO_2(s)$, Ni(s), and NiO(s) as the coexisting phases could be deduced. This was further confirmed by equilibrating these three phases together for extended periods at 773 K in sealed capsules. $Na_2NiO_2(s)$ is the only alkali rich sodium nickelate, and this suggests the

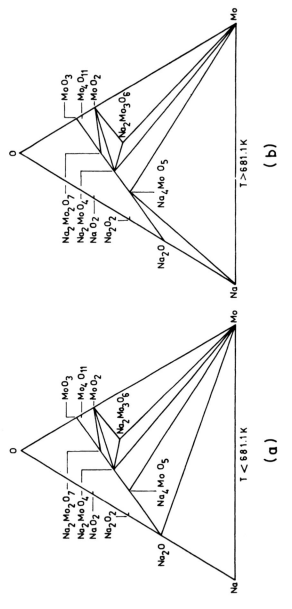

Fig. 2. Isothermal crossections of Na-M-O system (melting point of $Na_2Mo_2O_7(s)$ = 885.15 K).

presence of $Na_2NiO_2(s)$-$Ni(s)$-$Na_2O(s)$ phase field. This was also confirmed by equilibrating the equivolume mixture of these phases at 773 K. A similar approach was used to establish the $Na_2NiO_2(s)$-$NiO(s)$-$Na_2'NiO_2(s)$ phase field. Since no ternary oxygen compound of nickel is stable toward liquid sodium, and since $Na_2O(s)$ is the phase that coexists with liquid sodium, the presence of $Na_2O(s)$-$Ni(s)$-$Na(l)$ phase field could be deduced. These data were used to construct the isothermal crossection of Na-Ni-O phase diagram at 773, shown in Fig. 3 *(15)*.

## Na-Nb-O and Na-Ta-O Systems

Refractory metals, such as niobium and tantalum, are known to form ternary oxides with sodium. In-sodium equilibrations of $Nb_2O_5(s)$ and $Ta_2O_5(s)$ were carried out at 673–923 K. Formation of $Na_3NbO_4(s)$ and $Na_3TaO_4(s)$ were observed from these experimental results. Nb(s) was also identified along with $Na_3NbO_4(s)$ while equilibrating $Nb_2O_5(s)$ with sodium at 897 K, indicating the presence of $Na(l)$-$Nb(s)$-$Na_3NbO_4(s)$ phase field at this temperature. Thermochemical data in this system were also obtained by measuring the oxygen content in sodium, which exists in equilibrium with other phases in the appropriate ternary phase field as a function of temperature. The measured equilibrium oxygen potentials in sodium can be represented as

$$\Delta \overline{G}_{O_2} = 2\Delta G^o_{f,T} <Na_2O> + 2 \, RT \ln a_{[Na_2O]} \qquad (8)$$

Activity of sodium oxide in sodium, $a_{[Na_2O]_{Na}}$, can be related to the measured oxygen level, $C_{Na_2O}$, by applying Henry's law for the solution of oxygen in sodium as

$$a_{Na_2O} = C_{Na_2O}/C^S_{Na_2O} \qquad (9)$$

where $C^S_{Na_2O}$ is the saturated solubility of $Na_2O(s)$ in liquid sodium. Equilibrium oxygen concentrations in sodium refluxed with $Nb_2O_5(s)$ were measured in the temperature range 473–873 K. When $Na(l)$ -$Nb(s)$-$Na_3NbO_4(s)$ phase field exists at all temperatures, an expression for Gibbs energy of formation of $Na_3NbO_4(s)$ could be obtained from this data by considering the following reaction.

$$4 \, [Na_2O]_{Na} + Nb(s) \rightleftharpoons Na_3NbO_4(s) + 5 \, Na(l) \qquad (10)$$

The Gibbs energy of formation of $Na_3NbO_4(s)$ thus deduced can be represented as

$$\Delta G^o_{f,T} <Na_3NbO_4> = -1530572 + 254.694 \, T \text{ J/mol} \qquad (11)$$

The results are in good agreement with data derived from in-sodium equilibrium oxygen potential measurements using YDT based oxygen meters *(15)*.

## Na-Fe-O System

Iron is the major alloying element in structural steels employed in sodium circuits. The general corrosion rate of these steels would be

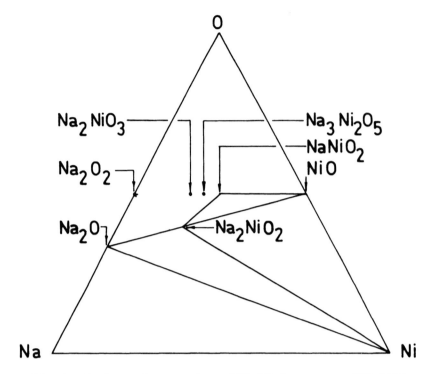

Fig. 3.   Isothermal crossection of Na-Ni-O system at 773.15 K.

determined mainly by the chemical reactions of iron in sodium containing dissolved oxygen. Only scanty data are available in the literature on
the Na-Fe-O system. $Na_4FeO_3(s)$ has been found to be the ternary compound formed by iron in sodium heavily doped with oxygen (*16*). The
required levels of oxygen were reported to be approximately 1000 ppm at
773 K (*17*). Detailed experiments involving in-sodium equilibrations of
oxides of iron in the temperature range 623–923 K were carried out in our
laboratory. Pseudoisopiestic equilibrations of oxides of iron were conducted at 923 and 773 K under sodium vapor of 880 and 4.7 Pa partial
pressures, respectively. The results of these experiments (*15*) indicated
that Fe(s) and Na(l) coexist with $Na_2O(s)$ at 623 K; $Na_4FeO_3(s)$ coexists
with these two metals at 923 K.

The invariant temperature at which $Na_2O(s)$ and $Na_4FeO_3(s)$ coexist
with Na(l) and Fe(s) was determined by differential thermal analysis.
Because of the high reactivity and high vapor pressures of alkali metals,
commercially available DTA equipments are not easily adaptable for this
purpose. A special DTA set up was fabricated and put into use (*18*). The
DTA runs were carried out with welded capsules containing sodium
along with oxides of iron. From the results, the invariant temperature
was deduced as 760 ± 6 K. In the absence of other ternary oxygen
compounds, stable in sodium at this temperature, the reaction occurring
can be represented as

$$3\ Na_2O(s)\ +\ Fe\ (s)\ \rightleftharpoons\ Na_4FeO_3(s)\ +\ 2\ Na\ (l) \tag{12}$$

$$\Delta G^{\circ}_{f,760} <Na_4FeO_3> = 3 \Delta G^{\circ}_{f,760} <Na_2O> \qquad (13)$$

Using the data on Gibbs energy of formation of $Na_2O(s)$ (19) and heat of formation of $Na_4FeO_3$ (20), Gibbs energy of formation of $Na_4FeO_3(s)$ was deduced as a function of temperature.

$$\Delta G^{\circ}_{f,T} <Na_4FeO_3> = -1211000 + 353.53 \, T \, J/mol \, (\pm 2550) \quad (14)$$

Equilibrium oxygen levels in sodium doped with iron and oxygen were measured in the temperature range 588–904 K (21). The invariant temperature determined by this technique is 626 K. Recently, Bhat and Borgstedt (22) measured thermodynamic data in this system by measuring equilibrium oxygen potential using YDT based oxygen meters. They reported the invariant temperature as 723 K. The Gibbs energy data on $Na_4FeO_3(s)$ reported in that work is in good agreement with the present DTA experimental results. The reasons for wide differences in the invariant temperatures determined by 3 different techniques is a matter of future research.

## K-O and K-Cr-O Systems

Unlike sodium systems, studies related to liquid potassium systems are not extensive. Even the 2 reported data (23,24) on the solubility of oxygen in potassium do not agree with each other. Experiments were conducted to determine these solubility values in the temperature range 343–675 K (25). From considerations of compatibility, a nickel lined SS vessel was used to equilibrate liquid potassium with $K_2O(s)$ and a nickel frit was used to obtain samples by filtration. From the experimental results, an expression for the solubility of oxygen in potassium was derived as follows.

$$\log (C/ppm) = 3.9702 - 420.4/TK^{-1} \, (343–675 \, K) \qquad (15)$$

Solubility measurements carried out in an all stainless steel apparatus led to extensive corrosion. However, the measured oxygen levels were interpreted as representing the $K(l)-KCrO_2(s)-K_4CrO_4(s)$ phase fields and used to derive the Gibbs energy of formation of $K_4CrO_4(s)$ (24).

## Threshold Oxygen Levels for the Formation of Ternary Oxygen Compounds in Na-316 SS System

The conditions that determine the formation of ternary oxygen compounds in sodium systems are

1. The relative thermodynamic stabilities of the ternary oxygen compounds involved;
2. The chemical activities of the individual alloying elements in the structural steels; and
3. The prevailing oxygen concentrations in sodium.

For the comparison of the stabilities of the ternary oxygen compounds of relevance in Na-316 SS system, the oxygen potentials of the following equilibria were calculated.

$$Na(l) + x[M]_{SS} + y/2\, O_2(g) \rightleftharpoons$$
$$Na/M_xO_y(s)\ [M = Fe,\ Cr,\ Mo,\ and\ Mn] \qquad (16)$$

The chemical activities of the alloying elements in 316 SS were measured recently by Azad et al. (27). Data on $NaMnO_2(s)$ were taken from (5). The data on other ternary oxygen compounds were taken from the present work. The calculated oxygen potentials are shown in Fig. 4. Oxygen potentials of sodium containing various levels of dissolved oxygen and the equilibrium oxygen potentials in $Na(l)$-$Nb(s)$-$Na_3NbO_4(s)$ are also plotted in this figure for purposes of comparison. It is seen that $NaCrO_2$ (s) is the ternary oxygen compound that would be formed in sodium containing 2–10 ppm oxygen, when it is in contact with 316 stainless steel systems. As seen in the figure, the equilibrium oxygen potentials in $Na(l)$-$Nb(s)$-$Na_3NbO_4(s)$ are comparable to that of $Na(l)$-$[Cr]_{ss}$-$NaCrO_2(s)$ at higher temperatures. In 347 SS systems (or in other niobium containing steels), the added niobium content is low and is bound by carbon. Hence, the activity of Nb would be considerably low in these steels, indicating that $NaCrO_2(s)$ alone would form in sodium of normal purity. $Na_4FeO_3(s)$, $Na_4MoO_5(s)$, and $NaMnO_2(s)$ would appear only when dissolved oxygen content is about 1000 ppm, a condition that would not be normally met. Influence of carbon on the formation of $NaCrO_2(s)$ should, however, be taken into account, as discussed in an earlier section.

# DEVELOPMENT AND APPLICATIONS
## OF ELECTROCHEMICAL ONLINE
## METERS FOR OXYGEN, CARBON,
## AND HYDROGEN IN ALKALI METALS

## *Electrochemical Oxygen Meter*

Monitoring the sodium coolant of fast breeder reactors for oxygen is important for controlling corrosion and activity transport and for detecting steam generator leaks at the very inception. The electrochemical meter for online monitoring of oxygen in liquid sodium makes use of an oxide-ion conducting solid electrolyte, Yttria Doped Thoria (YDT). The actual cell is an electrode concentration cell in which the sodium containing dissolved oxygen forms one electrode and another liquid metal coexisting with its oxide forms the reference electrode with the solid electrolyte separating the two liquids. The cells can be represented as

$$Na + [O]\ dissolved\ /\ YDT\ /\ In,\ In_2O_3$$

The EMF, E, generated by this cell is related to the oxygen potentials of the liquid metals by the relationship

$$E/Volts = (\Delta\overline{G}_{O_2}ref - \Delta\overline{G}_{O_2}sodium)/4F \qquad (17)$$

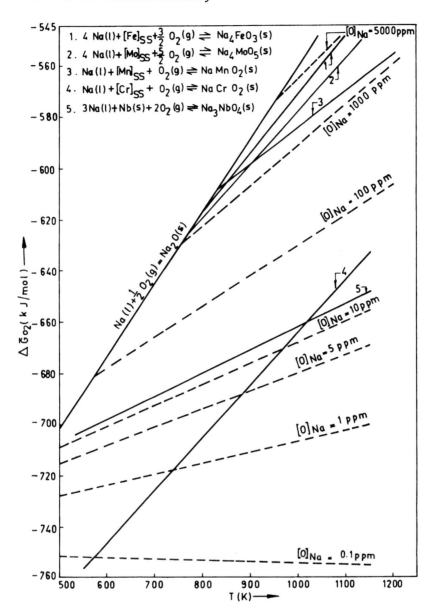

Fig. 4. Equilibrium oxygen potentials for the formation of ternary oxygen compounds in liquid sodium-316 SS system and Na-Nb-O system.

where $F$ is the Faraday constant. The $\Delta \bar{G}_{O_2}$ of sodium thus obtained can be related to oxygen concentration in sodium, $C_{Na_2O}$, assuming Henrian behavior for oxygen in sodium. Typical construction details of the meter are described elsewhere (2).

The YDT based meters suffer from unpredictable lifetimes (200–1000 h) and high cost (10,000 £/ea). These problems are caused by the poor thermal shock resistance of YDT and the difficulties in fabricating it in tube form. YDT is also attacked on exposure to high oxygen sodium. At IGCAR, we have succeeded in developing meters based on commonly available stabilized zirconia electrolytes by operating the cells at low temperatures when pure ionic conductivity extends to low oxygen potentials and sodium attack on the ceramic is minimum. They could be operated at low temperatures because of the choice of a new reference electrode K, $K_2O$. Sodium attack on the electrolyte was further reduced by coating it with calcium zirconate. These meters gave good performance when tested in sodium. Low temperature measurements of $\Delta\bar{G}_{O2}$ in sodium also showed that the oxygen behavior in sodium is nonHenrian at higher activity levels (*28*).

The electrochemical meters are simple and convenient devices and could be used to study phase equilibria in alkali metal systems. Applications of the YDT-based meters in Na-Cr-O and Na-Mo-O systems have been described in earlier sections.

### Electrochemical Carbon Meter

Measuring the carbon potential of sodium in fast reactor circuits using online meters is necessary to follow the carburization and decarburization processes occurring in sodium-steel systems and for detecting oil leaks into sodium through seals. The electrochemical carbon meters developed in our laboratory (*29*) are based on the principle of a concentration cell and are represented by

Reference electrode / Electrolyte / $[C]_{Na}$

where $[C]_{Na}$ represents carbon dissolved in sodium. An eutectic mixture of lithium and sodium carbonates is employed as the electrolyte. Since sodium and the electrolyte are not compatible, a thin-walled iron membrane having high permeability for carbon separates them. The reference electrode is pure graphite encapsulated in a nickel tube. The EMF, E, of this meter is given by

$$E/Volts = (RT/4F) \ln a_C^R/a_C^{Na} \qquad (18)$$

where $a_C^R$ and $a_C^{Na}$ represent the activities of carbon in the reference electrode and sodium, respectively. The other symbols have their usual significance.

In the actual design, the liquid electrolyte is contained in a high-purity iron cup (wall thickness at the sensing portion = 0.25 mm), which is welded to an SS tube. The reference electrode is positioned at the center of this cup. When the iron cup is immersed in sodium, the iron cup attains the carbon activity of sodium, and the EMF developed across this electrode and reference electrode is measured using a high impedance millivolt meter. When tested in distilled sodium, this cell gave a Nernstian response for different cell temperatures (*30*).

The electrochemical carbon meter was used to study the carbide phase equilibria and measure the carbon activity in structural steels. The material of interest was first equilibrated with liquid sodium and the carbon activity of sodium at equilibrium was measured by the electrochemical carbon meter. The attainment of equilibrium was indicated by the constancy of the meter output. The conventional gas equilibration technique involving $CH_4/H_2$ mixtures for carbon activity determinations would be affected by trace level moisture or oxygen in the gases. In the present technique,the sample is immersed in sodium, which does not form a thermodynamically stable carbide. Dissolved oxygen does not affect carbon chemistry in sodium (31), and, hence, oxygen-bearing impurities when present in sodium do not interfere with the experiment. Solubility of carbon in sodium is low and thus the time taken to attain equilibrium would be expected to be low.

### Measurement of Carbon Activity in Structural Steels

Carburization-decarburization phenomena that occur in fast breeder sodium coolant circuits would determine the mechanical properties of the structural steels employed. The extent and direction of carbon transport is determined by the carbon activities in sodium and steel, and their dependence on temperature (31). A detailed knowledge of carbon activities in structural steels in the temperature range of relevance to fast breeder reactor operation is, therefore, necessary. The earlier studies reported in literature on carbon activities in 18/8 type austenitic stainless steels were based on the gas equilibration technique (32,33).

Detailed studies were conducted in our laboratory to measure the carbon activities in commercial austenitic stainless steels by employing the electrochemical carbon meter. Annealed foils of type 304 stainless steel were used as samples. The measurements were carried out in the temperature range 860–960 K. The carbon activities were related to total carbon content (analyzed by combustion technique) and alloy element concentrations in the steel sample. Figure 5 shows the measured carbon activities of the 304 SS as a function of carbon content. The measured carbon activities are significantly higher than the data obtained by extrapolation of the high temperature data of Natesan et al. (33) for synthetic steel. This could be attributed to the presence of minor alloying elements in commercial steels. In the light of the present experimental results, the following expression to relate carbon activity and composition of 304 SS was proposed (34).

$$\ln a_C \text{ (SS)} = \ln (0.048\%C) + (0.525 - 300/T)\%C - 1.845 + 5100/T - (0.021 - 72.4/T)\% \text{ (Ni + Mn)} + (0.248 - 404/T) \% \text{ Cr} - (0.0102 - 9.422/T)\% \text{ Cr}^2 + 0.033\% \text{ Cr} \qquad (19)$$

The electrochemical carbon meter is being used to understand the behavior of carbon in 2¼ Cr − 1 Mo ferritic steel in sodium environment (35). These steels are characterized by high carbon activities and several competing carbide equilibria (31). Initial experiments have indicated that decarburization of foils of steel specimens (200 μm) at 878 K needed approximately 150 h. However, subsequent experiments showed decarburization was complete in about 20 h, whereas the process of attaining new carbon activitiy values in sodium is very slow, the exact reasons of which are presently not known. This behavior made it difficult to mea-

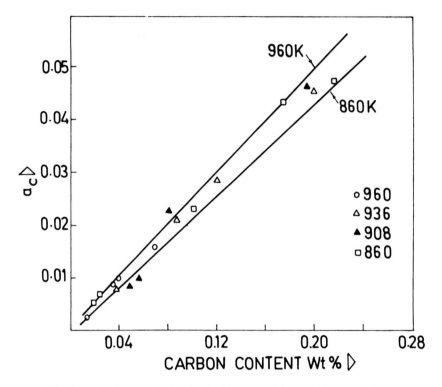

Fig. 5.   Carbon activity in 18/8 austenitic stainless steel.

sure the carbon activity of ferritic steels as a function of carbon concentration. The measured carbon activities were found to have significant scatter.

### Gibbs Energy of Formation
#### of Refractory Metal Carbides

The fast breeder test reactor (FBTR) at Kalpakkam uses a mixed uranium-plutonium carbide fuel clad in 316 SS tubes. The fuel consists of a two phase mixture of $U_{1-x}Pu_xC$ and $U_{1-y}Pu_6C_{1.5}$. The monocarbide phase dissolves significant quantities of oxygen and nitrogen impurities that are unavoidable as the fuel is produced by the carbothermic reaction of the mixed oxide. In order to assess the possibility of clad carburization by the fuel, a detailed knowledge of the carbon potential of the fuel is essential. The carbon potential of the fuel depends on the sesquicarbide content, oxygen and nitrogen levels, and, the U/Pu ratio. The measurement of carbon potential by conventional $CH_4/H_2$ gas equilibration technique could lead to erroneous data, if care is not taken to avoid trace level $O_2$ and $N_2$ impurities as otherwise the levels of these impurities would change during measurement. The present method involving equilibration in liquid sodium in conjunction with measurement of carbon activity by electrochemical carbon meter would avoid these interferences. This method was successfully used to measure the carbon potential of carbide fuel [Pu = 66.72 wt%; U = 27.48 wt%; C = 4.72 wt%; O = 3900 ppm; N = 728 ppm] in our laboratory (36). The measurements were carried out in the temperature range 847–913 K. The measured carbon potentials are shown in Fig 6.

Generation of Gibbs energy data of carbides of refractory metals such as Cr by conventional techniques are susceptible to be influenced by the formation of their oxides when trace level oxygen impurities are present in the equilibrating gas. The data in the literature on these metal carbides (37) show wide disparity, which could be largely attributed to this reason. Further, these measurements are generally carried out at high temperatures. The technique employing the electrochemical carbon meter was used to obtain data on $Cr/Cr_{23}C_6$ and W/WC couples. These data are also shown in Fig. 6.

### Electrochemical Hydrogen Meter

Monitoring of hydrogen in the sodium coolant of fast breeder reactors is important for detecting the leakage of steam into sodium in the steam generator as well as the ingress of hydrocarbon oil into sodium from centrifugal pump shaft seals. The electrochemical hydrogen meter developed at IGCAR makes use of $CaCl_2$ containing 5–7.5 mol% $CaH_2$ as a solid electrolyte. From considerations of compatibility with sodium, the electrolyte is encapsulated in an iron cup. A mixture of Li, LiH contained in another iron cup, which in turn is kept immersed into the electrolyte serves as the reference electrode. The cell can be represented as

$$\text{Li, LiH} \quad | \quad CaCl_2 - CaH_2 \quad | \quad [H]_{Na}$$
$$\text{Fe} \qquad\qquad\qquad \text{Fe}$$

The EMF, E, developed by the cell can be given as

$$E/\text{Volts} = (RT/2F) \cdot \ln (P^{Li}_{H_2}/P^{Na}_{H_2}) \tag{20}$$

where $P^{Li}_{H_2}$ and $P^{Na}_{H_2}$ represent the equilibrium hydrogen partial pressures in the Li-LiH reference electrode and in sodium sample, respectively. Other symbols have their usual significance. The details of construction and results of tests carried out with these meters in operating sodium loops have been described elsewhere (38,39).

The electrochemical meter is being used to measure equilibrium hydrogen partial pressures in ternary alkali metal-oxygen-hydrogen systems and deduce various equilibrium properties. Na-O-H system exhibits the following reaction (40).

$$\text{Na (solution 1) + NaOH (solution 2)} \rightleftharpoons Na_2O(s) + NaH (s) \tag{21}$$

The temperature at which the reaction occurs was determined by measuring equilibrium hydrogen partial pressures in liquid sodium containing large amounts of oxygen and hydrogen added as NaOH. The measurements were carried out in the temperature range 520–723 K. The reaction temperature was deduced as 673 K from the experimental results. This reaction temperature is in good agreement with the results obtained using the DTA technique (18,41).

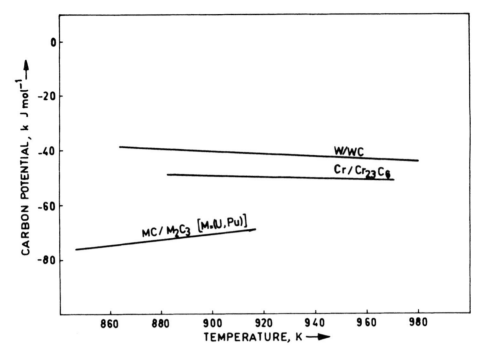

Fig. 6.  Carbon potential of metal-metal carbides measured by electrochemical carbon meter.

## ACKNOWLEDGMENT

The author wishes to acknowledge the contributions of several col-
leagues, especially G. Periaswami, T. Gnanasekaran, S. Rajendran Pillai,
and V. Ganesan, who carried out the studies discussed in this paper. T.
Gnanasekaran also assisted in preparing this manuscript.

## REFERENCES

1. Borgstedt, H.U. and Mathews C.K., *Applied Chemistry of the Alkali Metals*,
   Plenum, New York, 1987.
2. Gnanasekaran, T. and Mathews, C.K., *J. Nucl. Mater.* **140,** 202 (1986).
3. Jansson, S.A. and Berkey, E., *Corrosion by Liquid Metals*, Draley and Weeks,
   J.R., eds., Plenum, New York, 1970, p. 479.
4. Adamson, M.G., Aitkens, E.A., and Jeter, D.W., *Proc. Int. Conf. on Liquid
   Metal Technology in Energy Production*, Champion, USA, CONF-760503
   (1976), part II, p. 866.
5. Frankham, S.A., Ph. D. Thesis, University of Nottingham, UK, 1982.
6. Shaiu, B.J., Wu, P.C.S., and Chiotti, P., *J. Nucl. Mater.* **67,** 13 (1977).
7. Knights, C.F. and Phillips, B.A., *High Temperature Chemistry of Inorganic and
   Ceramic Materials*, Special Publication No. 30, Chemical Society, London,
   1977, p. 134.
8. Sreedharan, O.M., Madan, B.S., and Gnanamoorthy, J.B., *J. Nucl. Mater.*
   **119,** 296 (1983).

9. Venugopal, V., Iyer, V.S., Sundaresh, V., Ziley Singh, Prasad, R., and Sood, D.D., *J. Chem. Thermodynam.* **9**, 617 (1987).

10. Bhat, N.P., Swaminathan, K., Krishnamurthy, D., Sreedharan, O.M., and Sundaresan, M., *Proc. Third Int. Conf. on Liquid Metal Engineering and Technology*, The British Nuclear Energy Society, London, 1984, vol. I, p. 323.

11. Reau, J.-M., Fouassier, C., and Hagenmuller, P., *Bull. Soc. Chim. France* **11**, 3827 (1970).

12. Barker, M.G., *Re. Int. Hautes. Temper. Fr.* **16**, 237 (1979).

13. Gnanasekaran, T., Mahendran, K.H., Periaswami, G., Mathews, C.K., and Borgstedt, H.U., *J. Nucl. Mater.* **150**, 113 (1987).

14. Gnanasekaran, T., Mahendran, K.H., Kutty, K.V.G., and Mathews, C.K., *J. Nucl. Mater*, **165**, 210 (1989).

15. Ganesan, V., Sridharan, R., Gnanasekaran, T., and Mathews, C.K., *Proc. Fourth Int. Conf. on Liquid Metal in Engineering & Technology*, Avignon, France, 1988, Societe Francaise d'Energie Nuclearire, France, 1988, paper no. 533.

16. Horsley, G.W., *J. Iron and Steel Inst.* **183**, 43 (1956).

17. Addison, C.C., Barker, M.G., and Hooper, A.J., *J. Chem Soc. Dalton Trans.* 1017 (1972).

18. Gnanasekaran, T., Mahendran, K.H., Sridharan, R., Periaswami, G., and Mathews, C.K., in loc. cit., ref. *15*, Paper no. 521.

19. Fredrickson, D.R. and Chasanov, M.G., *J. Chem. Thermodynam.* **5**, 485 (1973).

20. Gross, P. and Wilson, G.L., *J. Chem. Soc.* (A), **11**, 1913 (1970).

21. Sridharan, R., Krishnamurthy, D., and Mathews, C.K., Paper paresented at STNM-7, Chicago, USA, Sept., 1988, *J. Nucl. Mater.* **167** (1989) (in press).

22. Bhat, N.P. and Borgstedt, H.U., *J. Nucl. Mater.* **158**, 7 (1988).

23. Williams, D.D., Grand, J.A., and Miller, R.R., *J. Phys. Chem.* **63**, 68 (1959).

24. Ganesan, V. and Borgstedt, H.U. *J. Less-Common Met.* **113**, 253 (1985).

25. Krishnamurthy, D., Thiruvengadasami, A., Bhat, N.P., and Mathews, C.K., *J. Less-Common Met.* **135**, 285 (1987).

26. Krishnamurthy, D., Bhat, N.P., and Mathews, C.K., *Proc. 10th Int. Congress on Metallic Corrosion*, Madras, India, Oxford & IBH, New Delhi, India, 1987, vol. IV, p. 3771.

27. Azad, A.M., Sreedharan, O.M., and Gnanamoorthy, J.B., *J. Nucl. Mater.* **144**, 94 (1987).

28. Periaswami, G., Rajan Babu, S., and Mathews, C.K., in loc. cit., ref. *15*, paper no. 607.

29. Rajendran Pillai, S. and Mathews, C.K., *J. Nucl. Mater.* **137**, 107 (1986).

30. Rajendran Pillai, S., Ranganathan, R., and Mathews, C.K., in loc. cit., ref *15*, paper no. 504.

31. Mathews, C.K., Gnanasekaran, T., and Rajendran Pillai, S., *Trans, Indian Institute of Metals* **40**, 89 (1987).

32. Tuma, V.H., Groebner, P., and Loebl, K., *Archiv. Eisenhuttenw.* **9**, 727 (1969).

33. Natesan, K., and Kassner, T.F., *Met. Trans.* **4**, 2557 (1973).

34. Rajendran Pillai, S. and Mathews, C.K., *J. Nucl. Mater.* **150**, 31 (1987).

35. Rajendran Pillai, S., Ranganathan, R., and Mathews, C.K. *Proc. National Symp. on Electrochemical Techniques and Instrumentation*, Kalpakkam, India, Dec. 1988, Society for Advancement of Electrochemical Science and Technology, Kalpakkam Chapter, India, 1988, p. 19.

36. Rajendaran Pillai, S., Anthonysamy, S., Prakashan, P.K., Ranganathan, R., Vasudeva Rao, P.R., and Mathews, C.K., Paper presented at STNM-7, Chicago, USA, Sept. 1988, *J. Nucl. Mater.* **167**, (1989), in press.

37. Coltters, R.G., *Materials Sci. Eng.* **76,** 1 (1985).
38. Ganesan, V., Gnanasekaran, T., Sridharan, R., Periaswami, G., and Mathews, C.K., in loc. cit., ref. *10*, vol I. p. 369.
39. Gnanasekaran, T., Sridharan, R., Periaswami, G., Mathews, C.K., Rajan, M., and Kale, R.D., in loc. cit., ref. *15*, paper no. 604.
40. Myles, K.M. and Cafasso, F.A., *J. Nucl. Mater.* **67,** 249 (1977).
41. Maupre, J.P., Centre' d'Etudes Nucleaires de Cadarache, France, Report CEA-R-4905 (1978).

# Knudsen Effusion
# Mass Spectrometric Determination
# of Metal Hydroxide Stabilities

L. N. GOROKHOV,* M. I. MILUSHIN, AND A. M. EMELYANOV

*Institute for High Temperatures, USSR Academy of Sciences,
Moscow 127412, USSR*

## ABSTRACT

Knudsen effusion mass spectrometric measurements of gaseous
equilibria in vapors of Cr and Fe oxides, with admission of hydrogen
or deuterium into an effusion cell, have been conducted. From the
third-law calculations, the values of dissociation energy $D_0$ (M-OH)
have been obtained (kJ mol$^{-1}$): CrOH, 372.7 $\pm$ 14; CrOD, 377.2 $\pm$ 14;
FeOH, 345.8 $\pm$ 17. Comparison of data on the monofluoride and
monohydroxide bond energies has led to a simple relation

$$D_0 \text{ (M-OH)} = D_0(\text{MF}) - 134$$

**Index Entries:** Knudsen effusion; mass spectrometry; high temperature equilibria; chromium monohydroxide; iron monohydroxide;
monohydroxide stabilities.

## INTRODUCTION

The interest in the investigation of the thermodynamic properties of
the metal–oxygen–hydrogen systems at high temperatures is a result of
the wide use of different metals, alloys, and ceramic materials in high-
temperature engineering. Gaseous metal hydroxides can play an important role in these systems. Interest in the thermochemical properties of
metal hydroxides is also accounted for by the fact that these molecules
can play an important role in the processes occurring in flames with
additions of metal compounds. In spite of their obvious importance,
gaseous metal hydroxides have been scantily studied. Reliable and comprehensive data are available only for hydroxides of alkali and alkaline-

*Author to whom all correspondence and reprint orders should be addressed.

High Temperature Science, Vol. 26     © 1990 by the Humana Press Inc.

earth metals. As regards transition metal hydroxides, the available literature data are scarce and for most metals are lacking in general.

The present paper is concerned with the investigation by Knudsen
effusion mass spectrometry of the interaction of the oxides $Cr_2O_3$ and
$Fe_2O_3$ with hydrogen at high temperatures. The enthalpies of formation
of CrOH and FeOH molecules have been determined as well as their
ionization energies. Earlier, the interaction of alumina with hydrogen
was studied and the enthalpy of formation of AlOH molecules was
determined (1).

The enthalpy of formation of FeOH molecules was determined in
two works. In Jensen and Jones' work (2) the equilibrium

$$Fe + H_2O = FeOH + H \tag{1}$$

was studied by flame spectrophotometry. From the temperature dependence of the equilibrium constant of the reaction in the temperature
range of 2035–2615 K using the second law treatment, the bond energy is
obtained $D_0(Fe–OH) = 382 \pm 20$ kJ mol$^{-1}$ (recalculated with the use of
auxiliary thermodynamic data from the IVTANTERMO data bank).

In Murad's work (3) the equilibrium (1) was studied by Knudsen
effusion mass spectrometry in the temperature range of 1723–1808 K.
Recalculation of Murad's results with the use of data from the IVTAN-
TERMO data bank gives the value $D_0(Fe–OH) = 343 \pm 17$ kJ mol$^{-1}$. For
this recalculation we have used atomic ionization cross sections from the
work of Mann (4). Molecular ionization cross sections were estimated
according to the additivity rule.

As regards the Cr–O–H system, the literature provides data only on
the thermochemistry of the molecules of chromium oxide–hydroxides
$CrO_2OH$ and $CrO_2(OH)_2$ in flames (5,6) and upon heating of chromium
oxides $Cr_2O_3$ and $CrO_3$ in steam and humid oxygen (7–9).

## EXPERIMENTAL

A magnetic mass-spectrometer MS 1301 (300 mm, 90°) was used in
the investigation. The molecular beam source was a molybdenum
Knudsen effusion cell. The inside diameter of the cell was 10 mm, the
height was 6.5 mm. A cylindrical molybdenum liner (inside diameter =
7 mm, height = 5.5 mm, wall thickness = 1 mm) was placed in the cell.
Hydrogen or deuterium entered the cell through a capillary. The molecular beam issued from the cell through an orifice, 1 mm in diameter. The
cell was heated by electron bombardment. Temperature was measured
with an optical pyrometer. The ion currents were measured by a secondary electron multiplier.

In the studies on the thermochemistry of CrOH molecules, the
samples used were chromium oxide of the natural isotopic composition
(sample No. 1) and chromium oxide with increased content of the heavy
isotope of chromium $^{54}Cr$ (94%, sample No. 2). Deuterium was admitted
to the cell with the first sample and hydrogen was introduced into the

cell with the second sample. When the thermochemistry of FeOH molecules was studied the cell contained a mixture of iron and chromium oxides (sample No. 2).

## RESULTS

### CrOH (g)

Without gas admission, the main ions in the mass spectrum of the vapor at the temperature of 2000–2300 K were: $O^+$, $Cr^+$, $CrO^+$, $CrO_2^+$, $MoO^+$, $MoO_2^+$, and $MoO_3^+$. The ionization energies of CrO and $CrO_2$ measured by the extrapolated voltage difference method, were found to be equal to $7.9 \pm 0.5$ and $9.8 \pm 0.5$ eV (the standard $Cr^+$). The ratios of the gain coefficients of the secondary electron multiplier for the ions $Cr^+$, $CrO^+$, $CrO_2^+$, and $H^+$ were found to be equal to 1.0:1.0:1.0:0.9, respectively. The ratio of the gain coefficients for $H^+$ and $D^+$ ions was taken to be equal to $\sqrt{2}$. A typical ratio of ion currents $Cr^+:CrO^+:CrO_2^+:O^+$ was 1000:100:20:3. The pressures of the vapor components were determined by the formula

$$p = kIT/\sigma$$

where $I$ is the ion current, $T$ is temperature, $k$ is the sensitivity constant of the instrument, and $\sigma$ is the ionization cross section of the atom or molecule. The ion currents were measured using ionization energies 3 eV greater than the ionization energies of the atoms or molecules. The atomic ionization cross sections were taken from Mann's work (4). The ionization cross sections of molecules were calculated according to the additivity rule. The treatment of experimental results is described in more detail in (1).

The sensitivity of the instrument was calibrated against the known equilibrium constant of the reaction

$$CrO = Cr + O \tag{2}$$

with the use of data from the reference book (10).

When hydrogen was admitted to the cell with sample No. 2, CrOH molecules were found in addition to the main components. The ionization energy of CrOH determined using the extrapolated voltage difference method was found to be $7.8 \pm 0.3$ eV using $Cr^+$ as a standard. The measured equilibrium constants for reactions in the gaseous phase

$$CrOH + O = CrO_2 + H \tag{3}$$

$$CrOH + Cr + O = 2CrO + H \tag{4}$$

and the enthalpies of the reactions calculated from these constants using the third law of thermodynamics are collected in Table 1.

The table also gives the atomic hydrogen pressures. When p(H) changes by a factor of 30 (T = 2113 K) the equilibrium constant of the

Table 1
Equilibrium Constants and Enthalpies of Reactions
$CrOH + O = CrO_2 + H$ (3),
and $CrOH + Cr + O = 2CrO + H$ (4)

| T,K | LgKp(3) | LgKp(4) | p(H), atm[b] |
|---|---|---|---|
| 1986 | 3.19 | — | $1,21.10^{-5}$ |
| 2113 | 3.18 | 3.52 | $2,97.10^{-5}$ |
| 2113 | 3.11 | 3.49 | $4,72.10^{-5}$ |
| 2113 | 3.11 | 3.50 | $1,33.10^{-5}$ |
| 2113 | 3.06 | 3.45 | $1,58.10^{-6}$ |
| 2171 | 2.81 | 3.14 | $1,32.10^{-6}$ |
| 2171 | 2.97 | 3.34 | $1,99.10^{-5}$ |
| 2218 | 2.98 | 3.31 | $1,58.10^{-5}$ |

[a]$\Delta_r H =$ $-204.85 \pm 15$  $-115.59 \pm 15$  hj.mol$^{-1}$
[b]latm $= 101325Pa$

reactions are the same within the experimental error. This shows that the vapor in the cell has attained equilibrium. The values of $\Delta_f H°(CrOH, g, 0)$ estimated from the enthalpies of reactions 3 and 4 are equal to $68.1 \pm 19$ and $59.4 + 15$ kJ mol$^{-1}$, respectively. The weighted mean of these values is $62.0 \pm 14$ kJ mol$^{-1}$. Similar measurements were carried out with deuterium admitted to the effusion cell (*see below*).

It should be noted that reactions (3) and (4) are isomolecular and absolute pressures of vapor components are not needed for these calculations. Thus, the error in the determination of the sensitivity constant of the instrument is not introduced in the calculations. Here and below, the thermodynamic functions of atoms and molecules and auxiliary thermochemical data used in calculations are taken from the reference book (*10*) and the IVTANTERMO data bank.

When deuterium was admitted to the cell with sample No. 1, in addition to the main components, the vapor was found to contain atomic and molecular deuterium and CrOD molecules. The ion currents $^{53}CrOD^+$ and $^{54}CrOD^+$ with mass numbers 71 and 72 were measured. The ionization energy of CrOD molecules determined by linear extrapolation (Cr$^+$ as standard) is equal to $7.3 \pm 0.5$ eV.

The equilibrium constants of reactions (5) and (6), which are similar to reactions (3) and (4)

$$CrOD + O = CrO_2 + D \qquad (5)$$

$$CrOD + Cr + O = 2CrO + D \qquad (6)$$

and the enthalpies of these reactions calculated from the latter with the use of the third law of thermodynamics are given in Table 2.

The enthalpies of formation of CrOD $\Delta_f H°(CrOD, g, 0)$ calculated from the enthalpies of reactions (5) and (6), are equal to $57.1 \pm 20$ and $47.4 \pm 17$ kJ mol$^{-1}$, respectively. The weighted mean of these values is $51.6 \pm 14$ kJ mol$^{-1}$. The errors do not include the error in the thermo-

Table 2
Equilibrium Constants and Enthalpies of Reactions
CrOD + O = CrO$_2$ + D (5),
and CrOD + Cr + O = 2CrO + D (6)

| T,K | LgKp(5) | LgKp(6) | p(D), atm |
|------|---------|---------|-----------|
| 2169 | 2.93 | 2.53 | $3,02.10^{-5}$ |
| 2263 | 3.01 | 2.67 | $2,29.10^{-5}$ |
| 2323 | 2.78 | 2.46 | $2,67.10^{-6}$ |
| 2323 | 2.87 | 2.52 | $1,27.10^{-5}$ |
| $^a\Delta H_r(o) =$ | $-190.08 \pm 18$ | $-99.80 \pm 16$ | |

dynamic functions of CrOD, which is taken into account, however, in the final error value. The value of $\Delta_f H°(CrOH, g, 0) - \Delta_f H°(CrOD, g, 0)$ was calculated as the difference of the energies of vibrational levels v = 0 of molecules and was found to be 6.8 kJ mol$^{-1}$, which leads to a value of $\Delta_f H°(CrOH, g, 0) = 58.4 \pm 15$ kJ mol$^{-1}$. The total error in the values of the enthalpies of formation was determined according to the method described in (1). The total error incorporates the reproducibility error, the error in the thermodynamic functions and the equilibrium constants (coefficient 1.5).

The final value of $\Delta_f H°(CrOH, g, 0)$ was taken to be $61.4 \pm 14$ kJ mol$^{-1}$, which was calculated as the weighted mean. It corresponds to the enthalpy of formation of CrOD molecules, $\Delta_f H°(CrOD, g, 0) = 54.6 \pm 14$ kJ mol$^{-1}$. The enthalpies of formation of CrOH and CrOD molecules correspond to D$_0$(Cr–OH) = $372.7 \pm 14$ and D$_0$(Cr–OD) = $377.2 \pm 14$ kJ mol$^{-1}$.

## FeOH (g)

In an experiment without admittance of hydrogen into the cell with the mixture of Cr$_2$O$_3$ and Fe$_2$O$_3$ (sample No. 2) at temperatures of 1700–1900 K, the mass spectrum contained the following ions formed by ionization of the molecular beam: Cr$^+$, Fe$^+$, CrO$^+$, CrO$_2$$^+$, and FeO$^+$ and also unidentified ions with the mass numbers 71 and 73. The ionization energy of FeO was determined using the extrapolated voltage difference method using Cr$^+$ and Fe$^+$ as standards: IE (FeO) = $8.6 \pm 0.3$ eV. The ion currents were measured for the ions containing the isotopes $^{54}$Cr and $^{56}$Fe. A typical ratio of the ion currents of Fe$^+$, Cr$^+$, FeO$^+$, and CrO$^+$ at 45 eV was 380:91:1:13 (T = 1923 K).

When hydrogen was admitted to the cell, chromium and iron monohydroxides were also detected in the vapor. The ion currents $^{54}$CrOH$^+$ and $^{56}$FeOH$^+$ were taken for quantitative measurements of the corresponding partial pressures; $^{56}$FeOH$^+$ was determined as the difference of the ion current with the mass number 73 after and before admittance of hydrogen into the cell. The fraction of the ion current $^{57}$FeO$^+$ in the total

Table 3
Equilibrium Constants
and Enthalpies of Reaction
CrOH + Fe = FeOH + Cr (7)

| T, K | | LgKp(7) | $\Delta_rH(0)$ |
|---|---|---|---|
| 1[a] | 1853 | −0.78 | 19.2 |
| | 1853 | −0.81 | 20.1 |
| | 1853 | −0.90 | 23.2 |
| | 1858 | −1.04 | 28.2 |
| | 1768 | −1.07 | 27.8 |
| | 1768 | −0.88 | 21.4 |
| | 1873 | −1.19 | 33.7 |
| 2[b] | 1858 | −0.89 | 22.9 |
| | 1858 | −1.05 | 28.6 |
| | 1908 | −1.19 | 34.4 |
| | 1923 | −1.05 | 29.6 |
| | 1923 | −0.89 | 23.7 |
| | 1923 | −1.19 | 34.6 |
| | 1923 | −1.10 | 31.4 |
| | | | Av. 26.8 ± 12 |

[a]Measured at electron energies 3 eV above ionization thresholds.
[b]Measured at electron energy 45 eV.

current of ions with the mass number 73 was 10–15% of the average. The fraction of the current of the unidentified ion with the mass number 73 (*see above*) was 10–15%. Thus, the ion current $^{56}FeOH^+$ was determined with a considerable degree of confidence. The values of the ion currents of ions with the mass number 71 included the ion currents, $^{54}CrOH^+$ ($\approx$80%), $^{54}FeOH^+$ ($\approx$10%), and the current of the unidentified ion with the mass number 71 ($\approx$10%). The ionization energy of FeOH was found by the extrapolated voltage difference method ($Fe^+$ and $Cr^+$ used as standards): IE (FeOH) = 7.6 ± 0.3 eV.

For the determination of FeOH stability the gas-phase reaction 7 was studied

$$CrOH + Fe = FeOH + Cr \qquad (7)$$

The equilibrium constants and the third-law enthalpy of reaction values are presented in Table 3.

In the first part of the table are given the results of measurements at 3 eV higher than the corresponding appearance energies of ions; the second part gives the data obtained at ionizing energies of 45 eV. From the enthalpy of reaction (7) the enthalpy of formation of FeOH(g) was calculated with the use of $\Delta_fH(CrOH, g, 0)$ determined in the present work: $\Delta_fH$ (FeOH, g, 0) = 106.3 ± 17 kJ mol$^{-1}$. This value corresponds to the bond energy $D_0(Fe-OH)$ = 345.8 ± 17 kJ mol$^{-1}$, in very good agreement with the data of Murad (3).

Table 4
Comparison of $D_o^*$(M–F) and $D_o^*$(M–OH) values, in kJ mol$^{-1}$

| M | $D_o$(M–F)[a] | $D_o$(M–OH)[a] | $D_o^*$(M–F)[c] | $D_o^*$(M–OH)[c] | Δ |
|---|---|---|---|---|---|
| Li | 578 | 440 | 250 | 264 | 14 |
| Na | 479 | 329 | 151 | 153 | 2 |
| K | 494 | 353 | 166 | 177 | 11 |
| Rb | 491 | 358 | 163 | 182 | 19 |
| Cs | 517 | 377 | 189 | 201 | 12 |
| Be | 575 | 459 | 247 | 283 | 36 |
| Mg | 455 | 330 | 127 | 154 | 27 |
| Ca | 530 | 398 | 202 | 222 | 20 |
| Sr | 540 | 390 | 212 | 214 | 2 |
| Ba | 580 | 464 | 252 | 288 | 36 |
| Al | 670 | 550[b] | 342 | 374 | 32 |
| Ga | 580 | 435 | 252 | 259 | 7 |
| In | 510 | 380 | 182 | 204 | 22 |
| Tl | 443 | 300 | 115 | 124 | 9 |
|  |  |  |  |  | Av. 18 ± 10 |

[a]Data from (10), except AlOH.
[b]Data from (1).
[c]EA(F) = 328 kJ mol$^{-1}$, EA(OH) = 176 kJ mol$^{-1}$ (10).

## DISCUSSION

Since hydroxides may be considered as pseudohalides it seems natural to compare the data on the stability of the hydroxides and halides. Hastie (11) proposed the following formula

$$D_o(M–OH) = 0.5 \ D_o(M–Cl) + 0.4 \ D_o(M–F) \qquad (8)$$

Later, Krikorian (12) used for the estimation of the stability of hydroxide molecules the relations

$$D_o(M–OH) = 0.83 \ D_o(M–F) \qquad (9a)$$

$$D_o(M–OH) = 1.05 \ D_o(M–Cl) \qquad (9b)$$

Both of these approaches are completely empirical. It would be useful to compare the data on the stability of gaseous monohydroxides and monohalides taking into consideration the information on their structure. First, it should be taken into account that monohydroxides are usually assumed to have the linear structure M–O–H, the distance M–O being close to M–F in monofluoride molecules. Therefore, in spite of a marked difference between $D_o$(M–OH) and $D_o$(M–F) values, one can try to compare these quantities rather than $D_o$(M–OH) and $D_o$(M–Cl). Further, it should be taken into account that in the molecules under consideration the ionic component of the bond energy can play an important role. In

Table 5
Experimental and Calculated $D_0$(M–OH) values (kJ mol$^{-1}$)

| MOH | $D_0$(M–OH) exptl | $D_0$(M–OH) calc. | $\Delta$ |
|---|---|---|---|
| LiOH | 440 | 444 | + 4 |
| NaOH | 329 | 345 | +16 |
| KOH | 353 | 360 | + 7 |
| RbOH | 358 | 357 | − 1 |
| CsOH | 377 | 383 | + 6 |
| BeOH | 459 | 441 | − 18 |
| MgOH | 330 | 321 | − 9 |
| CaOH | 398 | 396 | − 2 |
| SrOH | 390 | 406 | +16 |
| BaOH | 464 | 446 | − 18 |
| AlOH | 550 | 536 | − 14 |
| GaOH | 435 | 446 | +11 |
| InOH | 380 | 376 | − 4 |
| TlOH | 300 | 309 | + 9 |
| CuOH | 255 | 293 | +38 |
| CrOH | 373 | 366 | − 7 |
| MnOH | 305 | 310 | + 5 |
| FeOH | 346 | 343 | − 3 |

that case, one could compare the values of $D_0(M^+-X^-)$ rather than those of $D_0$(M–X), where

$$D_0(M^+-X^-) = D_0(M-X) + IE(M) - EA(X)$$

Since the ionization energy IE(M) enters in the same way into the calculations of $D_0(M^+-X^-)$ for halides and hydroxides, one may compare the values of

$$D_0^*(M-X) = D_0(M-X) - EA(X)$$

The results of the calculation of the values $D_0^*$(M–X) are given in Table 4. The table includes the data for monofluorides and monohydroxides of alkaline and alkaline–earth elements and also for the III group elements (Al, Ga, In, Tl).

It is clear from Table 4 that the values of $D_0^*$(M–F) and $D_0^*$(M–OH) are very close. Averaging gives

$$D_0^*(M-OH) = D_0^*(M-F) + 18 \text{ kJ mol}^{-1}$$

or

$$D_0(M-OH) = D_0(M-F) - 134 \text{ kJ mol}^{-1} \qquad (10)$$

It would be interesting to find out whether this relation can be used for estimation of transition metal monohydroxide stabilities. In Table 5 the experimental and calculated values of $D_0$(M–OH) for monohydroxides used in the derivation of relation (10), for CrOH and FeOH molecules investigated in the present work, and also for MnOH (*13*) and

CuOH (*14*) are compared. In all the cases, the experimental and calculated values show good agreement. It should be noted that for the CuOH molecule, relations (8) and (9) give a much larger value of $D_0(Cu-OH)$ than that obtained by experiment. Evidently, it is necessary to continue to accumulate experimental data on the stability of hydroxides of different groups and subgroups of the periodic system of elements and compare them with the stability of related compounds taking into account the structural data.

## REFERENCES

1. Milushin, M. I., Emelyanov, A. M., and Gorokhov, L. N., *Teplofiz. Vys. Temp.* **24**, 468 (1986).
2. Jensen, D. E. and Jones, G. A., *J. Chem. Soc. Faraday Trans.* Part 1, **8**, 1448 (1973).
3. Murad, E., *J. Chem. Phys.* **73**, 1381 (1980).
4. Mann, J. B., *Recent Developments in Mass Spectroscopy*, Ogata K. and Hayakawa T., eds., Tokyo University Press, 1970, p. 814.
5. Farber, M. and Srivastava, R. D., *Combust. and Flame* **20**, 43 (1973).
6. Bulewitz, E. M. and Padley, P. J., *Proc. Roy. Soc.* **A323**, 377 (1971).
7. Graham, H. C. and Davis, H. H., *J. Amer. Ceram. Soc.* **54**, 89 (1971).
8. Kim, Y.-W. and Belton, G. R., *Met. Trans.* **5**, 1811 (1974).
9. Glemser, O. von, and Mueller, A., *Z. Anorg. und Allg. Chem.* **334**, 150 (1964).
10. *Thermodynamic Properties of Individual Substances*, Glushko, V. P., ed. vols. 1–4, Nauka, Moscow, 1978–1982.
11. Hastie, J. W., *High Temperature Vapors*, Academic Press, NY, 1975.
12. Krikorian, O. H., *High Temp.–High Pressures* **14**, 387 (1982).
13. Padley, P. J. and Sugden, T. M., *Trans. Faraday Soc.* **55**, 2054 (1952).
14. Belyayev, V. N., Lebedeva, N. L. Krasnov, K. S., and Gurvich, L. V., *Izvestiya Vysshikx Uchebnykh Zavedenii. Khimiya i Khimicheskaya Tekhnologiya* **21**, 1698 (1978).

# High Temperature Mass Spectrometric Studies of the Thermodynamic Properties of Glass-Forming Systems

VALENTINA STOLYAROVA

*Institute of Silicate Chemistry, USSR Academy of Sciences, ul. Odoevskogo, 24, korp. 2, 199155, Leningrad, USSR*

## ABSTRACT

High temperature mass spectrometry was used to study the evaporation of oxide melts and glasses.

Vapor compositions, partial pressures, and chemical potentials of components, Gibbs free energies of formation, the partial and integral enthalpies of the formation of glasses, and melts in the $Na_2O-SiO_2$, $Na_2O-GeO_2$, $Na_2O-B_2O_3$, $Na_2O-B_2O_3-GeO_2$, $B_2O_3-SiO_2$, $GeO_2-SiO_2$ and $B_2O_3-GeO_2-SiO_2$ systems at 1200–1590 K have been determined.

It was shown that the thermodynamic properties of these systems within the whole concentration range can be described in terms of the generalized lattice theory of associated solutions. It was possible to predict the thermodynamic properties of melts for three glass-forming oxides based on the thermodynamic properties of the corresponding binary systems or by using data on the structure of individual oxide melts.

**Index Entries:** Mass spectrometric study; $Na_2O-B_2O_3$, $Na_2O-GeO_2$, $Na_2O-SiO_2$, $Na_2O-B_2O_3-GeO_2$, $B_2O_3-GeO_2$, $B_2O_3-SiO_2$, $B_2O_3-GeO_2-SiO_2$ systems; thermodynamic properties; theory of associated solutions.

## INTRODUCTION

Since the earlier work of Piacente and Matousek (1), which deals with the investigation of $Na_2O \cdot 2SiO_2$ evaporation with the Knudsen effusion method, high temperature mass spectrometry has been widely used to study the thermodynamic properties of glass-forming systems (2–9).

High Temperature Science, Vol. 26       © 1990 by the Humana Press Inc.

On the one hand, this is owing to the existing need for modern technology in high temperature oxide materials as well as to the application of glass-forming systems in various fields of technology such as metallurgy, nuclear waste disposal, and so on. On the other hand, the capability of the method in determining vapor compositions and thermodynamic properties of binary and multicomponent systems are more extensive than those of calorimetry and EMF measurements owing to the ability to measure the dimer-monomer equilibria and to apply the Belton-Fruehan or Neckel-Wagner methods (*10*).

The fields of application of glasses are often narrow owing to the selective evaporation of some components. This has to be taken into account in various applications of glass, such as solving problems of protection from gas corrosion, the use of glasses together with metals and ceramics in the electronics industry, the production of Pyrocerams and coatings on the basis of glasses, as well as in the use of glasses for containment of radioactive waste.

Besides traditional silicate glasses used for practical purposes of instrument-making and those used in the laser, medical, electronic, and communication industries, one searches for new applications of special glasses based on glass-forming oxides of boron, germanium, and so on. To choose among various glasses and their melts with regard to obtaining better characteristics and less volatility, one should pay greater attention to the prediction of their thermodynamic properties.

That is the goal of our investigations in studying thermodynamic properties of the $Na_2O-B_2O_3$ (*11*), $Na_2O-GeO_2$ (*12*), $Na_2O-SiO_2$ (*13*), $Na_2O-B_2O_3-GeO_2$ (*14–17*), $B_2O_3-GeO_2$ (*18–20*), $B_2O_3-SiO_2$ (*19–21*), $GeO_2-SiO_2$ (*22*), and $B_2O_3-GeO_2-SiO_2$ (*23–24*) systems. The present paper is a summary of the data obtained (*11–25*).

## METHODS

The investigations of vapor compositions and partial pressures of vapour components as well as the thermodynamic properties of glasses and glass-forming melts have been carried out using the traditional mass spectrometric Knudsen effusion method with the MS-1301 mass spectrometer at 1200–1590 K. The characteristics of the method are fully described in ref. *18*.

The evaporation has been carried out from platinum effusion cells with a 100:1 ratio of the cross-section area of the cell to the orifice area. The $Na_2O-SiO_2$ and $B_2O_3-SiO_2$ systems have been also studied using molybdenum cells with the above ratio of 400:1.

Partial pressures of components such as Na, $NaBO_2$, $(NaBO_2)_2$, $B_2O_3$, and GeO have been determined using the ion current comparison method. Individual oxides and silver (in the case of the $Na_2O-SiO_2$ system) have been taken as standards. The method has been significantly improved by the use of evaporation with two effusion cells: one for the investigated melt and the other for the standard.

For the $B_2O_3$–$SiO_2$ system, it has been found possible to determine partial pressures of $B_2O_3$ by the complete isothermal evaporation method (21) and to relate this value to the composition of the melt.

Several problems occurring during the investigation of glass-forming melts should be noted:

1. A very long time for equilibrium establishment for highly viscous melts during the increase and the following decrease of temperature, as has been observed in the $Al_2O_3$–$SiO_2$ system (4,5);
2. The creeping of germanate melts from platinum cells; the ways to prevent such phenomena are discussed in ref. 26; and
3. The appearance of diffusion barriers during the evaporation, for instance, in the $Na_2O$–$SiO_2$ and $GeO_2$–$SiO_2$ systems. Such a phenomenon has been observed by Hastie and Plante (27) in the $K_2O$–$Al_2O_3$–$SiO_2$ system owing to the appearance of glass on the surface of a sample as it loses potassium oxide.

High viscosity of the melts can be one of the causes of low evaporation coefficients of $SiO_2$ ($2 \cdot 10^{-2}$) (28) and $GeO_2$ ($6 \cdot 10^{-2}$) (29) and the decrease in the coefficient of $GeO_2$ from 0.06 for the individual oxide to 0.03 for the $B_2O_2$–$GeO_2$ system (29).

Enthalpies of evaporation in the systems have been determined using the Van't Hoff equation. The variety of vapor compositions in the systems studied enabled us to use various methods of data treatment to calculate the thermodynamic functions. These methods are the comparison method, the dimer-monomer method, the Belton-Fruehan and the Neckel-Wagner methods (10). The correlation of these methods has been demonstrated for the $Na_2O$–$B_2O_3$ and $Na_2O$–$GeO_2$ systems. Deviations do not exceed twice the experimental error (11,12,16).

Thermodynamic data obtained show negative deviations from ideality of the $Na_2O$–$GeO_2$ (Fig. 1), $Na_2O$–$B_2O_3$, $Na_2O$–$SiO_2$ (Fig. 2, and $Na_2O$–$B_2O_3$–$GeO_2$ systems; small negative deviations in the $B_2O_3$–$GeO_2$ system (Fig. 3a), small positive deviations in the $B_2O_3$–$SiO_2$ system, (Fig. 3b), a nearly ideal behavior of the $GeO_2$–$SiO_2$ system (Fig. 3c), and the variation of the sign from nonideality of the $B_2O_3$–$GeO_2$–$SiO_2$ system in various concentration ranges (Figs. 4–6).

## DISCUSSION

### Reliability of Experimental Data

The validity of the activities of components of glass-forming systems by the mass spectrometric data compared with the EMF data has been shown for the PbO–$SiO_2$ system using the dimer-monomer method (30) and for the $Na_2SiO_3$–$K_2SiO_3$ system using the Belton-Fruehan method (31).

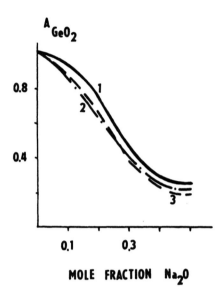

Fig. 1.   GeO$_2$ activities at 1543 K in the Na$_2$O–GeO$_2$ system (*12*). Results are based on (1) mass spectrometry data; (2) the Gibbs-Duhem equation; and (3) EMF data.

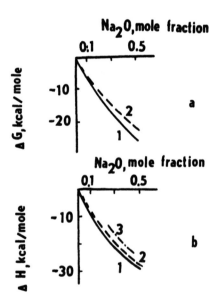

Fig. 2.   Thermodynamic functions for the Na$_2$O–SiO$_2$ system at 1473 K (*13*): Gibbs free energies (a) and integral enthalpies of formation (b) were determined by: (1) mass spectrometry; (2) EMF measurements; and (3) calorimetry, (1 cal = 4.1868 J).

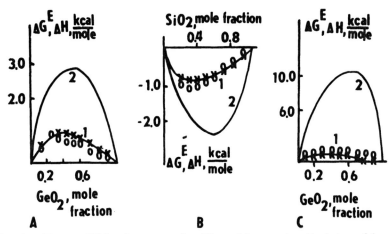

Fig. 3. Excess Gibbs free energies (1) and integral enthalpies of formation (2) for $B_2O_3$–$GeO_2$ at 1473 K (a), $B_2O_3$–$SiO_2$ at 1588 K (b), and $GeO_2$–$SiO_2$ at 1450 K (c) systems (solid line—experimental values; calculated values also given: x = the GLTAS with adjustable parameters; and o = the GLTAS without adjustable parameters).

We have observed (11–13,16) the correlation of the thermodynamic data obtained by mass spectrometry and the EMF method in the $Na_2O$–$GeO_2$ (Fig. 1), $Na_2O$–$B_2O_3$, and $Na_2O$–$SiO_2$ systems (Fig. 2). Enthalpies of mixing were analyzed for the first time. In the case of the $Na_2O$–$SiO_2$ system, a correlation has been also observed between these data and those obtained using high temperature solution calorimetry (32) and the ion-molecular equilibrium method (33).

On the basis of these results, one can conclude that in the case of the valid experiments there is little deviation in the thermodynamic functions obtained using the mass spectrometric and the EMF methods (i.e., the deviation is less than twice the experimental error) though, for instance, in ref. 34 such deviations were attributed to the difference in activities in the bulk and on the surface of samples.

However, our conclusion is supported by the correlation of the thermodynamic data, obtained by both mass spectrometry and the EMF method for the $MgO$–$SiO_2$ (35), $MnO$–$Al_2O_3$–$SiO_2$ (5), $FeO$–$SiO_2$, $FeO$–$CaO$–$SiO_2$ (4), and $PbO$–$B_2O_3$ (36) systems.

## Relative Volatility of Glass-Forming Melts

In principle, relative volatility of oxide melts can be evaluated using the data on evaporation of individual oxides, but the characteristics of the nature and structure of glass (37) may be the cause of errors in such an approach. The selective evaporation of glass components has been discussed as a method of study for the vitreous state (38).

The possibility of using the acid-base concept (39) to predict relative volatilities of glass-forming melts is considered (40). It is shown that relative volatilities in oxide melts can be quantitatively characterized by the partial pressure of oxygen and/or ionization potentials of elements

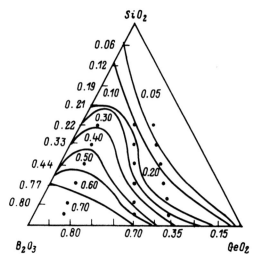

Fig. 4.   Constant $B_2O_3$ activity lines in the $B_2O_3$–$GeO_2$–$SiO_2$ system at 1453 K.

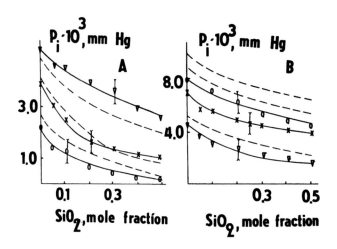

Fig. 5.   Experimental (solid line) and calculated (dotted line) values of partial pressures of $B_2O_3$ (A) and GeO (B) in the $B_2O_3$–$GeO_2$–$SiO_2$ system at 1453 K with $B_2O_3$:$GeO_2$ = 1:2 (o); $B_2O_3$:$GeO_2$ = 1:1 (x); and $B_2O_3$:$GeO_2$ = 1:0.17 ($\nabla$) (1 mm Hg = 133.3 Pa).

present in the systems. Such an approach accounts for the observed decrease of isothermal volatilities in the following sequence of systems: $Na_2O$–$SiO_2$ > $Na_2O$–$GeO_2$ > $B_2O_3$–$GeO_2$ ≥ $GeO_2$–$SiO_2$ > $B_2O_3$–$GeO_2$–$SiO_2$ > $Na_2O$–$B_2O_3$–$GeO_2$ > $Na_2O$–$B_2O_3$ > $B_2O_3$–$SiO_2$.

## Calculation of Thermodynamic Properties of Glass-Forming Systems

One should note two methods for prediction of thermodynamic properties: semiempirical methods of calculations for multicomponent

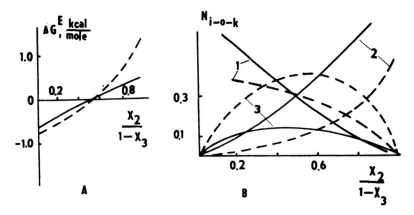

Fig. 6. Excess Gibbs free energies (A) and fractions (B) of B–O–Si (1), Ge–O–Si (2) and B–O–Ge (3) bonds in the $B_2O_3$–$GeO_2$–$SiO_2$ system at 1453 K: $x_{SiO_2} = 0.5$ (full line); $x_{SiO_2} = 0.2$ (dash line); $x_1 \simeq x_{B_2O_3}$; $x_2 \simeq x_{GeO_2}$; and $x_3 \simeq x_{SiO_2}$.

systems based on the binary system equilibrium data and statistical calculations using the information on individual oxide properties.

The first approach can be illustrated by the calculation of excess Gibbs free energies ($\Delta G^E$) in $Na_2O$–$B_2O_3$–$GeO_2$ melts using the Kohler method (15–17,41). The second type of calculation deals with the use of statistical models of associated solutions.

In the work of Hastie et al. (2,3,42) and Shultz and Shahmatkin (43), it has been convincingly shown that one can successfully use the theory of ideal mixing of complex phases (or the theory of ideal associated solutions) to calculate the thermodynamic properties of multicomponent silicate systems (42) as well as in the $Na_2O$–$B_2O_3$–$SiO_2$ system (43). But this theory cannot be applied to oxide systems formed by the network forming oxides owing to the formation of continuous network structures and the lack of information about chemical compounds. That is why Shultz et al. (19–24) proposed the use of the generalized lattice theory of associated solutions (GLTAS) (44,45) to calculate the thermodynamic properties of such systems. The theory is a generalization of the Guggenheim regular solution model for the systems with molecules of different size, in which the energy of interaction depends upon the orientation of molecules. The theory has been compared to those considering equilibria in solutions (45).

We have shown the applicability of the GLTAS to network forming oxide systems: $B_2O_3$–$GeO_2$, $B_2O_3$–$SiO_2$, $GeO_2$–$SiO_2$, and $B_2O_3$–$GeO_2$–$SiO_2$ (19–24,44,45). The potential energy of melts is the sum of the energies of interactions of contact site pairs, i.e., the bond energies, that differ owing to the type of atom in the second coordination sphere. Bond energies in the melt are the parameters of the theory. The calculation of thermodynamic properties has been made in two cases. The first (19,21) considers the energy parameters as adjustable. The second uses bond energies calculated by the Mulliken method (46). It should be noted that, in the latter case, only data on individual oxides have been used. The results of calculations of Gibbs excess free energies ($\Delta G^E$) in binary systems are given in Fig. 3. They correlate quite well with the experimental data. It is natural that, in the case where adjustable parameters were used, the correlation is much better.

Integral enthalpies of mixing have been calculated in the $B_2O_3$–$SiO_2$ and $B_2O_3$–$GeO_2$ systems using one more adjustable parameter describing the temperature dependence of bond energy (*19*). The thermodynamic functions of the ternary systems have been calculated in the same way. Partial pressures of $B_2O_3$ and GeO over the $B_2O_3$–$GeO_2$–$SiO_2$ melts at 1453 K have been calculated using excess chemical potentials of $B_2O_3$, $GeO_2$, $SiO_2$, and pressures of individual oxides (Fig. 5). One should note that, in this case, we try to predict partial pressures in the system with associated components based on the properties of individual oxides. This differs from the theory of ideal associated solutions dealing with the properties of compounds formed in a system. That is why we consider the correlation between experimental and calculated data on partial pressures of the same order of magnitude to be satisfactory.

## Correlation Between the Thermodynamic Properties and the Structure of Glass-Forming Systems

The concentration dependences of thermodynamic functions enable one to predict the phase structure of the systems, i.e., either the tendency to phase separate or the phase separation itself. The first case is represented, for instance, by the $B_2O_3$–$GeO_2$ system (*18*) and the second by the $Bi_2O_3$–$B_2O_3$ system (*47*).

The use of statistical approaches enables one to reveal the correlation between thermodynamic properties and the structure within the limitations of the model. Thus, one can obtain information upon the number of associates of a given type in a certain concentration range using the theory of ideal associated solutions (*3,27,42*). GLTAS enables one to calculate the ratio of bond numbers in the melt and reveal the correlation between the structure and properties in both binary (*20,22*) and ternary systems (*24*), as shown in Fig. 6.

The character of variation of Gibbs free energy with concentration correlates with the relative mixed-type bond number in the $B_2O_3$–$GeO_2$–$SiO_2$ system, (Fig. 6).

## SUMMARY

Vapor compositions, partial pressures of components, and the thermodynamic properties have been determined in binary and ternary glass-forming systems containing $Na_2O$, $B_2O_3$, $GeO_2$, and $SiO_2$ using high temperature mass spectrometry. The validity of the results is considered. The correlation between relative volatilities and acid-basic properties of the systems is discussed. Thermodynamic properties of the binary and ternary network-forming oxide systems are shown to be predictable on the basis of the generalized lattice theory of associated solutions, including the calculation of bond energies in the systems using the Mulliken method. The application of this theory to glass-forming systems allows one to consider, in detail, the system of interactions in a melt and the role of the size and form of structural fragments for which the ratio of mixed-type bond numbers may serve as a criterion.

## REFERENCES

1. Piacente, V. and Matousek, J., *Silikaty* **4**, 264 (1973).
2. Hastie, J. W., Horton, W. S., Plante, E. R., and Bonnell, D. W., *High Temp. Sci.* **14**, 669 (1982).
3. Hastie, J. W. and Bonnell, D. W., *High Temp. Sci.* **19**, 275 (1985).
4. Dhima, A., Stafa, B., and Allibert, M., *High Temp. Sci.* **21**, 143 (1986).
5. Stafa, B., *These présenteé é l'institut National polytechnique de Grenoble*, 1983.
6. Kambayashi, S. and Kato, E., *J. Chem. Thermodyn.* **15**, 701 (1983).
7. Ohara, N., Nunoue, S., and Kato, E., *J. Iron and Steel Inst. Japan* **73**, 1337 (1987).
8. Asano, M. and Yasue, J., *J. Nucl. Materials* **138**, 65 (1986).
9. Asano, M., Kou, T., and Yasue, J., *J. Non-Crystalline Solids* **92**, 245 (1987).
10. Raychaudhuri, P. K. and Stafford, F. E., *Mater. Sci. Eng.* **20**, 1 (1975).
11. Shultz, M. M., Stolyarova, V. L., and Semenov, G. A., *Fizika i Khimiya Stekla* **5**, 42 (1979).
12. Shultz, M. M., Stolyarova, V. L., Semenov, G. A., and Shahmatkin, B. A., *Fizika i Khimiya Stekla* **5**, 651 (1979).
13. Shultz, M. M., Stolyarova, V. L., and Ivanov, G. G., *Fizika i Khimiya Stekla.* **13**, 168 (1987).
14. Stolyarova, V. L. and Semenov, G. A., *Fizika i Khimiya Stekla* **5**, 127 (1979).
15. Shultz, M. M., Stolyarova, V. L., and Semenov, G. A., *Doklady Acad. Sci. SSSR* **246**, 154 (1979).
16. Shultz, M. M., Semenov, G. A., and Stolyarova, V. L., *Izvestiya Acad. Sci. SSSR, Neorganicheskiye Materialy.* **15**, 1002 (1979).
17. Shultz, M. M., Stolyarova, V. L., and Semenov, G. A., *J. Non-Crystalline Solids* **38/39**, 581 (1980).
18. Shultz, M. M., Stolyarova, V. L., and Semenov, G. A., *Fizika i Khimiya Stekla* **4**, 653 (1978).
19. Shultz, M. M., Ivanov, G. G., Stolyarova, V. L., and Shahmatkin, B. A., *Fizika i Khimiya Stekla* **12**, 285 (1986).
20. Shultz, M. M., Ivanov, G. G., and Stolyarova, V. L., *Doklady Acad. Sci. SSSR* **292**, 1198 (1987).
21. Shultz, M. M., Ivanov, G. G., Stolyarova, V. L., and Shahmatkin B. A., *Fizika i Khimiya Stekla* **12**, 385 (1986).
22. Shultz, M. M., Stolyarova, V. L., and Ivanov, G. G., *Fizika i Khimiya Stekla* **13**, 830 (1987).
23. Shultz, M. M., Ivanov, G. G., and Stolyarova, V. L., *XI Vsesoyuznaya Konferentia po Kalorimetrii i himicheskoi termodynamike*, Akademia Nauk SSSR, Novosibirsk, 1986, p. 130.
24. Shultz, M. M., Stolyarova, V. L., and Ivanov, G. G., *Vysokotemperaturnaya Khimiya Silikatov i Oksidov*, Nauka, Leningrad, 1988, p. 420.
25. Stolyarova, V. L. and Shultz, M. M., *Bazi fiziko-khimicheskih i technologicheskih dannih dlya optimizacii metallurgicheskih technologii*, Akad. Nauk SSSR, Dnepropetrovsk, 1988, p. 215.

26. Chatillon, C., Allibert, M., and Pattoret, A., *10th Materials Research Symposium on Characterization of High Temperature Vapors and Gases*, NBS Special Publication 561, vol. 1, 1979, p. 181.
27. Hastie, J. W., and Plante, E. R., NBS IR 81-2293, June 1981.
28. Nagai, S., Niwa, K., Shinmei, M., and Yokokawa, T., *J. Chem. Soc. Faraday Trans.* (Part I) **69**, 1628 (1973).
29. Stolyarova, V. L., Ambrock, A. G., Nikolaev, E. N., and Semenov, G. A., *Fizika i Khimiya Stekla* **3**, 635 (1977).
30. Kato, E. and Ohuchi, J. Chemical Metallurgy of Iron and Steel, *Proceedings of the International Symposium Metallurgical Chemistry-Application of ferrous metallurgy*, London, 1973, p. 26.
31. Belton, G. R., Choudary, U. V., and Gaskell, D. R., Physical Chemistry of Process Metallurgy, *Richardson Conference*, London, 1974, p. 247.
32. Ushakov, V. M., Borisova, N. V., Starodubzev, A. M., and Shultz, M. M., *Problemi Kalorimetrii i himicheskoi thermodynamiki*, vol. 2, Akademiya Nauk SSSR, Chernogolovka, 1984, p. 411.
33. Vovk, O. M., Rudnyi, E. D., Sidorov, L. N., Stolyarova, V. L., Rakhimov, V. I., and Shahmatkin, B, A., *Fizika i Khimiya Stekla* **14**, 218 (1988).
34. Okajima, K. and Sakao, H., *Trans. Jap. Inst. Met.* **14**, 75 (1973).
35. Kambayashi, S. and Kato, E., *J. Chem. Thermodyn.* **15**, 701 (1983).
36. Semenihin, V. I., Sorokin, I. D., Yurkov, L. F., and Sidorov, L. N., *Fizika i Khimiya Stekla* **13**, 542 (1987).
37. Mazurin, O. V., Porai-Koshitz, E. A., and Shultz, M. M., *Steklo: priroda i stroenie*, Znanie, Leningrad, 1985.
38. Kolykov, G. A., *Stroenie stekla*, Akademia Nauk SSSR, Leningrad, 1955, p. 234.
39. Shultz, M. M., *Fizika i Khimiya Stekla* **10**, 129 (1984).
40. Stolyarova, V. L., Rakhimiv, V. I., and Shultz, M. M., *10th IUPAC Conference on Chemical Thermodynamics*, Prague, Czechoslovakia, 1988, A 3.
41. Kohler, F., *Monatsh, Chem.* **91**, 738 (1960).
42. Hastie, J. W., Bonnell, D. W., and Plante, E. R., *Proceedings 23rd SEAM*, Somerset, PA, 1985, p. 701.
43. Shahmatkin, B. A., Kozhina, E. L., Shultz, M. M., *VIII Vsesouznoe sovechanie po sterkloobrasnomu sostoyaniyu*, Nauka, Leningrad, 1986, p. 419.
44. Barker, J. A., *J. Chem. Phys.* **20**, 1528 (1952).
45. Smirnova, N. A., *Molecular Theories of Solutions*, Chimia, Leningrad, 1987 (in Russian).
46. Mulliken, R. S., *J. Phys. Chem.* **56**, 295 (1952).
47. Minaeva, I. I., Karasev, N. M., Yurkov, L. F., Sharov, S. N., and Sidorov, L. N., *Fizika i Khimiya Stekla* **7**, 223 (1981).

# Comparison of Thermodynamic Data Obtained by Knudsen Vaporization Magnetic and Quadrupole Mass Spectrometric Techniques

GARY A. MURRAY,[1] ROBERT J. KEMATICK,[1]
CLIFFORD E. MYERS,*,[1] AND MARGARET A. FRISCH*,[2]

[1]Department of Chemistry, State University
of New York at Binghamton, Binghamton, NY 13901;
and [2]Thomas J. Watson Research Center,
IBM Corporation, Yorktown Heights, NY 10598

## ABSTRACT

The vaporization behavior of chromium silicides has been investigated by high temperature Knudsen effusion mass spectrometry in two laboratories, one employing a high resolution magnetic sector instrument, and the other a quadrupole instrument for mass analysis. The instrumentation differed also in cell design, furnace construction, temperature measurement and calibration, and experimental procedures. Chromium partial vapor pressures were based on calibrations with the pure element. The agreement of results from the two laboratories was found to be very good.

**Index Entries:** Mass spectrometry; Knudsen effusion; vapor pressure; chromium silicides.

## INTRODUCTION

Transition metal silicides have received attention owing to their refractory behavior, corrosion resistance, and electrical properties. The silicides of chromium have found application in evaporative resistor technology (1). In an attempt to achieve a better understanding of these systems and their manufacture, a collaborative research program was

---

*Authors to whom all correspondence and reprint orders should be addressed.

initiated between the Thomas J. Watson Research Center of IBM Corporation, in cooperation with the IBM facility at Endicott, NY, and the Department of Chemistry of the State University of New York at Binghamton. This research has involved a study of the vaporization behavior, phase equilibria, and thermodynamic stabilities of the chromium silicides. The existence in the two laboratories of different Knudsen effusion mass spectrometry systems provided an opportunity to compare results of experiments on samples covering a wide range of compositions in the chromium–silicon system. The instrumentation differed, not only in the manner of mass analysis, but also in effusion cell design, furnace construction, temperature measurement and calibration, and experimental procedures. This paper presents and compares chromium vapor pressure data from the two laboratories. A preliminary account of results on phase equilibria and thermodynamic stabilities, primarily from the SUNY-Binghamton laboratory, has been published previously (2). A complete treatment of thermodynamic activities, phase equilibria, and free energies and enthalpies of phase formation will be forthcoming.

A review of the early work on the system has been published by Chart (3) whose phase diagram is presented in Fig. 1. Intermediate phases that have been reported (4) for the system are $Cr_3Si$ (which exhibits a significant range of homogeneity), $Cr_5Si_3$, $CrSi$, and $CrSi_2$ (which also may exhibit a homogeneity range). Previous studies of thermodynamic stabilities include combustion calorimetry (5), EMF measurements that employed a fused salt electrolyte (6), effusion studies on both solid (7) and liquid (8) samples, and, more recently, an effusion study of the reaction of respective silicides with $SiO_2(s)$ to give $SiO(g)$ (3). None of the effusion studies employed mass spectrometry. In addition, the several published results do not show good agreement.

## EXPERIMENTAL

As described previously (2), samples were prepared by arc melting on a water cooled copper hearth under argon and were characterized by X-ray powder diffraction and chemical analysis. Samples used for effusion studies were in the form of chunks with typical dimensions of a few millimeters; sample weights were generally 50–80 mg. A description of the two sets of apparatus for Knudsen cell mass spectrometry, together with experimental procedures, is given in the following sections. A comparison of the significant differences between the two is given in Table 1. Both instruments employed electron impact ionization, and it was necessary to use a low electron energy in order to avoid interference of CO and $N_2$ with Si at mass 28. Inasmuch as the first ionization potential of Si is about 8.1 eV, whereas those of CO and $N_2$ are, respectively, about 14.0 and 15.6 eV, an electron energy of 16 to 12 eV, depending on the background pressure in the source, was found to be sufficiently low to suppress the signals from these impurities.

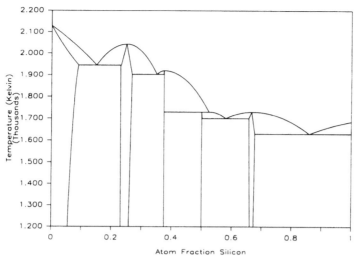

Fig. 1. Chromium-silicon phase diagram [after Chart (3)].

## SUNY-Binghamton

### High Temperature Magnetic Sector Mass Spectrometry (HT-MMS)

The vaporization studies in the SUNY-Binghamton laboratory made use of a 90° sector, single focusing, high resolution mass spectrometer, manufactured by Nuclide Corporation (Model 12-90-HT), equipped with a computer controlled data acquisition system. This apparatus has been described previously (9). Ion current signals were obtained by means of a Galileo channeltron multiplier operated in the current output mode. Parameters under the control of an IBM-PC computer included

1. Mass position;
2. Ion current signal;
3. Temperature setpoint;
4. Thermocouple readout; and
5. Pyrometer readout.

The instrument was fitted with a high temperature Knudsen cell furnace assembly that consisted of a resistively heated tungsten element contained within a set of five concentric cylindrical tungsten and tantalum radiation shields. The effusion cell was positioned within the furnace so that the orifice in the center of the lid was aligned with the entrance to the ion source. A shutter could be interposed to prevent the molecular beam from entering the source. The mass spectrometer configuration was such that the molecular beam from the effusion cell, the path of the ionizing electrons, and the path of the resultant ions were mutually perpendicular. The furnace, source, and analyzer were pumped separately; typical vacuums were $1 \times 10^{-6}$, $1 \times 10^{-6}$, and $3 \times 10^{-8}$ Torr, respectively. The analyzer section of the apparatus was kept under vacuum during change of samples. A scale drawing of the effusion cell is

Table 1
System Comparison[a]

|  | Magnetic Sector | Quadrupole |
|---|---|---|
| Orifice location | Center of lid | Upper cylindrical wall |
| Temperature measurement and control | Thermocouple (95%W-5%Re vs 74%W-26%Re) calibrated *in situ* with optical pyrometer | Thermocouple (94%W-5%Re vs 74%W-26%Re) |
| Heating program | Cooling stepwise 10° drop with 5 min. delay before measurement | Heating 2°/min., 30 min. anneal at highest T, cooling at 2°/min. |
| Signal Measurement | Ion current, multiplier, with Cr and Si signals measured in sequential cooling ramps | Pulse counting, with Cr and Si signals measured in same ramp: rapid peak shifting |
| Background correction | High resolution, baseline correction sufficient | Ultra high vacuum, modulated molecular beam |
| Ionizing electron energies | Cr: 30.0 eV Si: 12.5 eV | Cr: 16.0 eV Si: 16.0 eV |

[a]Signal calibration: Known vapor pressures of the pure elements Cr and Si.

given in Fig. 2. The cell body and the sample cup were both fabricated from tungsten metal and were obtained from Rembar, Inc. The temperature of the cell was measured by a thermocouple (95%W,5%Re vs 74%W,26%Re) mounted in a well located in the center of the base of the effusion cell. The thermocouple was calibrated *in situ* by means of an optical pyrometer sighted through a right-angle prism and shuttered window into the effusion cell orifice. The pyrometer has a certificate of calibration traceable to the US National Bureau of Standards. The pyrometer calibration was checked against the freezing temperature of gold by observation of an arrest in the ion current signal on cooling through the gold point temperature. Observed arrest temperatures were within ±5° of the gold point. Silicon signals were obtained with an ionizing electron energy of 12.5 eV in order to minimize the effect of background at mass 28 (9). Chromium signals were obtained with ionizing energies of 30.0 eV. In a typical run, the sample in a fresh sample cup was placed in the effusion cell and, after obtaining vacuum, was heated to the upper limit of the temperature range. The sample was maintained at this temperature until the ion current signal became constant; typically this required an hour or more. The temperature was then lowered in 10° increments with a 5-min delay at each temperature before recording of the ion current. For the first run in a cycle, ion current signals were obtained as a function of temperature for one of the pure components

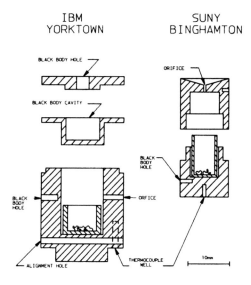

Fig. 2.   Effusion cells.

followed by a change of samples and measurement of signals for that component in a silicide sample. Following this second run and without breaking vacuum, data were collected for the other component. The final run was on the second pure component. This sequence of calibration and measurement runs was usually completed within a 24-h period.

## IBM-Yorktown Heights

### High Temperature Quadrupole Mass Spectrometry (HT-QMS)

The vaporization studies at IBM were carried out by means of an automated quadrupole mass spectrometer coupled to a vacuum furnace. This system has been described elsewhere (10), but the important details are given here. The following operating parameters of the system were under the control of an IBM Series/1 computer

1. Mass position;
2. Ionizing energy;
3. Integration time (modulation cycles);
4. Scaler readout;
5. Temperature setpoint; and
6. Temperature readout.

The mass spectrometer consisted of 6 mm quadrupole rods and electron impact ionizer from Finnigan Corporation; the electronics were obtained from Extranuclear, Inc. The ion detector was a Galileo channeltron operated in the pulse counting mode. The furnace was a resistively heated tungsten mesh element surrounded by 13 concentric tantalum radiation shields which provided an excellent approximation to a black body cavity. The mass spectrometer and furnace were in separately pumped UHV chambers that were isolated by a tube 5 mm in diameter by 30 mm long. The base pressure of the mass spectrometer chamber was about $10^{-10}$ Torr, whereas the furnace chamber was usually less than $10^{-8}$ Torr. Both the ion source and analyzer were kept under vacuum during change of samples. The mass spectrometer was mounted at a right angle to the furnace such that the molecular beam entered the ion source in the crossed electron beam mode. The molecular beam was modulated at 20 Hz. Only the center 80% of each half of the modulation cycle was used for gating two channels of a LeCroy 550B Quad 100 MHz scaler. A scale drawing of the effusion cell is given in Fig. 2. The cell body and sample cup were fabricated from tungsten metal and were obtained from Rembar, Inc. The effusion orifice was 1 mm in diameter by 5 mm in length. This orifice can be accurately aligned with the ion source entrance. Temperatures were measured by means of a reference grade thermocouple (95%W,5%Re vs 74%W,26%Re), secured in the cell wall, which was used for both temperature measurement and control. The thermal emf was measured with a HP3456A digital voltmeter. Temperature control was achieved by establishing the reference setpoint with a D/A converter. The overall temperature control stability of 0.2° was readily obtained. An ionizing electron energy of 16.0 eV was used for obtaining both chromium and silicon signals. In each run, following an initial heating to about 1300K, the cell temperature was increased at 2°/min. The temperature was then held constant at the selected maximum temperature for at least 30 min, following which it was cooled at a rate of 2°/min. Both the Cr and Si ion intensities were measured in alternating sequence over the entire temperature program with three measurements of each being made per minute. However, owing to the 80% gating period to the scalers, this 10-s integration time is equivalent to 4 s of actual counting during the open portion of the chopper cycle. Over the 7-h run in a typical experiment, the very large number of data points collected was reduced to a reasonable number for plotting by a simple average of 15 sequential measurements. Each data point plotted in the accompanying figures is thus equivalent to 60 s of total integration time. Since the temperature was slowly varying over this time interval, the temperature for each data point was taken as the average for the interval. This 15 point average is equivalent to a temperature interval of 10°, which is comparable to the SUNY data being presented. Using this simple averaging technique, we find that there is no significant effect on the second law or intercept over a wide range in binning intervals. A typical plot of Cr ion intensity versus time for a pure chromium sample is shown in Fig. 3; the temperature program for this run is also presented. Standardizing runs with pure chromium and silicon were generally made within one week of each silicide run.

## RESULTS

In this paper, we report data on chromium pressures which were obtained from ion current data in the following manner. For runs on pure

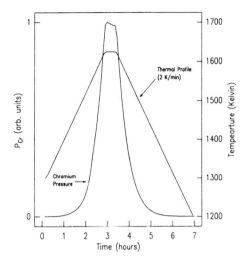

Fig. 3. Chromium signal and thermal program (HT-QMS).

chromium, the ion current-temperature product at the highest temperature, together with vapor pressure data from the JANAF Tables (11), were used to obtain an instrument sensitivity factor. This sensitivity factor was used to calculate vapor pressures in the pure chromium run and partial pressures in the silicide run. Second Law enthalpies were calculated in the usual manner.

It was shown to be necessary to equilibrate the interior walls of the cell with the sample. This was crucial in the HT-QMS experiments in which none of the effusing vapor came directly from an encounter with the surface of the sample. This equilibration was most effectively accomplished at temperatures where the chromium vapor pressure was above $10^{-5}$ atm; 30 min at $10^{-6}$ atm was insufficient to establish equilibrium. It was somewhat less problematic in the HT-MMS experiments in which most of the effusing vapor, particularly that entering the ion source, came directly from an encounter with the sample. As noted above, samples in the HT-MMS experiments were equilibrated for an hour or more at peak temperature. Figure 4 is a plot, from the HT-QMS experiments, of the pressure of pure chromium in a cell which had previously been heated to 2000K for approximately 2 h. As the temperature is increased, the points for the heating ramp approach the straight line defined by the cooling ramp points. In Fig. 5, data for pure chromium obtained in the HT-MMS experiments are compared with the cooling ramp points from the HT-QMS experiments. Agreement is excellent, which, given the differing configurations of the two effusion cell assemblies, supports the assumption that equilibrium was obtained in both experiments. Well equilibrated samples of pure chromium gave second law enthalpies of sublimation at 1500K of 379.6 ± 12.5 kJ mol⁻¹ in the HT-MMS experiments and 379.2 ± 8.3 kJ mol⁻¹ in the HT-QMS experiments, where the uncertainty limits represent the range of values in each

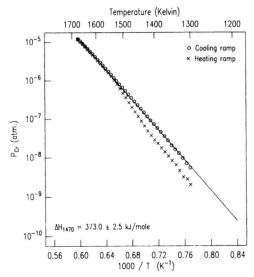

Fig. 4.  Cell equilibration (HT-QMS): chromium vapor pressure for heating and cooling ramps, pure chromium sample.

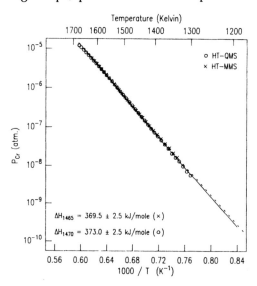

Fig. 5.  Comparison HT-QMS cooling ramp data with HT-MMS data: pure chromium sample.

case. These enthalpies may be compared to the third law value of 384.8 kJ mol$^{-1}$ given in the JANAF Tables (*11*).

Figure 6 shows data from the HT-QMS experiments for a sample with composition $X_{Si} = 0.39$, a two-phase mixture of $Cr_5Si_3$ and CrSi. The immediately previous run had been on a sample with a significantly higher chromium content. The heating ramp points again show equilibration of the cell walls with the sample, in this case from a higher chromium activity. Following the run described in Fig. 6, the sample was again taken through a heating and cooling cycle, the heating ramp data for which are given in Fig. 7 in comparison with the cooling ramp data from Fig. 6. It should be noted that in this run the data for the heating

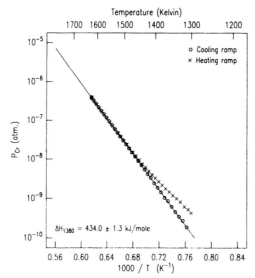

Fig. 6.   Cell equilibration (HT-QMS): chromium vapor pressure for heating and cooling ramps, chromium silicide sample ($X_{Si}$ = 0.39).

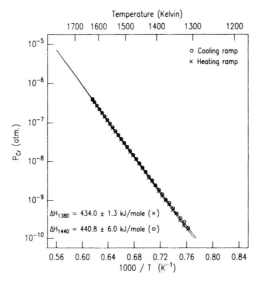

Fig. 7.   Cell equilibration (HT-QMS): cooling ramp from Figure 6 and a repeat heating ramp, chromium silicide sample ($X_{Si}$ = 0.39).

and cooling ramps coincide and agree with the cooling ramp data in Fig. 6 that is interpreted as implying equilibration of the sample with the cell walls throughout the cycle. Figure 8 is a comparison of HT-MMS data on the $X_{Si}$ = 0.39 sample with the HT-QMS cooling ramp data in Fig. 6. Agreement is excellent. For both the pure chromium and the silicide samples, it was found to be necessary to carry the heating ramp to a temperature sufficiently high that a linear plot of log P vs 1/T was obtained. This was necessary to ensure that equilibration was achieved, and failure to satisfy this condition gave data which did not agree in the interlaboratory comparison.

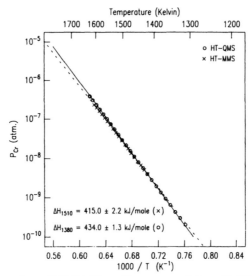

Fig. 8.  Comparison HT-QMS cooling ramp data with HT-MMS data: chromium silicide sample ($X_{Si} = 0.39$).

## SUMMARY

This paper has reported a comparison of vaporization data obtained by Knudsen effusion mass spectrometry on pure chromium and chromium-silicon samples. The data were obtained with two sets of apparatus which differed in the manner of mass analysis, effusion cell design, furnace construction, temperature measurement and calibration, and experimental procedures. It was possible to obtain very good agreement in results by adherence to the following practice.

1. Proper temperature calibration and measurement (although different in the two laboratories);
2. Use of the vapor pressures of the pure elements to establish instrumental sensitivity; and
3. Exercising care that the cell walls were in equilibrium with the sample.

## ACKNOWLEDGMENT

This research was supported in part by the US Office of Naval Research.

## REFERENCES

*1.* (a) Fronz, V., Rosner, B., and Storch, W., *Thin Solid Films,* **65,** 33 (1980); (b) Weiss, B. Z., Smith, D. A., and Tu, K. N., *Acta Met.,* **34,** 1491 (1986); (c) Narizuka, Y., Kawahito, T., Kamei, T., and Kobayashi, S., *IEEE Trans. Components, Hybrids, Manuf. Technol.,* **11,** 433 (1988); (d) Bedeker, C. J., Kritzinger, S., and Lombaard, J. C., *Thin Solid Films,* **141,** 117 (1986); (e)

Mazzega, E., Michelini, M., Nava, F., *J. Phys. F: Met. Phys.*, **17**, 1135 (1987).

2. Myers, C. E., Murray, G. A., Kematick, R. J., and Frisch, M. A., *High Temperature Materials Chemistry-III*, Munir, Z. A., Cubicciotti, D., eds., Electrochem. Soc. *Proceedings Vol.* **86-2**, 54 (1986).
3. Chart, T. G., *Met. Sci.*, **9**, 504 (1975).
4. Goldschmidt, H. J. and Brand, J. A., *J. Less-Common Metals*, **3**, 34 (1961).
5. Golutvin, Yu. M. and Chin-k'uei, L., *Russ. J. Phys. Chem.*, **35**, 62 (1961).
6. Eremenko, V. N., Lukashenko, G. M., Sidorko, V. R., and Kar'kova, A. M., *Sov. Powder Metall. Met. Ceram.*, **1971**, 563.
7. Bolgar, A. S., Gordienko, S. P., Lysenko, A. A., and Fesenko, V. V., *Russ. J. Phys. Chem.* **45**, 1154 (1971).
8. Riegert, J.-P., Vermande, V., and Ansara, I., *High Temp.-High Pressures*, **5**, 231 (1973).
9. Myers, C. E. and Kematick, R. J., *J. Electrochem. Soc.*, **134**, 720 (1987).
10. (a) Frisch, M. A. and Reuter, W., *J. Vac. Sci. Tech.* **16**, 1020 (1979); (b) Frisch, M. A., and Reuter, W., *Proc. 23rd Conf. Mass Spectr. and Allied Topics*, Houston, TX, May 1975, p. 560.
11. Chase, M. W., Curnutt, J. L., Prophet, H., McDonald, R. A., and Syverud, A. N., *J. Phys. Chem. Ref. Data*, **4**, 1 (1975).

# Mechanistic Aspects of Metal Sulfate Decomposition Processes

D. L. Hildenbrand,* K. H. Lau, and R. D. Brittain

*Materials Research Laboratory, SRI International,
Menlo Park, CA 94025*

## ABSTRACT

The decomposition reactions of the crystalline sulfate phases $CaSO_4$, $MgSO_4$, $ZnSO_4$, $ZnO \cdot 2ZnSO_4$, $CuSO_4$, and $CuO \cdot CuSO_4$ were studied by the torsion-effusion method with simultaneous mass-loss measurement, and by mass spectrometry. Decomposition pressures and vapor compositions were derived from the results. In all instances, these phases decompose to the corresponding oxides or oxysulfates, plus a gas phase that would be predominantly $SO_2$ and $O_2$ at equilibrium. However, most of the processes are severely limited kinetically because of failure to attain $SO_3$–$SO_2$ equilibrium in the gas phase, and direct desorption of $SO_3$ is frequently observed. Certain noble metals and $p$-type semiconducting oxides are observed to catalyze the $SO_3$–$SO_2$ conversion and in some instances lead to dramatic increases in the observed effusion pressure. In a few cases, the catalytic additives convert the gas phase to the equilibrium $SO_2$ + $O_2$ composition, but have no detectable effect on decomposition pressure, suggesting that the $SO_3$ to $SO_2$ conversion may or may not be closely coupled to the initial sulfate ion decomposition step. An attempt is made to correlate this wide range of observed behavior. In addition, decomposition pressures derived by extrapolation to zero orifice size are compared with values calculated from thermochemical data to test the suitability of the effusion method for thermodynamic studies of these materials.

**Index Entries:** Metal sulfate decomposition; effusion studies; thermodynamics; sulfur oxides; kinetics.

## INTRODUCTION

Over the last decade or more, we have studied a number of metal sulfate decomposition reactions by the effusion method, and very div-

*Author to whom all correspondence and reprint orders should be addressed.
High Temperature Science, Vol. 26      © 1990 by the Humana Press Inc.

erse behavior has been observed (1–4). Our goal initially was to characterize these processes thermodynamically. Since they were expected to follow the equilibrium pathway

$$MSO_4(s) = MO(s) + SO_2(g) + \tfrac{1}{2} O_2(g) \tag{1}$$

information about the reaction thermodynamics and the properties of the $MSO_4(s)$ phases could be obtained from measured decomposition pressures and the established properties of the products. Early on, however, effusion studies on $MgSO_4(s)$ showed that only $SO_3$ was observed in the effusing gas under some conditions, rather than the dominant equilibrium products $SO_2$ and $O_2$ (1). The composition of the effusate from $MgSO_4$ (s) was dependent not only on orifice size, but also on the fractional degree of decomposition (1), making the interpretation of the results somewhat complex. In accord with the results of Pechkovsky (5) on the kinetics of sulfate decompositions, it was also found that some transition metal oxides catalyzed the decomposition process and dramatically increased the observed effusion pressure (2); certain noble metals were later found to show similar catalytic properties. Because of this complex and unexpected behavior, we extended our effusion studies to other metal sulfate systems and redirected our interest to clarification of the decomposition process itself, including the underlying chemical and kinetic factors. The results of these systematic effusion studies on $CaSO_4$, $MgSO_4$, $ZnSO_4$, $ZnO \cdot 2ZnSO_4$, $CuSO_4$, and $CuO \cdot CuSO_4$ are presented here in a unified way, so that trends and correlations in behavior patterns may be more easily discerned. Furthermore, the results provide an interesting test of the effusion method for systems with severe kinetic limitations.

Effusion pressures and vapor molecular weights were determined by means of the torsion method with simultaneous mass loss determination. This is a rapid and accurate method that permits one to monitor pressure and molecular weight continuously. In many instances, supplemental information about vapor composition was obtained from mass spectrometric sampling of the effusion beam.

## EXPERIMENTAL

The torsion-effusion apparatus and associated microbalance, along with details of the experimental technique, have been described in a previous publication (1). This assembly is basically an effusion manometer, a device with which pressure is determined in terms of the geometrical and mechanical constants of the system; vapor molecular weights can also be derived if mass effusion rates are measured concurrently. Alumina effusion cells of the type described previously (1) were used throughout. Both pressures and vapor molecular weights are estimated to be accurate within 5%, as corroborated by frequent checks on laboratory vapor pressure standards. For an effusing vapor containing more than one gaseous species, the molecular weight $M$ is a weight average value defined by the relation

$$M = (\Sigma m_i M_i^{-1/2})^{-2} \qquad (2)$$

where $m_i$ and $M_i$ are the mass fraction and molecular weight of the $i$ gaseous species. The magnetic mass spectrometer (6) was used to examine the effusate from the solid sulfates.

Metal sulfate samples were prepared by dehydration of reagent grade starting materials. Details of the individual sample preparations are given in earlier papers (1–4).

## RESULTS

For reference in interpreting the measured vapor molecular weights $M$, reaction (1) yielding $SO_2$ and $O_2$ yields a calculated value of 54.6, whereas $SO_3$ release by the process

$$MSO_4(s) = MO(s) + SO_3(g) \qquad (3)$$

gives $M = 80.1$. Mixtures of $SO_3$, $SO_2$, and $O_2$ will yield intermediate values. For each system, the experimental data are summarized in the earlier publications on $CaSO_4$ (1), $MgSO_4$ (1,2), $ZnSO_4$ and $ZnO \cdot 2ZnSO_4$ (3), and CuO and $CuO \cdot CuSO_4$ (4). Only derived results are presented here.

### CaSO₄

Figure 1 shows a Whitman-Motzfeldt (7,8) plot of the observed effusion pressure data on $CaSO_4$ at 1200 K for several orifice sizes (1). According to the W-M model (7,8), reciprocal observed pressure $P$ shows a linear correlation with effective orifice area, $\Sigma fa$, where $f$ and $a$ are orifice force factor and area, respectively; the intercept of such a plot at zero effective orifice area is the reciprocal of equilibrium pressure, $P_{eq}$. Over time, a great deal of evidence has accumulated in support of the W-M model for interpreting effusion data, and particularly for deriving equilibrium pressures. The pressure data for $CaSO_4$ in Fig. 1, show a significant hole-size variation but a satisfactory linear correlation. In addition, the average vapor molecular weights of 53.4, 54.8, and 53.9 are in close accord with the $SO_2 + \frac{1}{2} O_2$ vapor composition, indicating the decomposition process to be

$$CaSO_4(s) = CaO(s) + SO_2(g) + \frac{1}{2} O_2(g) \qquad (4)$$

Vapor pressures and molecular weights were independent of time and fractional degree of decomposition.

From the intercept of Fig. 1, we derive an equilibrium pressure of $1.2 \times 10^{-5}$ atm for reaction (4) at 1200 K, in good agreement with the equilibrium total effusion pressure of $1.5 \times 10^{-5}$ atm calculated from tabulated thermodynamic data (9,10). Thus the interpretation of the

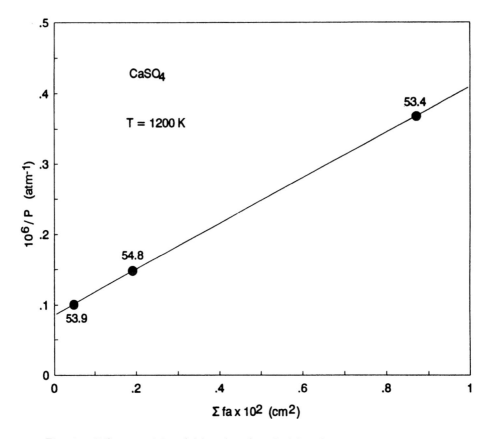

Fig. 1.   Whitman-Motzfeldt plot for $CaSO_4$ decomposition: numerical values are vapor molecular weights.

$CaSO_4$ effusion measurements appears to be entirely straightforward and more or less predictable. Mohazzabi and Searcy (11) found rather similar behavior in effusion studies of $BaSO_4$ decomposition near 1500 K. Note that the equilibrium $SO_3$ pressure is several orders of magnitude lower than that of $SO_2$ under these conditions.

## $MgSO_4$

Unexpectedly, the effusion results on $MgSO_4$ (1) at 950–1050 K proved to be quite different from those on $CaSO_4$. As seen in Fig. 2, the Whitman-Motzfeldt plot for the pure material shows two quasilinear regions extrapolating to different zero hole-size pressures. For the larger effusion orifices with A/a (A = cell cross sectional area) ratios of 10–30, the average M values were 80.8, 78.6, and 81.5, with an extrapolated "$P_{eq}$" of $7.5 \times 10^{-7}$ atm at 1000 K. These M values indicate clearly that decomposition proceeds by the reaction

$$MgSO_4(s) = MgO(s) + SO_3(g) \tag{5}$$

with the larger orifices, even though the equilibrium ratio $P(SO_2)/P(SO_3)$ $\cong 75$ at 1000 K under these conditions. Furthermore, the extrapolated

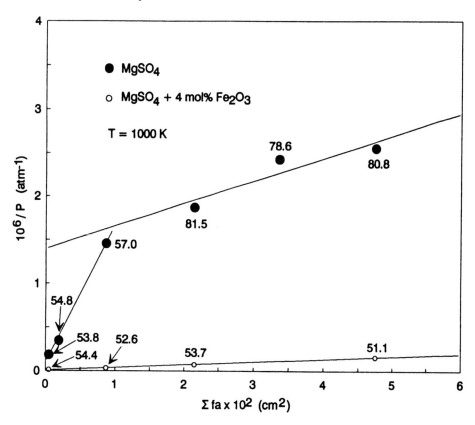

Fig. 2. Whitman-Motzfeldt plot for $MgSO_4$ decomposition: numerical values are vapor molecular weights.

pressure of $7.5 \times 10^{-7}$ atm is comparable to the calculated equilibrium pressure of $3.2 \times 10^{-6}$ atm for reaction (5); there are reasons to be discussed later for believing that the calculated pressure should be somewhat lower. Nevertheless, the long extrapolation makes the derived zero hole-size pressure for the larger orifices relatively uncertain.

Effusion-beam mass spectrometry showed $SO_3$ to be the only detectable vapor species under the above conditions, verifying the measured $M$ values. All of these results were obtained during the initial phase, when the percentage of decomposition was less than 5%. As the decomposition was extended to levels of 10% or greater, however, $M$ values with the larger orifices fell into the 55–60 range, whereas pressures decreased only slightly. Conversion of $SO_3$ to $SO_2$ and $O_2$ is evidently occurring during passage of the gas through the accumulating $MgO(s)$ product layer, but this apparently has little influence on observed pressure.

With the smaller effusion orifices, where $A/a = 50$–500, the measured M values of 57.0, 54.8, and 53.8 show unambiguously that the effusing gas has the $SO_2 + \frac{1}{2} O_2$ stoichiometry, suggesting that the system is attempting to approach the equilibrium process

$$MgSO_4(s) = MgO(s) + SO_2(g) + \tfrac{1}{2} O_2(g) \qquad (6)$$

with decreasing hole size. This is also evident from the sharply increased slope of the W-M plot, although the zero hole-size pressure of $1.7 \times 10^{-5}$ atm is a factor of ten lower than the calculated equilibrium pressure for reaction (6). This is believed to be the first clear example of a W-M plot showing two distinct linear regions of different slope, each one extrapolating to a different limiting pressure at zero orifice size. Evidently, there are severe kinetic limitations to attainment of $SO_3$–$SO_2$ equilibrium and the system is trying to adjust as zero hole size is approached.

Following Pechkovsky's report (5) that small amounts of $Fe_2O_3$ and CuO markedly accelerate the overall decomposition of $MgSO_4$ at 1200–1300 K, we investigated the effect of a few mol% $Fe_2O_3$ addition on the effusion measurements. The results were very dramatic. As seen in Fig. 2, measured pressures were higher than those of the pure sulfate by a factor of ten or more and the $M$ values for all orifice sizes were in the region $53 \pm 2$, with the plot linear over the entire range. This behavior indicates that the $Fe_2O_3$ additive effectively catalyzes the $SO_3$–$SO_2$ conversion and leads to sharply increased pressures, approaching equilibrium values. The extrapolated zero hole size pressure of $1.1 \times 10^{-4}$ atm at 1000 K is close to the equilibrium value of $2.8 \times 10^{-4}$ atm for reaction (6), calculated from tabulated thermodynamic data (9,10).

Subsequently, the effects of other catalytic additives were studied (2), some of which were known to be effective oxidation catalysts. It was found that the metals Pt, Ir, and Ru, along with the $p$-type semiconducting oxides $Cr_2O_3$, $Mn_2O_3$, CuO, CoO, and NiO all yielded effusion pressures a factor of ten or more higher than pure $MgSO_4$, with corresponding $M$ values of $54 \pm 2$ for all orifices. Conversely, the $n$-type and insulator oxides MgO, $Al_2O_3$, $TiO_2$, $Y_2O_3$, and ZnO were found to have little or no effect. In an auxiliary study in which catalysts of the $SO_3$–$SO_2$ conversion was investigated in the absence of solid sulfate phases (12), the same metals and $p$-type oxides were found to be highly effective, although there was a spread in the onset temperature of catalytic activity. Pt and Ru were found to have the lowest threshold temperatures, and yielded near equilibrium compositions. Again, $Al_2O_3$ and ZnO had no effect.

Our interpretation of all this is that $SO_3$ is the initial gaseous decomposition product of the lattice sulfate ion, and that subsequent conversion to the equilibrium $SO_2 + \frac{1}{2} O_2$ composition is a critical slow step in the overall process. Direct desorption of $SO_3$ into the effusing vapor occurs for relatively large orifices, but the system begins to approach the equilibrium condition with decreasing orifice size. In the presence of an effective catalyst, $SO_3$–$SO_2$ equilibrium is apparently achieved close to the decomposition site, and both pressure and gas composition are shifted toward that expected for reaction (6). Further discussion of this point is given later.

### $ZnSO_4$ and $ZnO \cdot 2ZnSO_4$

As seen in the Whitman-Motzfeldt plots in Figs. 3 and 4, the zinc sulfate phases $ZnSO_4$ and $ZnO \cdot 2ZnSO_4$ exhibited yet a different type of

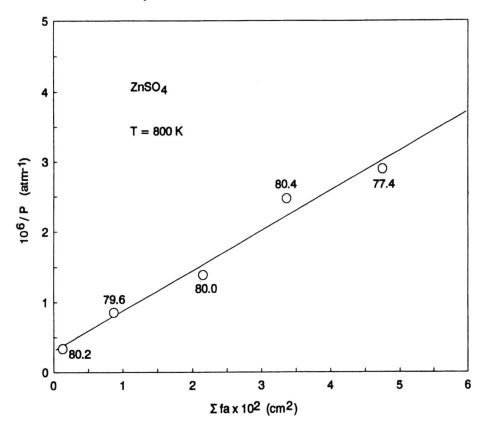

Fig. 3. Whitman-Motzfeldt plot for $ZnSO_4$ decomposition: numerical values are vapor molecular weights.

effusion behavior at 800–850 K. For the pure phases, both vapor molecular weights and effusion-beam mass spectrometry showed clearly that $SO_3$ is the sole gaseous decomposition product with all orifice sizes, although at gaseous equilibrium $P(SO_2)/P(SO_3) > 10$. The W-M plots show a smooth orifice–size correlation, and yield limiting zero hole-size pressures of $3.4 \times 10^{-6}$ atm for $ZnSO_4$ at 800 K, and $1.3 \times 10^{-6}$ atm for $ZnO \cdot 2ZnSO_4$ at 850 K. Analysis of cell residues showed the processes to be

$$3\ ZnSO_4(s) = ZnO \cdot 2ZnSO_4(s) + SO_3(g) \qquad (7)$$

$$\tfrac{1}{2}\ ZnO \cdot 2ZnSO_4(s) = \tfrac{3}{2}\ ZnO(s) + SO_3(g) \qquad (8)$$

The extrapolated pressures from the effusion data noted above are each about a factor of five lower than corresponding equilibrium pressures of $10.5 \times 10^{-6}$ atm at 800 K for reaction (7) and $6.7 \times 10^{-6}$ atm for reaction (8) at 850 K, calculated from accurate thermodynamic data (*13*). Formation of the $ZnO \cdot 2ZnSO_4$ and ZnO decomposition product phases in a finally divided state could be responsible for the low extrapolated effusion pressures.

Interestingly enough, with $ZnO \cdot 2ZnSO_4$ and the smallest effusion orifice size, the torsion pressures did show a 50% increase and the

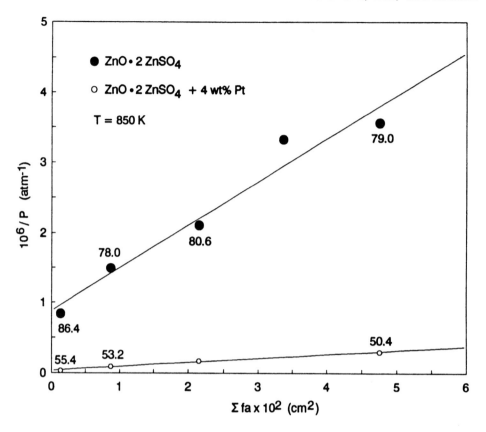

Fig. 4. Whitman-Motzfeldt plot for $ZnO \cdot 2\,ZnSO_4$ decomposition: numerical values are vapor molecular weights.

molecular weights a corresponding decrease with time, as a larger fraction of the sample was decomposed. This is somewhat similar to the small orifice behavior of $MgSO_4$, and indicates the onset of a trend toward equilibrium. Presumably, a W-M plot similar to that of pure $MgSO_4$ would result if measurements could be extended to significantly smaller orifice dimensions, although this might not be possible practically.

Of the metal and oxide powder additives found to catalyze $MgSO_4$ decomposition, only Pt had a significant effect on the zinc sulfates. For $ZnO \cdot 2ZnSO_4$, addition of a few mole percent of Pt powder led to striking increases in decomposition pressure by factors of 12 to 25, depending on orifice size, and $M$ values fell to the $53 \pm 3$ level characteristic of $SO_2 + O_2$. For the smallest orifice, this is the pressure increase expected for transition to the process

$$\tfrac{1}{2}ZnO \cdot 2\,ZnSO_4(s) = \tfrac{3}{2}\,ZnO(s) + SO_2(g) + \tfrac{1}{2}\,O_2(g) \qquad (9)$$

The W-M plot in Fig. 4 shows the Pt-catalyzed pressure data on $ZnO \cdot 2ZnSO_4$ to be very well behaved.

With $ZnSO_4(s)$, however, the observed pressure increase was substantially smaller and unreproducible, although molecular weights dropped into the $52 \pm 3$ range. Repeated attempts failed to yield consistent results on $ZnSO_4$, suggesting that, for whatever physical and chemical reasons, this system is on the threshold of catalytic activity but cannot sustain such behavior.

### $CuSO_4$ and $CuO \cdot CuSO_4$

The copper sulfates have still higher decomposition pressures than the magnesium or zinc sulfates and might, therefore, show still different effusion pressure behavior. In addition, the known catalytic behavior of $CuO(s)$ for $SO_3$–$SO_2$ conversion (2,12) suggests the possibility of a "built-in" catalyst in the oxide decomposition product. The W-M plots in Fig. 5 and 6 summarize the effusion pressure results on $CuSO_4$ and $CuO \cdot CuSO_4$ at 720 and 750 K, respectively. For both substances, the pressures show a good hole–size correlation, but the measured $M$ values of 65–75 are intermediate to those expected for $SO_2 + \frac{1}{2} O_2$ (54.6) and $SO_3$ (80.1). However, the molecular weights clearly showed a trend with time, decreasing from the initial values that were near 80. Mass spectrometric examination corroborated this increasing $SO_2$ content of the effusing vapor with larger fraction of decomposition; although only $SO_3$ was detected initially, $SO_2$ levels rose gradually, even for less than 1% sample decomposition. Pressures remained relatively steady during this falloff in the $M$ values.

We interpret this behavior to indicate that $SO_3$ is the primary gaseous decomposition product of both $CuSO_4$ and $CuO \cdot CuSO_4$, and that the accumulating oxide product partially catalyzes the conversion to $SO_2$ and $O_2$; $SO_2$–$SO_3$ equilibrium is not attained, however, since $P(SO_2)/P(SO_3) > 5$ at equilibrium at these temperatures and pressures, whereas the $M$ values of 65–75 indicate only a 30–40% $SO_2$ content. Addition of a few mole% Pt powder to both sulfates effected an immediate drop to $M$ values of 56–58, about that expected for gaseous equilibrium, but the effusion pressures remained unchanged. Because of the invariance of pressure with increasing $SO_2$ level, we conclude that the measured pressures are associated with $SO_3$ release, just as with the zinc sulfates. Subsequent partial conversion to $SO_2$ and $O_2$ by the oxide decomposition product apparently is uncoupled from the initial decomposition step so that effusion rate and pressure are unaffected by this change in effusing gas composition.

Analysis of cell residues showed CuO to be the solid decomposition product of both sulfates. Therefore, the observed processes are

$$CuSO_4(s) = CuO(s) + SO_3(g) \qquad (10)$$

and

$$CuO \cdot CuSO_4(s) = 2CuO(s) + SO_3(g) \qquad (11)$$

$CuO \cdot CuSO_4(s)$ is only 1.5 kJ mol$^{-1}$ more stable than $CuO(s)$ and $CuSO_4(s)$ at 700 K, and this marginal stability is presumably insufficient to force nucleation of the oxysulfate phase as the product of reaction (10).

The extrapolated "equilibrium" pressure of reaction (10) is $2.2 \times 10^{-6}$ atm at 720 K, compared to a value of $6.3 \times 10^{-7}$ atm calculated from

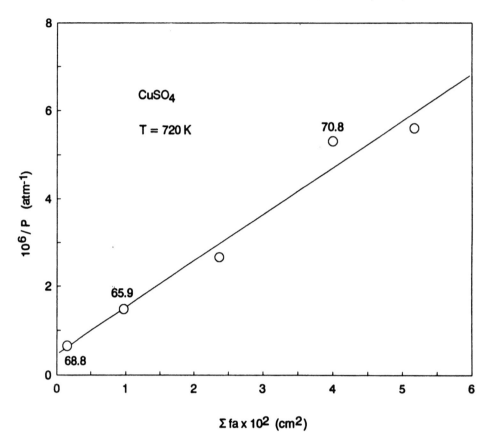

Fig. 5.   Whitman-Motzfeldt plot for $CuSO_4$ decomposition: numerical values are vapor molecular weights.

thermodynamic data (14). Note that the effusion pressure is higher than the calculated value, rather than lower as found with the zinc sulfates. Static pressure (15) and EMF (16) measurements also yielded an equilibrium $SO_3$ pressure of $2.0 \times 10^{-6}$ atm for reaction (10) at 720 K, close to our extrapolated value, suggesting a possible error in the thermodynamic data (14). For reaction (11), the W-M extrapolation of our effusion data yields a limiting pressure of $1.8 \times 10^{-6}$ atm at 750 K, whereas the recommended value extrapolated from static pressure measurements (15) is $1.5 \times 10^{-6}$ atm, and a value of $2.0 \times 10^{-6}$ atm is calculated from calorimetric and related thermodynamic data (14). In contrast to the disparities in the results for $CuSO_4$, the equilibrium $SO_3$ pressure derived from effusion measurements on $CuO \cdot CuSO_4$ is in remarkably good agreement with the value calculated from thermodynamic data, which should normally be accurate to within 20% when the properties are well established.

    In summary, the measured effusion pressures for both $CuSO_4$ and $CuO \cdot CuSO_4$ are associated solely with the $SO_3$ release step, and effective catalytic additives for $SO_3 - SO_2$ conversion have no detectable influence on effusion rate or pressure, unlike the corresponding results on $MgSO_4$, $ZnSO_4$, and $ZnO \cdot 2 ZnSO_4$. Although partial or near-complete

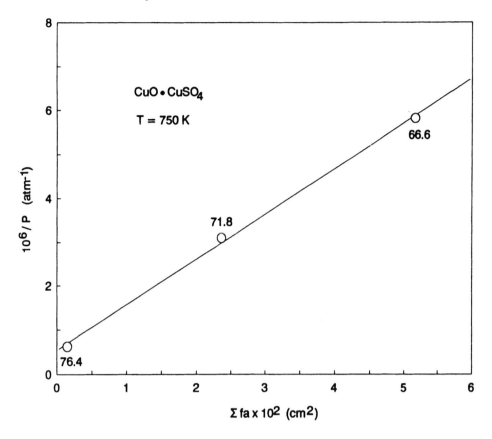

Fig. 6.   Whitman-Motzfeldt plot for CuO · CuSO₄ decomposition: numerical values are vapor molecular weights.

conversion to $SO_3$–$SO_2$ equilibrium in the effusing gas can occur by contact with the CuO(s) decomposition product or Pt powder additive, the pressures must be interpreted in terms of reactions (10) and (11).

## DISCUSSION

It is evident that there is a substantial kinetic barrier to the gaseous equilibrium

$$SO_3(g) = SO_2(g) + \tfrac{1}{2} O_2(g) \tag{12}$$

and that this barrier dominates the effusion behavior of the Ca, Mg, Zn, and Cu sulfates reported here. Taken as a whole, the results indicate that release of $SO_3$ from the sulfate ion is the initial step in the decomposition process, and that subsequent equilibration of $SO_3$, $SO_2$, and $O_2$, reaction (12), is a critical slow step in the overall decomposition process. The magnitude of the observed effusion pressure then strongly depends on the coupling between the initial $SO_3$ release step and reaction (12). For $CaSO_4$ at about 1200 K, the kinetics of reaction (12) are sufficiently fast that final equilibrium apparently occurs on or adjacent to the decomposition site.

With $MgSO_4$ at 1000 K, the situation is much more complex but nevertheless revealing as the various pieces of the puzzle fit together. For the larger orifice dimensions, direct release of $SO_3$ to the effusing vapor occurs without vapor equilibration. Although vapor phase equilibration gradually becomes detectable as fractional decomposition increases, probably during passage of $SO_3$ through the accumulating MgO product layer, this conversion is too far removed from the actual decomposition site, and effusion rate and pressure are essentially unaffected. With decreasing orifice size, there is a crossover point at which the system attempts to approach full equilibration, as evidenced by a shift to the $SO_2$ + ½ $O_2$ composition and a rapidly rising pressure; the latter manifests itself in the unusual W-M plot with sharply increased slope at the smallest effective orifice areas. For the larger collision numbers and residence times associated with the smallest orifices, the vapor equilibrates closer to the decomposition site and the limiting pressure approaches but cannot quite reach the equilibrium value for reaction (6). Small additions of certain noble metals and $p$-type semiconducting oxides catalyze reaction (12) adjacent to the decomposition site, promoting full equilibration of reaction (6). The common feature responsible for the catalytic activity of the metals and $p$-type oxides is apparently their affinity for oxygen chemisorption (2).

At temperatures in the range 800–900 K, $ZnSO_4$ and $ZnO \cdot 2ZnO_4$ yield $SO_3$ only, for all orifice sizes, although there was a weak hint of increasing pressure and $SO_2$ level with $ZnO \cdot 2ZnSO_4$ at the smallest orifice size. Only Pt was found to catalyze the decomposition rate. At this lower temperature, the reduced rate of reaction (12) decreases the coupling between $SO_3$ release and $SO_3$–$SO_2$ equilibration steps, and there are fewer options for attaining equilibrium under effusion conditions.

At still lower temperatures of 700–780 K, the $SO_3$ desorption step alone dominates the effusion behavior of $CuSO_4$ and $CuO \cdot CuSO_4$ under all conditions, with and without catalytic additions. Although the CuO solid decomposition product and Pt powder promote partial or perhaps full equilibration of reaction (12) in the effusing gas, this process must be essentially completely decoupled from the initial $SO_3$ release step, and no enhancement of decomposition rate or effusion pressure is observed under any conditions. Since reaction (1) is a thermally activated process with apparent activation energy of about 170 kJ mol$^{-1}$, this diverse effusion behavior over the range 700–1200 K can be accounted for on a purely thermal basis.

Each sulfate decomposition process showed a substantial orifice size dependence, as seen in the W-M plots of Figs. 1-6. This is, of course, characteristic of kinetically retarded processes of the type described here. It is likely that this effect arises largely from the major structural changes associated with transformation of the tetrahedral sulfate ion with r(S–O) ~ 0.155 nm to the planar symmetric $SO_3$ molecule with r(S–O) = 0.143 nm. Upper limit values to the vaporization coefficients of these processes are all approximately $10^{-3}$ (1–4).

The Whitman-Motzfeldt model (7,8) provides a very satisfactory interpretation of the effusion pressures, and in all cases yields derived "equilibrium" pressures at zero hole–size that are in reasonable accord with calculated and measured values. In a few instances ($MgSO_4$, $ZnSO_4$, $ZnO \cdot 2ZnSO_4$), the extrapolated effusion pressures are lower than calculated equilibrium values by factors of 4 to 6. It seems likely that such effects could arise from formation of the solid decomposition products in a finely divided metastable state of higher energy content. This type of behavior was recognized by Giauque (17) in connection with the MgO(s) product of $Mg(OH)_2$ decomposition, and Beruto et al. (18) have verified that surface–area effects can readily account for $H_2O$ effusion decomposition pressures over $Mg(OH)_2$ that are low by about $10^4$. The importance of particle size effects on dynamic effusion measurements should always be kept in mind.

Perhaps the lesson to be learned from all this is that things are not always what they seem, that in the study of complex processes there are often many misleading signs. But with care and caution, the pieces of evidence can be put in place and the mystery unraveled. If the $MgSO_4$ measurements were not begun until more than 5% of the sample was decomposed, as is often the case when samples are outgassed at temperature to eliminate volatile impurities, one would be forced to conclude from $M$ values in the 55–60 range that reaction (6) was the process observed, and that the very low pressures indicated some fault with the method or W-M model. Just as with the copper sulfates, the molecular weights indicated an effusing vapor rich in $SO_2$ and $O_2$, whereas the effusion rates and derived pressures are actually dominated by the $SO_3$ desorption rate, and the observed processes are reactions (5), (10), and (11). Even the vapor mass spectrum can be misleading; more detailed examination showed that a spectrum indicating equilibrium abundances of $SO_3$ and $SO_2$ over $Al_2(SO_4)_3$ (19) was probably influenced by secondary ion source reactions yielding $SO_2$, whereas the actual effusion beam was purely $SO_3$ (3). The effusion method and the related Whitman-Motzfeldt model are certainly applicable to the study of these complex decomposition reactions, provided that we clearly define the chemical process being studied.

## ACKNOWLEDGMENT

This research was supported by the Division of Chemical Sciences, Office of Basic Energy Sciences, Office of Energy Research, US Department of Energy.

## REFERENCES

1. Lau, K. H., Cubicciotti, D., and Hildenbrand, D. L., *J. Chem. Phys.* **66,** 4532 (1977).

2. Knittel, D. R., Lau, K. H., and Hildenbrand, D. L., *J. Phys. Chem.* **84,** 1899 (1980).
3. Brittain, R. D., Lau, K. H., Knittel, D. R., and Hildenbrand, D. L., *J. Phys. Chem.* **90,** 2259 (1986).
4. Brittain, R. D., Lau, K. H., and Hildenbrand, D. L., *J. Phys. Chem.,* **93,** 5316 (1989).
5. Pechkovsky, V. V., *J. Appl. Chem. USSR* **29,** 1229 (1956).
6. Hildenbrand, D. L., *J. Chem. Phys.* **48,** 3657 (1968); **52,** 5751 (1970).
7. Whitman, C. I., *J. Chem. Phys.* **20,** 161 (1952).
8. Motzfeldt, K., *J. Phys. Chem.* **59,** 139 (1955).
9. JANAF Thermochemical Tables, 3rd ed., J. Phys. Chem. Ref. Data **14,** Suppl. No. 1 (1985).
10. DeKock, C. W., *US Bu. Mines Rept. IC 9081* (1986).
11. Mohozzabi, P. and Searcy, A. W., *J. Chem. Soc. Faraday Trans. 1* **72,** 290 (1976).
12. Brittain, R. D. and Hildenbrand, D. L., *J. Phys. Chem.* **87,** 3713 (1983).
13. Beyer, R. P., *J. Chem. Thermodyn.* **15,** 835 (1983).
14. DeKock, C. W., *US Bu. Mines Rept. IC 8910* (1982).
15. Kellogg, H. H., ITrans. Met. Soc., AIME **230,** 1622 (1964).
16. Skeaff, J. M. and Espelund, A. W., *Can. Metall. Q.* **12,** 445 (1973).
17. Giauque, W. F., *J. Am. Chem. Soc.* **71,** 3192 (1949).
18. Beruto, D., Rossi, P. F., and Searcy, A. W., *J. Phys. Chem.* **89,** 1695 (1985).
19. Knutsen, G. F. and Searcy, A. W., *J. Electrochem. Soc.* **125,** 327 (1978).

# Author Index

442

# Subject Index

CPSIA information can be obtained at www.ICGtesting.com
Printed in the USA
LVOW131602260212

270490LV00009B/13/P